Stability in
Coal Mining

Stability in Coal Mining

Proceedings of the first International Symposium on
Stability in Coal Mining
Vancouver, British Columbia, Canada, 1978

Edited by C. O. Brawner
Department of Mineral Engineering
University of British Columbia

and Ian P. F. Dorling
North American Editor
WORLD COAL magazine

MILLER FREEMAN PUBLICATIONS
San Francisco

Technical Advisors to the First International Symposium on Stability in Coal Mining

Golder, Brawner and Associates, Ltd.

Society of Mining Engineers of the American
Institute of Mining, Metallurgical, and Petroleum Engineers, Inc.

The Canadian Institute of Mining and Metallurgy

The University of British Columbia Mineral Engineering Department

British Columbia Ministry of Mines and Petroleum Resources

British Columbia and Yukon Chamber of Mines

Copyright © 1979 by MILLER FREEMAN PUBLICATIONS, INC.,
500 Howard Street, San Francisco, California 94105 USA. Printed in the
United States of America.

Library of Congress Catalog Card Number: 78-73842
International Standard Book Number: 0-87930-108-2

Contents

Underground Mining

Preparation Plant Refuse and Waste Disposal

Reclamation

Foreword

The recent oil and energy crisis has focused attention on the potential use of coal for the development of power and heat. The available coal reserves in North America that can be practically mined using existing technology are estimated to be adequate for at least 200 years. This presupposes that the costs of mining and power development using coal will be competitive with gas and oil and that the environment can be returned to at least its original state.

It will not be easy to maintain long-term price competitiveness and environmental approval. Coal mining involves changes in surface and subsurface soil, rock, and groundwater conditions. In the past it has been common to look at each of these geotechnical and environmental changes more or less independently. The organizers of the first International Symposium on Stability in Coal Mining considered that there was a great need for a single meeting to provide a wide overview of the interplay of geotechnical and environmental concerns.

Accordingly, a program was developed to provide a review, including case examples, of geotechnical stability and environmental control for surface and subsurface mining and for disposal of waste materials. Concerns included the influence of geology, surface and subsurface water, blasting, climate, and mining method and geometry on physical stability. In recognition of the concern for safety at waste disposal facilities, major attention was paid to mine waste dump and sludge pond stability. Since coal mines have a finite life a review of typical experience with reclamation programs was included.

Major attention was directed to the inclusion of many case examples. Experience is the best teacher and it is gratifying to see mining organizations recognizing the need to share geotechnical and environmental experience. This sharing of data will improve stability, safety, and, subsequently, mining economics. At this stage in our technical development there is still a great need for experience-related investigation and research rather than theoretical research.

The program was grouped into four sections: surface coal mining, underground coal mining, preparation plant refuse and waste disposal, and reclamation.

To highlight each section of the program a keynote speaker was selected to provide an overview of the concerns and experience relating to surface and underground stability, waste disposal, and reclamation. The first chapter in each section of this book presents the keynote addresses. The organizers are indebted to H. D. Dahl, R. Fujimoto, and I. H. Reiss for their contributions.

The symposium developed broad interest with a total of 36 speakers from the United States, 13 from Canada, 7 from Australia, 2 from India, and one each from the Federal Republic of Germany, Turkey, and Hungary.

The organizers wish to thank the authors for the high quality of presentations and the Technical Advisory groups for their contributions. Special thanks are extended to the Honorable James C. Chabot, Minister of Mines and Petroleum Resources, Province of British Columbia for the opening address and to Mr. M. N. Anderson, President, Cominco Ltd., Mr. G. H. Roman, Publisher, *World Coal* magazine, and Mr. F. G. Higgs, Manager, British Columbia and Yukon Chamber of Mines, as guest luncheon speakers.

The organizational success of the symposium was greatly enhanced by the contributions of Mr. Ray Chow of Golder Associates, Vancouver, British Columbia.

Thanks are also given to my co-editor Mr. Ian Dorling and other members of the staff of Miller Freeman Publications, including Ms. Jann Donnenwirth, Ms. Peggy Boyer, and Mr. George Roman, for their major contribution to the publication of the symposium proceedings.

With papers from many countries and the standardization of units not yet international, the editors have attempted to use both metric and English measurement systems and American spelling wherever practical. Unless otherwise noted, all tonnage figures are given in metric tons.

C. O. Brawner
Symposium Chairman and Co-Editor
Vancouver, British Columbia, Canada
December 1978

About the Editors

C. O. Brawner has been associated with stability studies on over 200 mining projects around the world, including some 20 studies for major coal mining operations.

Mr. Brawner spent 15 years as a consultant, ultimately becoming president of Golder, Brawner & Associates, Ltd. He resigned in 1978 to join the faculty of the University of British Columbia and give specialized instruction in geomechanics. He continues with some private, specialized consulting.

A native of Canada and graduate of the University of Manitoba and Nova Scotia Technical College in soil mechanics, foundations, and geology, Mr. Brawner worked with the Province of British Columbia Department of Highways prior to joining Golder Associates in 1963. He is a registered Professional Engineer.

Among other awards, Mr. Brawner received the 1978 B. T. A. Bell commemorative medallion for service to the Canadian mining industry. His advisory activities include work with the Canadian Department of Mines as a specialist consultant for development of the "Design Guide of Mine Waste Embankments in Canada." He is a past member of the National Research Council of Canada Associate Committee on Geotechnical Research and served as vice chairman of the Canadian Advisory Committee on Rock Mechanics.

Mr. Brawner is the co-editor of two other books on stability in mining: *Stability in Open Pit Mining* and *Geotechnical Practice for Stability in Open Pit Mining.* He has written over 60 technical papers on soil and rock mechanics engineering.

Ian P. F. Dorling, North American editor of *World Coal* magazine, has worked with and observed slope instability in many countries.

Born in England, he graduated from Leeds University in 1971 with a degree in mining engineering. He joined the National Coal Board that year and worked at the Lofthouse and Ackton Hall collieries as overman on development work and fully mechanized longwall production faces and safety engineer. He became assistant to the manager at Kellingley colliery, one of the newest and largest underground mines in the United Kingdom. Control of surface subsidence and underground strata was an integral part of his work.

Mr. Dorling joined *World Coal* in 1978 and covers industry activities in the United States, Canada, Latin America, and Australasia.

He holds a First Class Certificate of Competency (manager's certificate) from the United Kingdom Mining Qualifications Board and is studying for registration as a Professional Engineer in the state of California. He is a member of the British Institution of Mining Engineers.

Strip and Open Pit Mining

Stability in Open Pit and Strip Mining Coal Projects

1

C. O. Brawner, Department of Mineral Engineering,
University of British Columbia,
Vancouver, British Columbia, Canada

Introduction

The cost of oil has increased dramatically over the last few years. As a result, electric power generated from oil and gas has placed other alternative means of power development in greater cost effective competitiveness. Accordingly, increasing emphasis is now being placed on the use of coal to generate electricity. Since many of the free-world countries have extensive coal deposits, major economic pressures are developing to exploit these deposits for the generation of power. These include a number of large lignite deposits which will be developed by open pit mining, and bituminous and anthracite deposits which are being developed by strip mining methods.

To minimize the cost of power and maximize the economy of mining it is essential that mining progresses with control of ground stability. Engineering and design must include the evaluation of the stability of slopes for open pit projects, the stability of the highwall in strip mine projects, the control of pit bottom heave which could develop where high water pressures exist below the base of excavations, and the stability of spoil piles.

This chapter provides an overview of factors that must be considered to maintain the stability and illustrates these factors with a number of typical examples.

Factors that Influence Stability

A great many factors influence stability in coal mining projects (Brawner, 1968; Brawner, 1974; Hoek, 1977). These include the following eight factors:

1. Geology.
2. Shear strength of the materials in the slopes and the base of the pit.
3. Topography.
4. Climate.
5. Hydrology.

6. Mining method and the rate of mining.
7. Need for, and program of blasting.
8. Mine economies.

The more important of the geotechnical factors that influence stability are reviewed in the following text.

Geology

An assessment of the geology must take into account the extent and characteristics of any overburden which may overlay the deposit, and the influence on stability of the rock which may lie above it. The assessment must also consider the characteristics of the coal and the influence of the geologic conditions that lie beneath the coal.

It has been common in the past to drill down to the coal before good samples are taken. It must be emphasized that, since all materials can be involved in potential failure, it is absolutely essential that in a number of the boreholes as complete a core recovery as possible should be obtained from the surface to a depth of 6.1 to 15.2 meters (20 to 50 feet) below the base of the coal deposit.

In order to evaluate the stability of any overburden, the principles of soil mechanics should be used. Extensive experience is available to obtain and test good samples of overburden soil, and to assess the shear strength, permeability, and influence of groundwater to determine stable slope angles in overburden materials. Particular attention should be paid to the potential existence of bentonitic clay layers. Where overburden material rests directly on the coal, the contact zone between these two materials will frequently contain the weakest material within the system. Figure 1.1 shows a failure along the silty overburden-coal contact which almost hit a bucket wheel excavator.

Figure 1.1. Failure of silt overburden soil along silt-coal contact.

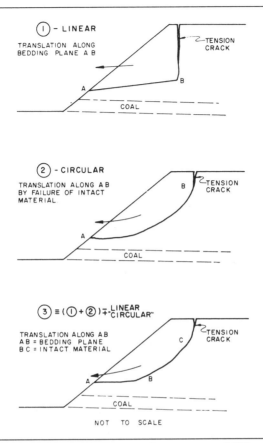

Figure 1.2. Potential highwall failure modes (Brawner, 1977).

The geology of the bedrock is extremely important. The major factors that influence stability will be the existence of discontinuities such as bedding, cross-bedding, contact zones, jointing, faulting, steeply dipping beds, the existence of fire holes, etc. Where the rock is approximately horizontally bedded, stress relief due to the excavation of material can result in the opening of steeply dipping joints or develop tension cracks behind the face. If these tension cracks become filled or partially filled with water, high hydrostatic pressures can develop which can result in block failures in the highwall. In many deposits there will be discrete cross-bedding. Where these dip into the excavation they provide potential surfaces of weakness along which failure can occur.

Figure 1.2 illustrates types of failures that may be influenced by primary or cross-bedding.

Extensive experience has shown that weak contact zone material frequently exists in the rock directly at the top and base of the coal deposit in which block-type failures may occur.

Figure 1.3. Block failure along the mudstone-coal contact which partially undermined a dragline.

Figure 1.4. Failure of spoil pile into strip excavation along weak shale layer below coal, which greatly reduced available spoil volume for the next strip cut.

Figure 1.3 shows a block failure in the highwall which occurred within the contact zone of weak mudstone lying on top of the coal seam. This type of instability can extend back far enough from the highwall face to undercut and endanger drilling or excavating equipment.

Where weak rock exists at the base of the coal, this material can lead to potential failures under spoil piles which are placed on it after the coal has been excavated. Figure 1.4 shows a spoil pile that has slid across the base of the excavation greatly reducing the available spoil volume for the next strip cut. In some instances failure occurs before all of the coal has been excavated from the base of the excavation, which results in double handling of spoil or loss of coal.

In some instances, interbedded clay layers may exist within the coal measures, such as in the open pit mine at Centralia, Washington, discussed in the conference. It is essential that the existence of such layers be predetermined, their shear characteristics be evaluated, and their influence on stability be assessed prior to development or extension of the project.

The existence of jointing and faulting within the deposit should be determined prior to the design of the mine. Figure 1.5 shows a steeply dipping, through-going joint which has been exposed by an open pit excavation and along which movement has occurred. These joints can be expected to open up as a result of stress relief, or if they dip steeper than the angle of friction of the joint and are undercut, failure will likely occur along the joint.

Figure 1.5. Failure along steeply dipping, through-going joint, which damaged the bucket wheel excavator.

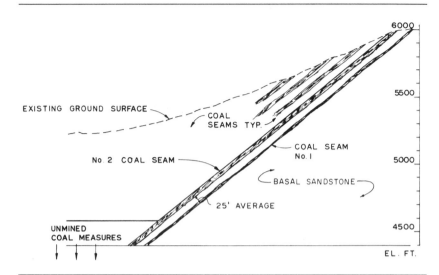

Figure 1.6. Steeply dipping coal seams in Rocky Mountains. Shear strength along seams is critical to stability.

In mountainous areas, coal deposits are frequently encountered where the dip of the deposit and the bedding may be relatively steep. Figure 1.6 is a cross section through a series of coal seams which dip at 39°. This steep dip will obviously have a major influence on slope stability and on the method of mining that is selected. For this example it is absolutely essential that the shear characteristics along the bedding and along the coal-rock contacts be determined and assessed. Where the shear strength along the contact is reasonably low, it is likely that the excavation will have to be developed along the angle of the bedding. It will also be very likely that one will not be able to use the standard bench construction that is so common in normal mining. If steeply dipping smooth slopes are planned in northern climates where snow occurs, it is essential that the control of potential snow avalanche slides, particularly in the spring time, be evaluated. In order to develop stability it may be necessary to install special drainage within the slope to reduce water pressures behind the steeply dipping contacts (Brawner, 1968). One of the most important factors to determine and assess is whether weak layers exist below the lowermost layer of coal that will be excavated.

In a number of coal mining projects, underground fires have developed in past geologic history, and these fires have led to shrinkage within the coal deposit. Cracks that have subsequently developed usually become infilled with clay or silty materials. In other instances, there may have been created a weak ash zone. If the dip of burn hole contact zones is out of the slope, there is a potential of block-type failures that can occur into the excavation. Figure 1.7 shows a typical burn hole failure.

Important factors that should be determined are the thickness of the coal horizons, and the topography of the surface and base of the coal deposit.

Figure 1.7. Burn hole which failed when undercut.

Shear strength

In order to assess stability in quantitative terms, it is necessary to determine the shear strength parameters of the various materials that could be involved in failure. At existing mines, it is possible to obtain undisturbed samples by excavating block samples of the material from the slopes and testing them to determine the shear strength characteristics.

Where new projects are being considered the most important factor is to obtain good samples of the material from the ground surface down to below the base of the deposit. The best samples can be obtained by excavating test cuts from which block samples can then be taken. If this is not possible, careful drilling procedures using triple tube core barrels with as large a diameter as economically practical are strongly recommended. Since the dip and strike of the contact layers are important, it is necessary that the discontinuities in the core be oriented. As soon as the core is recovered and exposed it should be photographed in color. This provides a permanent record of the drill core. Where weak materials are recognized in the core, samples of these weak materials should be subjected to strength tests and these tests should be performed as quickly as possible. Sedimentary materials, particularly coal, have the notorious habit of breaking down very quickly on exposure to air. It is recommended that at least one out of every ten boreholes drilled on the coal deposit be sampled continuously from the surface of the ground to well below the base of the coal deposit. It must be emphasized that it is the core which is not recovered that is usually the culprit in stability. Therefore a major effort should be made to obtain 100 percent core recovery.

Where it is obvious that the geologic discontinuities can have a major influence on stability, it is essential that the shear strength along these discontinuities, particularly bedding, joints, faults, and contacts, be determined. The most effective test for this is the direct shear test. This should be performed on samples in excess of 152 millimeters (6 inches) in width. When the bedding is steeply dipping

Figure 1.8. In situ shear test in steeply dipping coal bed.

in the field, it may be necessary to determine shear strength along these discontinuities in situ. Figure 1.8 shows a typical direct shear test being performed in an adit in a coal seam which dips at 39°. If there is any indication of slickensides or past movement on any of these discontinuities, it is essential that both the peak strength and the residual strength be determined for use in the analysis. If the assessment of potential instability indicates that failure could occur across or through intact material, then the triaxial test will be necessary to assess shear strength parameters. A typical large scale triaxial cell is shown in Figure 1.9.

If it is likely that failure could occur along discontinuities, and these discontinuities have an irregular surface, the surface roughness of the contact materials must be assessed, as it can have a significant influence on the effective angle of friction along any potential zone of failure (Patton, 1966). It is particularly important that the direction of the testing be oriented in the direction of failure that may occur in the field.

Experience has shown that many rock slopes will remain stable for some time after excavation prior to failure. This indicates that time-dependency of slope movement can be extremely important in sequencing mining operations, particularly in strip mining projects. Special laboratory tests or test box cuts in the field can be performed to assess this possibility. If the rock has very low permeability, and excavation is rapid, negative pore pressures develop which, for a period of time, increase the shear strength until the groundwater conditions reach equilibrium. For short periods of time, this will allow steeper slopes than normal to be utilized in the mining operation.

Figure 1.9. Large size high-pressure triaxial cell.

Groundwater

The existence of groundwater influences stability in a number of ways:

1. It reduces the shear strength by reducing the normal load on the potential shear surface due to buoyancy.
2. Seepage pressures may exist which act in the direction of potential movement to impose a friction head loss.
3. Tension cracks or existing joints may open up as a result of stress relief, which allow water to build up hydrostatic pressure at the back of potential failure blocks.
4. Where blasting is involved the seismic acceleration forces create a hydrodynamic shock in the water system which in turn creates sudden, very high water pressures. These high pressures can be sufficient to liquefy uncemented silty or sand layers within the slope, or repetitive hydrodynamic shock can induce failure along discontinuities in the rock.

As part of any geotechnical investigation it is advisable to obtain as much information from each borehole as possible. After the core has been recovered, it is usually feasible to perform water packer permeability tests throughout the borehole to determine the average permeability, or the rate at which water will flow through the rock at various horizons. Following this program it is recommended that a number of piezometers be installed in each borehole to measure the water pressure at different elevations. Because of the normal occurrence of different rock types with depth, it is unlikely that a uniform hydrostatic head exists from the groundwater surface to the base of the mining excavation.

Experience has shown that bituminous and anthracite coal seams may often be the most pervious of the rock types in the slope. As a result of this, these seams

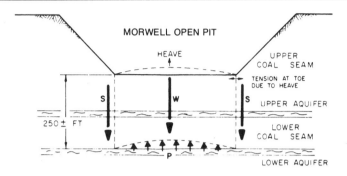

P   WATER PRESSURE IN LOWER AQUIFER
W   WEIGHT OF STRATA
S   SHEAR RESISTANCE AGAINST UPLIFT

Figure 1.10. Excess water pressure will cause pit bottom heave if the pressure exceeds the weight of the coal and shear strength of the resisting block.

may be used as drainage layers. Lignite coals, on the other hand, frequently have very low permeability and behave differently in a groundwater regime. In any drilling program it is strongly recommended that a number of the boreholes be carried down well below the base of the coal, in order to determine the potential existence of weak layers which may be conducive to instability, or the existence of high water pressures which may lead to heaving of the base of the coal excavation. In any situation where the water pressure exceeds the weight of the coal remaining in the pit, there is a significant potential for heave, as can be seen in Figure 1.10.

There are a number of techniques that can be used to reduce water pressures if they are found to be detrimental to stability. If the major problem is high water pressures within the highwall slope or within the open pit slope, the use of wells within or around the pit to lower the water table can be effective as shown in Figure 1.11. In other instances the use of horizontal drains, which may be up to 122 to 131 meters (400 or 500 feet) long to lower the water table, can be utilized. This last procedure normally is the least expensive of the techniques available. As a rule of thumb, the drains should not be longer than about one-half of the height of the slope, and to reduce the cost of the system a number of horizontal drains can be installed for one setup. If the holes will not cave, no casing is required. If caving may occur plastic perforated pipe is recommended. The inner portion of the drain should have the perforations down to let the water in, and the outer half of the drain should have the perforations up or no perforations at all. Figure 1.12 shows a typical horizontal drain.

If there are a large number of discontinuities within the highwall it may be desirable to construct vertical drain wells 30.5 to 46 centimeters (12 to 18 inches) in diameter, backfill these with a well-graded filter material, and then construct horizontal drains from the toe of the pit to intercept these wells. This results in a gravity drainage system in the slope.

Figure 1.11. In-pit pumping to control pit bottom heave.

Figure 1.12. Perforated horizontal drain to improve slope stability.

Where there is a potential that the base of the excavation may heave due to underlying water pressures, pressure relief wells, commonly used in civil engineering to stabilize the toe area of earth dams, are recommended. These comprise holes 30.5 to 38 centimeters (12 to 15 inches) in diameter drilled into the underlying aquifer. Assessment of the water pressure, the weight of the rock above the aquifer, the depth to the aquifer, and width of excavation will determine whether it is necessary to install pumps, or whether the relief wells may simply be allowed to flow freely.

Blasting

Blasting can cause extensive damage for some considerable distance into the final slopes or within the highwall (Brawner, 1968). The damage may be due to the seismic acceleration forces generated by the blast, the expansion of gases along bedding or joints, or it may result from water penetrating cracks after the cracks have been opened by blasting. Figure 1.13 shows a mine site where blasting has induced large cracks behind the highwall. It is important that blasting be delayed so that very large charges are not utilized to break the rock near final slopes.

Figure 1.13. Overblasting and resulting cracking beyond highwall.

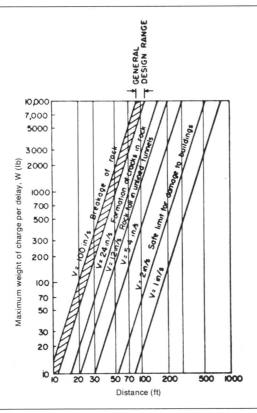

Figure 1.14. Variation of maximum particle velocity with distance and weight of charge per delay.

Figure 1.14 illustrates the relationship between distance from the center of gravity of the blast, the pounds per delay of each charge, and the particle velocity as a measure of the potential damage to the rock.

The influence of blasting on stability can be reduced by minimizing the amount of explosive per delay and yet maintaining large total blasts to provide the productivity and the fragmentation that is required. In strip operations, in order to develop a better highwall slope, the use of angle drilling along the line at which the highwall is proposed to be cut will usually be effective. There may be instances, however, where this may be considered impractical. Close to the face of the highwall, it is recommended that a line drilling concept with closer spaced holes and lighter charges be utilized.

Fire potential

Figure 1.15 illustrates a recent major fire that developed in an open pit mine in Australia. In this instance it was believed that sparks from vehicles caused the fire. A major problem that can develop as a result of fires is a need for massive

Figure 1.15. Fire in an open pit coal mine which required massive volumes of water to control.

amounts of water which must be sprayed on the slopes to put the fire out. This large amount of water can raise groundwater levels conducive to instability. If dewatering wells exist below the pit floor to control heave, fire can burn out power lines and cut off the dewatering system. In this example, about two-thirds of the pumps were shut down due to the fire and greatly reduced the stability of the bottom of the pit against heave.

In order to minimize the fire potential on permanent slopes they can often be covered with overburden.

Evaluation of Stability

Following the evaluation of the geologic conditions, the groundwater conditions, influence of blasting, the assessment of the mining procedures and costs, the mechanics of potential stability, and the geometry of potential failure can be assessed so that an evaluation of safety, usually in the form of a safety factor, can be determined. It is essential that the shear strength parameters and groundwater conditions in particular be very carefully assessed.

In sedimentary conditions the geometric configurations of potential failure will likely be relatively few, and so the use of probability theory, as is being applied to random structural geologic conditions in some open pit mining projects, is not necessary.

If any failures should occur it is strongly recommended that the field geology, groundwater conditions and geometry be determined. With this information and the knowledge that the safety factor can be accurately assumed to be 1.0 at failure, back analysis can be used to determine the shear parameters that existed at the time of the failure. This type of analysis can be used to assess quantitatively the influence of any stabilization program, and to reassess and redesign pit slope angles where site conditions are similar.

No matter how much care is taken to obtain good samples, to perform good shear tests, and to obtain good geologic data, the determination of any safety factor will be subject to some error. It is, therefore, absolutely essential than an observational record of movements and potential movements be maintained throughout the operation of any mine. If any cracks or movement develop it is recommended that the cumulative movement be plotted against time. If any acceleration in this movement develops it is obvious that stability is reducing. A continued plot of this type of movement will enable one to decide whether mining can be safely continued, whether it should be discontinued, or whether stabilization measures should be instituted (Brawner, 1977).

In large open pit operations one of the most efficient methods of measuring movement is to use electronic distance measuring devices. This type of equipment, such as shown in Figure 1.16, is capable of measuring distances up to 3.2 to 4.8 kilometers (2 to 3 miles) by sighting on prisms with an accuracy of about plus

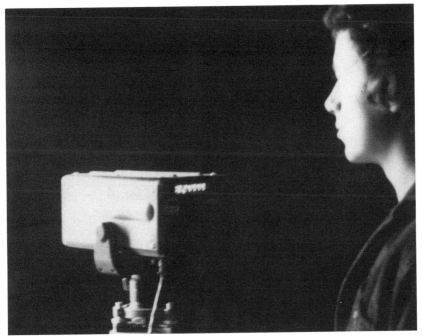

Figure 1.16. Direct distance measuring instrument used to monitor slope movements can measure distances of 3.2 to 4.8 kilometers (2 to 3 miles).

or minus 5 millimeters (0.2 inch). This is sufficiently accurate for most surface mine monitoring requirements.

It is also recommended that a number of piezometers be installed in and around the pit or the strip operation. If the groundwater conditions are found to be rising this also is an indication that the stability is reducing.

It will frequently be desirable to monitor the influence of blasting. For this purpose a number of mines are purchasing and installing seismographs to monitor the particle acceleration forces due to blasting. If these forces are indicated to be excessive, modification of the blasting procedures can be developed.

References

Brawner, C. O., 1968, "The Three Major Problems in Rock Slope Stability in Canada." Second International Conference on Surface Mining, Minneapolis.

Brawner, C. O., 1974, "Rock Mechanics in Open Pit Mining." Supplementary Report, 3rd International Rock Mechanics Conference, Denver.

Hoek, E. and Bray, J. W., 1977, "Rock Slope Engineering." Revised second edition, Institution of Mining and Metallurgy, London.

Brawner, C. O., 1968, "The Influence and Control of Groundwater in Open Pit Mining." Fifth Canadian Symposium on Rock Mechanics, Toronto.

Patton, F. D., 1966, "Multiple Modes of Shear Failure in Rock." Proceedings, First International Conference on Rock Mechanics.

Brawner, C. O., 1977, "Open Pit Slope Stability Around the World." Journal of Canadian Petroleum Technology, Montreal.

The Effects of Geologic Structures on Slope Stability at the Centralia Coal Mine

2

R. A. Paul, Engineering Geologist,
Washington Irrigation and Development Company,
Centralia, Washington, USA

Introduction

The geology of the Pacific Northwest presents what may be some unique stability problems to those anticipating developing the coal seams of the region. The purpose of this chapter is to alert potential developers to the geologic features and associated stability problems found at the Centralia coal mine as typical of those which might be encountered elsewhere in the region.

General Geologic Setting

The Centralia coal mine is located in western Washington midway between Seattle, Washington, and Portland, Oregon. It lies in the western foothills of the Cascade Mountains and along the eastern edge of the Willamette-Puget lowland.

The coal reserves occur within the Skookumchuck Formation, a geologically young sequence of shallow marine, estuarine, and lacustrine sediments. This sequence has been warped into a series of northwest to southeast trending folds and broken by a series of major parallel faults. Several series of secondary faults further break and disorient the rock sequence. After subsequent erosion had developed considerable relief on the Skookumchuck rocks, the area was blanketed by glacial outwash debris presently known as the Logan Hill Formation. Erosional remnants of these materials cap the uplands of the area. The present valleys are filled with thick deposits of peat and soft sediments.

Geologic Setting of the Centralia Coal Mine

The Centralia coal mine is located on the western flank of a major downfold or syncline in the Skookumchuck rocks. Mining commenced at the outcrop of the coal-producing seams and has progressed eastward down the flank of the structure. Although eastward dips range up to 22°, the maximum dip experienced in the mining to date is 11°.

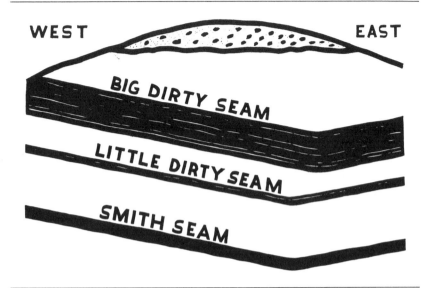

Figure 2.1. Generalized cross section at the Centralia coal mine.

As shown in Figure 2.1, three coal seams, the Big Dirty, Little Dirty, and Smith are mined at Centralia. The seams range in thickness from a maximum of 15.2 meters (50 feet) for the Big Dirty to a minimum of 0.61 meter (2 feet) for the Little Dirty Seam. The thickness of each seam varies considerably over the property, apparently dependent on the location within the coal basin.

The Logan Hill Formation is composed of approximately equal proportions of silts and clays, sands and gravels. These materials are deeply weathered and poorly indurated. The ancient erosion surface or unconformity separating the Logan Hill and underlying Skookumchuck Formations exhibits considerable relief.

The valley sediments are typically very fine-grained, weak, plastic, and saturated. A shear strength of 976 kilograms per square meter (200 pounds per square foot), bearing capacity of 641 kilograms per cubic meter (40 pounds per cubic foot), and water content of 500 percent are typical characteristics.

The Big Dirty and Little Dirty Seams derive their names from the numerous partings which can comprise up to 30 percent of the seams. The partings represent layers of volcanic ash deposited contemporaneously with the organic matter comprising the coal. These ash layers have altered to soft, weak, expensive, plastic bentonitic clays. The primary mineral is montmorillonite.

Slope Stability Problems

The effect of partings within the coal seams
The partings are composed of the weakest materials within the entire rock sequence, and probably represent the single most important factor affecting

slope stability at Centralia. When highwall failures occur, invariably the slip planes are within the parting materials. Slip planes occur in partings apparently irrespective of their thickness. With movement, the parting materials are extruded from the coal seam and can be found in large sheets at the toe of the slide.

Highwall failures are generally of the block-glide type. Typically, portions of a coal seam and the overlying rock materials move laterally on one or more partings into the open excavation. The slide blocks are invariably wholly or partially bounded by fault planes. The fault represents a plane of weakness across which there is little, if any, cohesion. This type of movement occurs along very shallow components of dip of only a few degrees.

Occasionally block-glide type slides develop on bedding planes. Invariably these planes are clay-coated and lubricated by groundwater. To date, slope failures of this type have only occurred in areas where the dips exceed 12°.

The effect of dip & strike
From a slope stability standpoint, one critical factor in surface mining pitching seams is pit orientation in relation to the rock structure. At Centralia the strips were initially orientated parallel to the strike of the rocks, as shown in Figure 2.2, providing for a maximum dip into the highwall for highwall stability. This requires that the spoils rest on the sloping pit bottom updip of the open excavation. The result of this pit configuration was massive spoil slides involving several million cubic meters (or yards) of material. Stable strips have been developed parallel to the strike in areas of flatter dips, in the order of 8°, by benching the pit floor.

Subsequent to the spoil slides at Centralia, consideration was given to orientating the strips parallel to the dip of the rock. In this configuration, spoils would

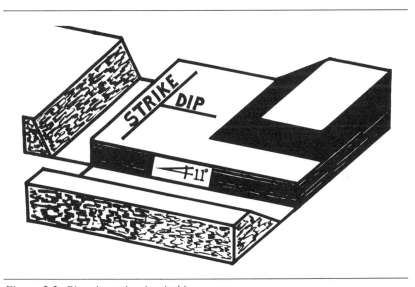

Figure 2.2. Pit orientation in pitching seams.

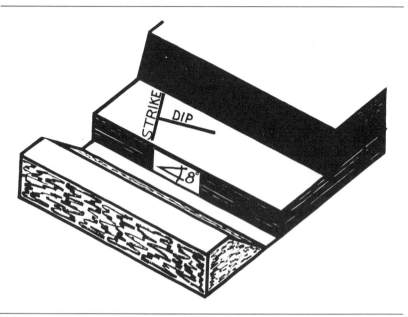

Figure 2.3. Compromise pit orientation.

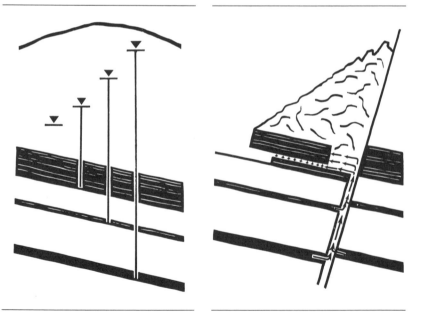

Figure 2.4. Hydrostatic heads in coal seams and overburden.

Figure 2.5. Effect of fault-transmitted hydrostatic heads.

not lie directly updip of the open excavation as shown in Figure 2.2. However, the coal hauling equipment could not negotiate the resulting 20 percent grade of the pit bottom.

Orientating the pit on an apparent dip of 14 percent, the maximum grade negotiable by the coal haulers, provided sufficient dip into the highwall for stability and a relatively stable spoil situation. This orientation is shown in Figure 2.3.

Effect of groundwater

Both the horizontal and vertical permeability within the Skookumchuck Formation is generally very low. The notable exceptions are the coal seams which act as artesian aquifers. The lowest seams have the highest hydrostatic head as shown in Figure 2.4.

Faults can transmit this head vertically to the vicinity of the open pit excavation. If the strip is orientated roughly parallel to the fault, block-glide slides develop as the hydrostatic head provides an uplift and driving force on the highwall block between the fault and the pit face. Movements of this type have been experienced up dip slopes of several degrees as illustrated in Figure 2.5.

Effect of local changes in dip

Although care is taken to orient a pit in relation to the geologic structure to achieve optimum highwall and spoil stability, there are a number of factors which can locally change the orientation of the rock structures and, therefore, alter the stability conditions. One such factor shown in Figure 2.6 is dragfolding, or deformation associated with faulting. This phenomenon is common in weak, poorly indurated sediments such as those of the Skookumchuck Formation. Dragfolding often produces components of dip into or parallel to the open excavation, creating a potential for landslides. A hazard is also created for excavating machines working on the deformed pit bottom.

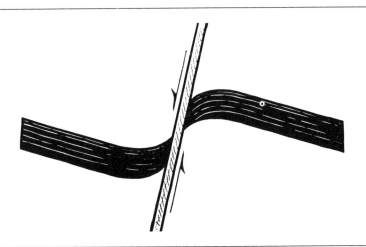

Figure 2.6. Typical dragfolding associated with faulting.

The rock sequence in the Centralia area has, at some time, been subjected to substantial compressional forces. The result is a large number of complicated and varied compressional features, developed on a horizontal slip plane within the mining sequence. These features include various types of folds and flexures in combination with reverse faults, upthrusts, and overthrusts. Each of these features produces crushed, weakened rock conditions; permeable zones with hydrated, weakened partings, and components of dip into or parallel to the open excavation. Slope stability is, of course, greatly reduced in the area of these structures. In addition, stability is reduced over a wide area by the presence of the horizontal slip plane.

Effect of valley sediments

If the open pit excavation encroaches upon the sediment-filled valleys, these weak materials tend to flow toward the excavation. In some instances this situation is worsened by the combination of the mining activity and the quick condition of the sediments. Spoiling on these weak materials causes them to flow, further reducing the inherent strength and creating conditions that are difficult to stabilize.

Conclusion

The geologic conditions along the western flanks of the Cascade Mountains are complicated and varied. Their combination may produce stability conditions somewhat unusual in the coal mining industry. Those anticipating developing the regions' coal reserves should be aware that commonly used plans, methods, and costs may not be directly applicable to this area.

Assessment of the Stability of Open Pit Mine Slopes in the Rhenish Brown Coal District

3

Karl-Josef Pierschke, Head of the Geomechanical Department, Rheinische Braunkohlenwerke AG, Rheinbraun-Consulting GmbH, Cologne, Federal Republic of Germany

Introduction

The Rhenish brown coal district of West Germany shown in Figure 3.1 is situated west of Cologne and covers an area of approximately 2,500 square kilometers (965 square miles). The deposit is the largest in Europe and contains reserves of 55 billion tons of brown coal. At the present time, five open pit mines are operated by the Rheinische Braunkohlenwerke AG which contain approximately 3 billion tons of brown coal. Considering present mining technology, about 35 billion tons of coal are thought to be economically minable. Extensive mining in the southern open pits has resulted in the relocation of operations to the north and west. The depth of mining has increased to 320 meters (1,050 feet) in the Fortuna-Garsdorf open pit, and to 500 meters (1,641 feet) in the Hambach open pit, because of the geology of the deposit. The Hambach mine is expected to be opened in the autumn of 1978.

An assessment of the stability of the open pit mine slopes e.g., at a depth of 270 meters (886 feet) and inclined at 1 in 2.5 (see Figure 3.2), is of considerable importance. The aim of such an assessment based on investigations is to establish slopes stable enough at an optimum inclination for maximum exploitation of the deposit.

The safety and economic interests of a favorable overburden to coal ratio must be considered. The criteria resulting from geological, hydrological, and operating conditions as well as stability investigations upon which the assessment of stability by the Rheinische Braunkohlenwerke and Rheinbraun-Consulting is based, are shown below.

Parameters Influencing Stability

The stability of a slope depends on the parameters given in the list that appears on the following text page.

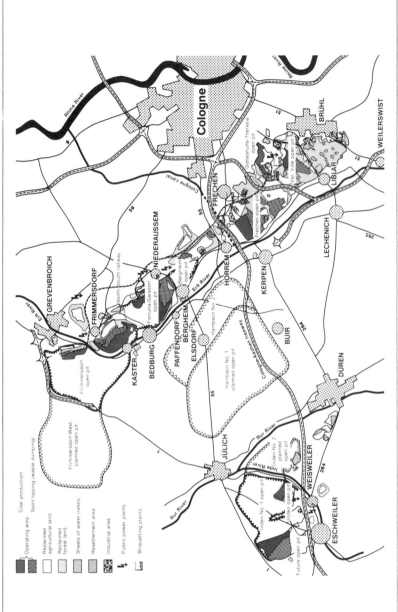

Figure 3.1. Map of the Rhenish brown coal district.

Figure 3.2. Rheinbraun open pit mine, Fortuna.

1. Structure and characteristics of the overburden.
2. Effect of water.
3. Slope height.
4. Slope angle.
5. Slope length.
6. Period during which slopes are kept open (standing time).

Structure and characteristics of overburden

The geology of the overburden in the Rhenish brown coal district is marked by an alternating sequence of gravels, sands, clay, and silts with different strength characteristics (Figure 3.3). These individual layers were formed during Tertiary and Quaternary times and have been disturbed from their original flat disposition by tectonic movements. Blocks were formed in which layers are exposed between faults at different angles of inclination. The inclination of layers up to a maximum of 20°, and the angle of dislocation (according to the slope position) can have positive effects on stability for opposite angle, and negative effects for equal slope angle.

Besides these parameters of influence and the different stability characteristics of the individual strata, there are also weak zones in the overburden exhibiting low strength. Weak zones are really dislocations which have weakened the layers of overburden because they were subjected to movements. These faults are marked by so-called filling-in faults containing different materials of low strength. In addition layers with reduced strength exist up to 100 meters (328 feet) in front of and behind the faults.

Effect of water

When considering hydrological conditions it can be assumed that, as a rule, the groundwater level of the strata lies below the lowest open pit bottom. Groundwater, therefore, does not influence stability. Studying the geological conditions and the economics of dewatering open pit mines, total dewatering of the overburden is not possible so that in certain areas of the strata there is residual groundwater at various heights.

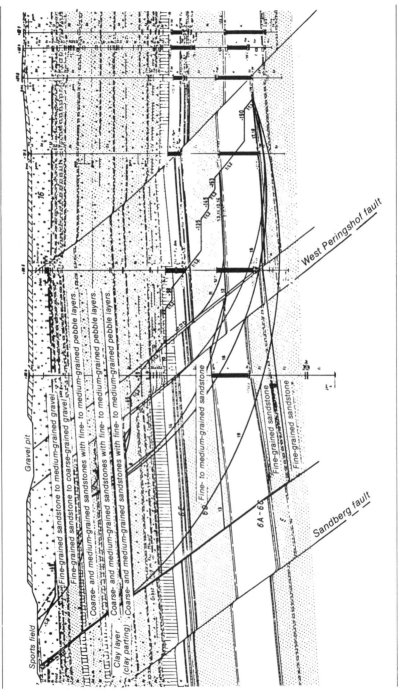

Figure 3.3. Geological cross section of the Rhenish brown coal deposit.

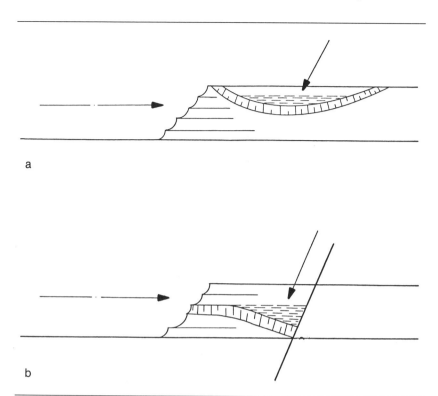

Figure 3.4. Residual groundwater is found (a) *in synclines which cannot be dewatered and* (b) *in the vicinity of faults.*

Quantities of residual groundwater are frequently found in the vicinity of faults. Wells are not sunk because they would be destroyed by strata movements. Residual groundwater may also be found in synclines which cannot be dewatered because of their indeterminate positions (Figure 3.4).

Slope geometry
Slope height and angle influence stability i.e., the slope geometry must be adjusted to the geological, geomechanical, and hydrological conditions. Slope height and angle are limited by the geometry of bucket wheel excavators and spreaders used in the open pits.

Slope length
The progression of the production slope, and the dumping slope which follows it, is conditioned by open pit mine technology, so that the mine slopes are limited in their length (Figure 3.5). The final slope is supported by the production and dumping slopes. The length L of the final slope is mainly controlled by the equipment used in the mines. The less the distance L, the greater is the stability of the final slope.

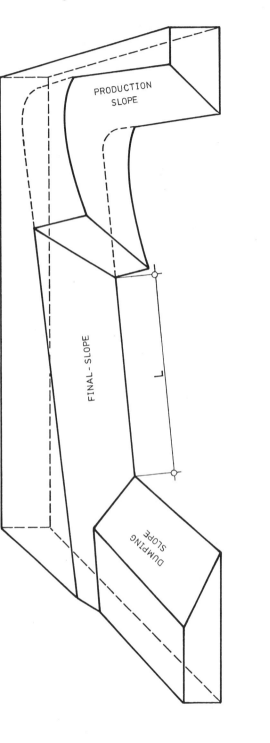

Figure 3.5. The final slope is supported by the production and dumping slopes.

Standing time

The strata in the slope areas are relieved by mining, and strata movements are produced which are noticeable as vertical and horizontal displacements. Connected with this time-dependent decompression is a decrease in the rock strength. Because stability is largely dependent on rock strength, it is therefore influenced by the standing time (the period during which the slope is kept open).

Stability Investigation

The actual basis for assessment of the stability of a slope is the stability investigation. A stability investigation is divided into three parts:

1. Geological investigations.
2. Geomechanical investigations.
3. Stability calculations.

Geological investigations

Geological investigations include a description of the geology and the expected tectonics of the overburden and base rocks, and are shown in the form of geological sections. In the case of dumping slopes, the overburden and base rocks are described. In addition, the present and future hydrological conditions in the slope areas are indicated.

Geomechanical investigations

The properties of the overburden material expected to be dumped are determined by geomechanical investigations in the field and in the laboratory. Special importance is attached to the reference values describing the strength. The strength characteristics are determined in Rheinische Braunkohlenwerke's own laboratory by direct shear box tests (Casagrande) and triaxial shear tests. The test equipment is adjusted to the preconsolidation pressure at up to a depth of 600 meters (1,969 feet). The geomechanical core values are statistically evaluated and stored in the data bank "Geomechanik" (Winter, 1977).

Stability calculations

Different procedures are applied for stability calculations according to the slope type and scope. It must be pointed out that it is necessary to make these stability calculations at short notice depending on the situation of operations in the open pits.

Slice methods such as those introduced by Janbu, Neuber, the Bureau of Reclamation, and Bishop are used for making the calculations (Janbu, 1954; Neuber, 1968; Düro, 1968).

Janbu

$$\eta = \frac{1}{\Sigma \Delta G \cdot \sin a} \cdot \Sigma \frac{c \cdot \Delta X + (\Delta G - p \cdot \Delta X) \cdot \tan\varphi}{\cos a + \dfrac{\sin a \cdot \tan\varphi}{\eta}} \tag{1}$$

Neuber for $\epsilon = 0$

$$\eta = \frac{(\text{Ho} - \text{Hn}) + \Sigma \dfrac{\Delta G \cdot \tan\varphi \cdot \cos a + (c - p \cdot \tan\varphi) \cdot \dfrac{\Delta X}{\cos a}}{-\tan\varphi \cdot \sin a - \eta \cdot \cos a}}{\Sigma \dfrac{\Delta G \cdot \sin a}{-\tan\varphi \cdot \sin a - \eta \cdot \cos a}} \qquad (2)$$

Bureau of Reclamation

$$\eta = \frac{\Sigma \Delta G \cdot \tan\varphi \cdot \cos a + (c - p \cdot \tan\varphi) \cdot \dfrac{\Delta X}{\cos a}}{\Sigma \Delta G \cdot \sin a} \qquad (3)$$

Bishop

$$\eta = \frac{\Sigma \left[\dfrac{c \cdot \Delta X + \tan\varphi \cdot (\Delta G - p \cdot \Delta X)}{\eta \cdot \cos a + \tan\varphi \cdot \sin a} \right] \cdot \eta}{\Sigma \Delta G \cdot \sin a} \qquad (4)$$

At first the calculation methods are applied to investigate the open pit boundary slopes, final slopes, and the residual pit slopes. In the case of final slopes up to a depth of 500 meters (1,641 feet) with various geological strata and hydrological conditions, a manual calculation would take several weeks to complete. Therefore stability calculations are made with electronic data processing systems.

The BETAJA program developed by Rheinische Braunkohlenwerke does not only consist of the calculation formulas, but also comprises the entire data preparation for carrying out and controlling the calculation (Dermietzel, Metzmacher, Pierschke, 1977). The electronic data processing system is a UNIVAC 1110 (1 × 1).

For making stability calculations the following data are recorded (Figure 3.6):

1. Geological strata.
2. Faults.
3. Groundwater level.
4. Slope system.
5. Slip planes.

Geological strata and faults are registered by a Gradicon digitizer and passed onto a data carrier. The boundaries of the geology are selected starting from the determined open pit upper boundary. Depending on the calculated stability coefficient η a modification of the slope geometry is possible without the acquisition of new geological data. In this case the geomechanical reference values are assigned to the geological layers. A possible slope system is listed on the section, digitalized, and passed onto the data carrier.

Considering the weak zones in the overburden, slip planes in any form are kinematically possible and their coordinates are punched into the data carrier.

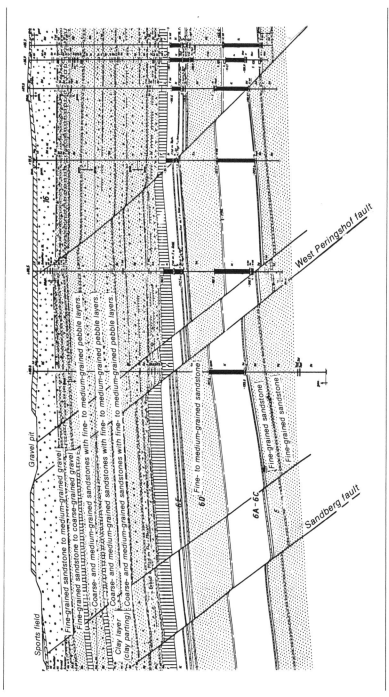

Figure 3.6. Geological cross section of the Rhenish brown coal deposit showing the outline of an open pit slope system and major slip planes.

SLIP SURFACE	JANBU	NEUBER	B.O.R.	BISHOP
1	1,08	1,08	1,05	1,08
2	1,33	1,36	1,36	1,33
3	,96	,89	,88	,96
4	1,15	1,13	1,14	1,15
5	2,38	1,45	2,12	2,38
6	1,68	1,34	1,59	1,68
7	1,86	1,16	1,58	1,86
8	1,93	1,53	1,65	1,93
9	1,58	1,18	1,45	1,58
10	1,37	,97	1,18	1,37
11	1,60	1,39	1,52	1,60
12	2,10	1,36	1,84	2,10
13	1,80	1,17	1,54	1,80
14	1,48	1,41	1,43	1,48
15	1,51	1,40	1,50	1,51
16	1,53	1,48	1,51	1,53
17	1,45	,99	1,21	1,45
18	1,82	1,12	1,49	1,82
19	1,26	1,16	1,22	1,26

Figure 3.7. Results of a typical stability calculation (η-value).

By means of the designed slope polygon points the BETAJA program selects the digitalized calculation and calculates the intersection points at the top of the slope. Then it divides the block of failures in the slope into slices according to the criteria stated in subprograms, and determines the specific slice values. These are substituted in the formulas and the stability coefficients are calculated. The stability coefficient η of each slip plane to be investigated is calculated using the formulas of Janbu, Neuber, the Bureau of Reclamation, and Bishop, and the results are tabulated as shown in Figure 3.7.

The time required by the electronic calculator to make calculations from the geological section takes on average one minute. For the entire processing duty i.e., the acquisition of data including the determination of the slip planes, necessary controls and corrections, and the calculations, takes about two to three days for one section.

Assessment of Stability

After the parameters influencing the stability of a slope and the calculations have been made, the assessment of stability is carried out. The result of the stability calculation for each slip plane studied is given as the stability coefficient η. A slope is supposed to be stable at $\eta = 1.0$. The view of staff at Rheinische Braunkohlenwerke AG is that the size of the stability coefficient for open pit slopes should generally not be given a numerical limit value e.g., 1.2, 1.5. The η-value necessary for the stability of a slope i.e., a sufficiently stable slope, must be determined and proven taking into account the following:

1. The important factors and interests to be protected including:
   a. Personnel safety.
   b. Security of the mine operation.
   c. Security of nearby inhabited land.
   d. Security of public traffic.
2. The reliability of geological maps and sections.
3. The geomechanical coefficients indicated by calculation.
4. The optimum extraction of the deposit.
5. The period during which the slopes are kept open.
6. The measures for conservation or increase in the stability e.g., special dewatering measures immediately following dumping.

Because dumping immediately follows mining, the open pit slope is limited in its length and standing time. It is possible, therefore, to considerably improve the overburden to coal ratio; the slopes can be made steeper than those slopes at creep resistance limit. The calculation relating to the stability of open pit boundary slopes, taking into account the supporting effect, is carried out according to a process developed by Rheinbraun. This calculation allows an assessment of the distance between dumping and production slopes to be made in order to assure sufficient stability.

Conclusions

A thorough application of this method contributes to a considerable improvement in the overburden to coal ratio in the open pit mines. This results in the additional extraction of coal, or a smaller extraction of overburden. In one mine an additional 3.0 million tons of coal could be mined along an open pit mine boundary slope with direct dumping (Figure 3.8).

Figure 3.8. Open pit mine boundary slope showing direct dumping.

Sufficiently stable slopes already exist in the mines with a calculated η-value of 1.1. Slopes with an η-value of only 1.5 and a little higher are also considered to be sufficiently stable. The determination of the η-value for a sufficiently stable slope is dependent on the assessment engineer having a large knowledge of the operational components in order to make decisions on both aspects of safety and economy.

References

Dermietzel, E., Metzmacher, H., Pierschke, K.-J., 1977, "Standfestigkeitsberechnung von Tagebauböschungen auf einer elektronischen Rechenanlage mit dem Programm BETAJA." Braunkohle, Heft 8, August 1977.

Düro, F., 1968, "Elektronische Berechnung der Standsicherheit hoher Böschungen nach der sogenannten Streifenmethode nach im Geologischen Landesamt Nordrhein-Westfalen entwickelten Formeln." Fortschritte Geologie Rheinland und Westfalen, 15, Krefeld.

Henning, D., 1977, "Zur Standfestigkeit von Schaufelradbaggerböschungen." Braunkohle, Heft 8, August 1977.

Janbu, N., 1954, "Application of composite slip surfaces for Stability analysis." Proceedings Conference Stability, 3, Stockholm 1954.

Neuber, H., 1968, "Untersuchung der Standsicherheit hoher Böschungen nach der sogenannten Steifenmethode." Fortschritte Geologie Rheinland und Westfalen, 15, Krefeld.

Winter, K., 1977, "INDAG-System zur Integrierten Datenverarbeitung in der Geomechanik." Braunkohle, Heft 8, August 1977.

Considerations in the Stability Analyses of Highwalls in Tertiary Rocks

4

Dr. Walter E. Jaworski,
Assistant Professor of Civil Engineering,
Northeastern University, Boston, Massachusetts, USA,
and Robert L. Zook,
Corporate Geotechnical Engineer,
North American Coal Corporation, Cleveland, Ohio, USA

Introduction

The importance of assessing the stability of deep open pit highwalls and its relationship to the efficient, safe, and profitable operation of a mine has been well presented in the literature (Steward and Seegmiller, 1972; Moffitt, Friese-Greene, and Lillico, 1972; and Kim and Cassun, 1977). In the case of strip mining operations where it is planned to cut highwalls greater than 30.5 meters (100 feet), the slope angle of the highwalls has a great impact on the economics of the mine operation. The slope affects the type and size of mining equipment which will operate most efficiently, as well as the amount of rehandling of spoil that will be necessary during the stripping operation. In addition, where high groundwater tables exist, the amount of dewatering necessary to maintain safe, economical slopes becomes a major consideration. All of these factors have a major effect on the annual operating cost for the mine. It is, therefore, necessary to have a reasonably accurate assessment of the safe slope angles during the early planning stages, not only to verify the economic feasibility of the mine, but to permit ordering of the proper mining equipment.

This chapter presents the approach used to predict design highwalls for developing the mining plan and assessing the economic feasibility of a lignite strip mine in east Central Texas, shown in Figure 4.1. Of concern was the determination of strength parameters used for the stability analysis and the assessment of dewatering needs. The study included field reconnaissance of existing strip mines in the area, an extensive laboratory test program, and an in-depth stability analysis of typical highwall profiles. The objective was to establish safe slopes for the highwalls, particularly for those which were 30.5 meters (100 feet) or more in height, and to establish them without being overly conservative in the stability analysis, because at these depths the economics of mining the coal becomes quite sensitive to slope angle.

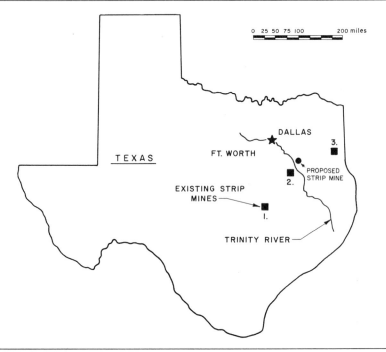

Figure 4.1. Map showing the location of the proposed lignite mine in Texas.

Site Geology

The proposed lignite mine lies along the Trinity River, and stratigraphically occurs in the upper 30 percent of the Wilcox Group of Eocene age. Here the Wilcox rocks consist of alternating and interbedded friable sandstones, claystones, and lignite.

Figure 4.2 shows typical columnar logs of generalized profiles found at the site. Section 1 is representative of lowland areas immediately adjacent to the Trinity River. Here the overburden consists of 6.1 to 9.1 meters (20 to 30 feet) of alluvial clays and sands underlaid by either claystones with thin interbeds of sandstone or thick bedded channel sandstones. Section 2 is representative of the upland areas where alluvium is absent. The profile consists of claystones underlaid primarily by sandstones. It is in this latter area where the cuts to the lignite will be deepest.

The Tertiary rocks at the site, being weakly cemented, exhibit the properties of both soft rock and heavily overconsolidated soil. The sandstones consist of fine-grained, very friable to weakly cemented silty sands. The claystones are generally very plastic and are either intact with slickenslides or contain horizontal thin seams of silt or carbonaceous materials. Joints in the sandstones and claystones are essentially vertical, and bedding planes are essentially horizontal. The lignite contains vertical and horizontal fractures, and is locally soft.

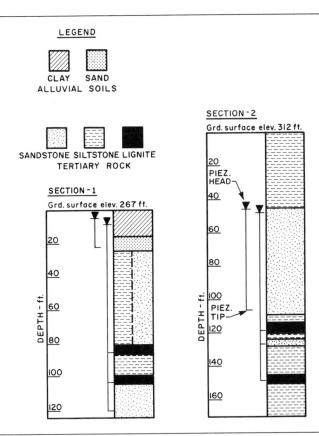

Figure 4.2. Two generalized stratigraphic columnar sections at the project site.

The groundwater levels monitored by piezometers installed at several depths in the sandstones and lignite are shown in Figure 4.2. In the lowland area (Section 1), the groundwater level is approximately 2.7 meters (9 feet) below ground surface. In the upland area, the groundwater level lies slightly below the claystones, approximately 13.7 meters (45 feet) below ground surface. Throughout the site area, the claystone layers create localized perched and/or artesian water conditions. The groundwater levels fluctuate seasonally and are primarily controlled by the elevations of the Trinity River and smaller streams in the site area.

Local Mining Experience

Three surface mines are currently in operation near the proposed mine. These are identified as sites 1 through 3 on Figure 4.1. At site 3, the mining is limited to depths not exceeding 19.8 meters (65 feet). At sites 1 and 2, the mining depths have presently reached 29 meters (95 feet) which is comparable in depth to the proposed early stage development of the subject mine.

Figure 4.3. Ninety-five-foot highwall in claystone.

Figure 4.3 shows a 29-meter (95-foot) highwall at site 1, which consists of 6.1 meters (20 feet) of alluvium underlaid by claystone down to the mined lignite bed. This highwall is similar to one at the proposed mine. The slope in the claystone averages 60°, whereas the slope in the alluvium is laid back to approximately 50°. No stability problems were encountered with this highwall. Of particular interest is the seepage pattern of the groundwater along the highwall. The seepage line, defined by the light-dark zone interface, showed a marked drop in the claystone only a few days after exposure. The prime reason for the rapid drawdown appears to be related to the flow along more permeable bedding planes where seepage was observed. This suggests that steady-state seepage conditions may develop rapidly in the claystones and that the stability analysis should consider this condition. The importance of this will be discussed later.

Figure 4.4 shows a 24.4-meter (80-foot) predominantly sandstone highwall at site 1. The highwall slope averages 60° and, when first cut, had a groundwater level within 9.1 meters (30 feet) of the ground surface. The mining operation was shut down because of large inflows of groundwater along the sandstone-lignite contact. The only instability of the highwall at this location was the toppling, or

Figure 4.4. Eighty-foot highwall in sandstone.

sloughing off, of the sandstone along a vertical joint pattern, which is evident in the extreme right of the figure. A mining engineer at the site reported that a 24.4-meter (80-foot) highwall in similar sandstone failed with a dragline operating on it, when the wall was cut at a slope of approximately 65°. There was no evidence of gross instability along the highwall shown in the left portion of Figure 4.4, where overburden had been placed on top and the wall formed a slope approaching 70°. The sandstones here appear to be stronger than at the locations where failures had occurred, and they indicate that one might expect significant variations in the strength of the sandstone within a given mine site.

At site 2 there were problems with highwall stability in cuts as shallow as 19.8 meters (65 feet). Figure 4.5 shows a local failure in a 55° highwall, 19.8 meters (65 feet) in height, which consisted of 15.2 meters (50 feet) of sandstone underlaid by claystone. The failure occurred approximately three days after exposing the wall. The account of the failure by mine personnel indicates that the failure was initiated by seepage along the sandstone-claystone contact. Apparently, there was some erosion of the sandstone initiating a tension crack in the top of the highwall, followed by a gross failure through the claystone. The mining in this area continued, but with the slopes laid back to a 45° to 50° slope. This seemed to solve the immediate problem, although some tension cracks were observed along the top of the highwall, and it was feared that rain would fill these cracks and initiate further slope failures. For this reason, precautions were taken to drain rainwater away from the highwall.

Overall, the limited experience in surface mining in this part of Texas suggests that, for the most part, 60° highwalls of up to 30.5 meters (100 feet) in either claystone or sandstone would be stable. This is governed, however, by local variations in the groundwater conditions and in the strength of the Tertiary rock. The

FAILED ZONE

Figure 4.5. Slope failure in 65-foot highwall.

experiences here also point out the need to evaluate the necessity for ground-water control, not only from the stability standpoint, but also from the aspect of providing access to the coal resource after stripping.

Stability Analysis

General considerations

As shown in Figure 4.6, the prime factors controlling the stability of a highwall are:

1. The strength of overburden material $\bar{c}$, $\bar{\phi}$.
2. The pore water pressures within and below the highwall.
3. The presence and orientation of weak zones such as joints, shear zones, and slickenslides.

The ability to identify these factors controls the confidence level one can have when predicting the factor of safety against a slope failure.

Strength parameters

In determining the strength parameters for use in a stability analysis, consideration must be given to the potential mode of failure and the strength tests conducted accordingly. Hoek, 1971, suggests that for cases of horizontally bedded

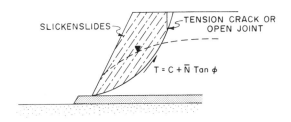

Figure 4.6. Factors influencing highwall stability.

intact rock, failure will occur along a circular surface. Since the bedding at the proposed mine site was horizontal, and the claystone beds with steeply dipping slickenslides are randomly interbedded with intact claystone beds, a circular failure surface was considered appropriate for the stability analysis.

Figure 4.7 shows a circular failure surface for the highwall. In the upper region of the slope, the failure surface cuts across the bedding planes. In the lower regions the surface is nearly parallel to the bedding. Hence, to determine the possible variations in strength parameters along the failure zone surface, tests were conducted which develop shear planes in an orientation similar to those shown. The upper zone was modeled by triaxial compression tests and the lower zone by direct shear tests parallel to the bedding.

Figure 4.8 shows a summary of the strength envelopes for the predominant rock types at the mine site. The envelopes are based on the peak shear resistance obtained from both direct shear tests and from triaxial tests with pore pressure measurements. The specimens tested were obtained from either continuous PQ wireline cores or Pitcher tube samples. Shown for comparison are the results from direct shear tests obtained by the U.S. Army Corps of Engineers, Texas, for the proposed Tennessee Colony dam, which is approximately 24.2 kilometers (15 miles) south of the site.

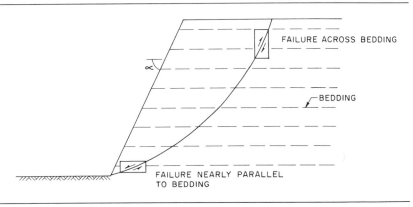

Figure 4.7. Failure surface in horizontally bedded rock.

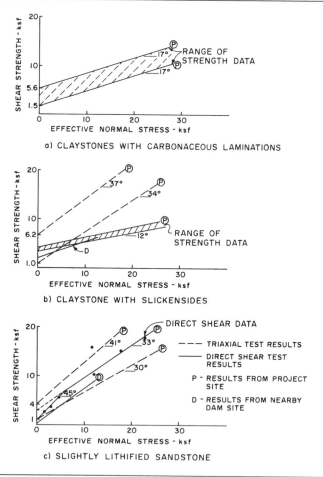

a) CLAYSTONES WITH CARBONACEOUS LAMINATIONS

b) CLAYSTONE WITH SLICKENSIDES

c) SLIGHTLY LITHIFIED SANDSTONE

Figure 4.8. Results of strength tests.

The results from triaxial tests and direct shear tests on the claystone with slick-enslides show a significant difference in cohesion intercept: 47.9 kilonewtons per square meter (kN/m²) versus 167.7 kN/m² (1 kip per square foot [ksf] versus 3.5 ksf) and in friction angle — 12° versus 34° to 37°. On the other hand, direct shear tests on the sandstones gave results which fell within the range of strength envelopes obtained from the triaxial tests. This suggests that anisotropic strength behavior is of limited concern in the sandstones. It is highly significant in claystones, however, and should be considered when analyzing the stability of claystone highwalls.

For the three rock types tested, the results show a significant range of cohesion intercepts. Visual observations of the test specimens show the range was, in part, linked to slight variations in the degree of lithification of the rock. Since the cohesion has a significant impact on the predicted safety of the highwall, its value must be cautiously selected for the stability analysis. This is particularly true where the friction angles are low, such as for the claystones, and the cohesion makes up the major portion of the shear resistance.

Overall, it is quite difficult, if not impossible, to select a unique set of strength parameters for each rock type. One may resort to using either average parameters and high factors of safety or the lower bounds of the strength envelopes with lesser factors of safety. Either approach leads to some inescapable conservatism in the design of the highwall slopes during the planning stages. It is considered prudent, at least in the planning stage, to select lower values of cohesion unless field observation of existing slopes indicates otherwise.

Pore water pressures
Pore pressure changes occur with time, and when analyzing their effects on the highwall stability, the short-term case (immediately after excavation) and the long-term case (when steady-state seepage occurs in the slope) are important.

Two prime reasons exist for evaluating changes in pore water pressure with time. First, it allows an assessment of how the factor of safety will change with time, and hence identifies the most critical time for a failure. Second, when evaluating the performance of existing highwalls for purposes of calculating strength parameters, it is necessary to have some idea of the magnitude of the pore pressures existing within the highwall.

For claystone highwalls, the pore pressures immediately following excavation are primarily a function of the stress changes along the highwall. Bishop and Bjerrum, 1960, showed that the change in pore pressure at a point from this stress relief can be evaluated using the equation:

$$\Delta\mu = B \left[\Delta\sigma_3 + A \left(\Delta\sigma_1 - \Delta\sigma_3 \right) \right]$$

where:
- $\Delta\mu$ is the change in pore pressure
- $\Delta\sigma_1$ is the change in major principle stress
- $\Delta\sigma_3$ is the change in minor principle stress
- A and B are pore pressure factors determined from triaxial strength tests

Figure 4.9 shows typical results from triaxial tests on claystone from the mine site. The results show that the A-factor continually decreases during shear, achieving a value of approximately 0.4 at peak shear stress. The B-factor was 1.0. Using the data in the above equation, and recognizing that during excavation the vertical stresses in the highwall will experience little change ($\Delta\sigma_1 = 0$), and lateral stresses essentially decrease to zero ($\Delta\sigma_3$ is negative), the predicted pore pressures immediately following excavation will be negative. In fact, negative pore pressures will develop for these stress conditions as long as the A-factor is less than 1.0.

The magnitude of the negative pore pressures developed will be a function of the depth and the existing lateral stresses in the ground. These values are extremely difficult, if not impossible, to ascertain. Fortunately, it is not necessary to

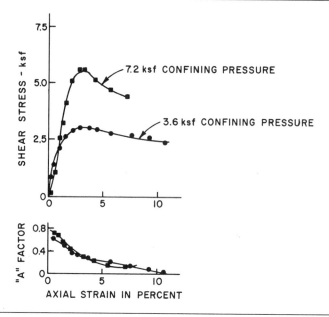

Figure 4.9. Typical stress-strain behavior of claystones in triaxial shear.

determine these values, since the stability analysis immediately following excavation may be evaluated using the strength parameters from undrained triaxial shear tests, or unconfined compression tests (undrained triaxial tests with confining pressures equal to the rock overburden pressure at the sample depth is the preferred test). These tests give shear strengths which inherently account for the negative pore water pressure buildup. Ultimately, the effect of the negative pore pressures is to increase the frictional component of the claystone's shear resistance, i.e., increase the effective normal pressures on the failure surface. This yields higher factors of safety than the long-term, steady-state seepage case. For the steady-state case, the pore pressures are positive or zero, depending on the location of the seepage line. Thus, the effective normal stresses are lower on the failure surface, resulting in a lower factor of safety (Figure 4.10).

The decision on what condition should be used for analyzing the stability of claystone slopes is dependent on how quickly the negative pore pressure will dissipate. This is a function of the permeability of the claystone mass. If the claystone is intact, without joints or multiple sand or silt seams, pore water cannot readily migrate to the failure surface and relieve the negative pore pressures. Thus, the undrained conditions will exist for some time, and the slope may be analyzed using the undrained shear strength parameters. On the other hand, if the claystone is jointed, fractured, and/or contains many thin, silty fine sand seams, the negative pore pressures will dissipate quickly. For this case, the effective strength parameters should be used for analyzing the highwall stability along with an appropriate assumption for the seepage line.

At the existing strip mines in Texas, field observations suggest that the seepage line dropped quickly within the slopes (see Figure 4.3). The claystone did have vertical joints and silt or fine sand seams parallel to the bedding planes among which seeps were seen. Thus, it would appear that the horizontal permeability of the rock mass was significantly higher than its vertical permeability. Based on these observations, it was decided that using effective strength parameters with horizontal seepage lines, breaking out on the slope was the appropriate condition for stability analysis. The approach was deemed prudent, since it appeared that steady-state seepage conditions developed within a few days after stripping.

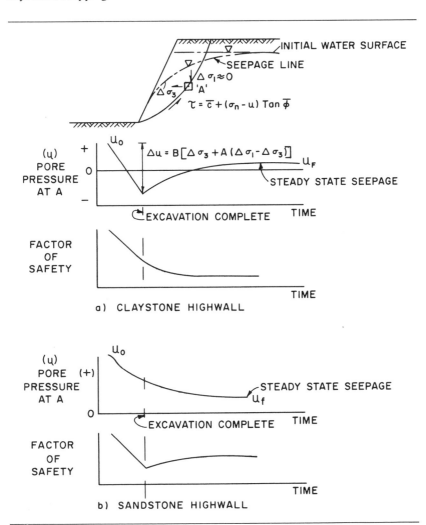

Figure 4.10. Variation in factor of safety with time.

For the sandstone highwall, the short-term stability will depend on how rapidly the groundwater level will drop in the highwall during excavation. When the highwall first reaches maximum height, the pore pressures will still be higher than the steady-state seepage condition. Therefore, the factor of safety for the slope will be lowest at this time. As the groundwater level decreases with time, the factor of safety of the slope will increase. In either case, stability analysis using effective stress parameters is appropriate.

Predicted Highwall Stability

Figures 4.11 and 4.12 show a summary of results of the stability analysis on typical sections of highwalls for the mine site. The stability analysis used was the simplified Bishop (Bishop, 1955). Because of the possible range of strengths as defined by the laboratory tests, and the possible range of drawdown of the phreatic surface in the highwalls, these *sensitivity plots* were developed to see what effect these factors have on the predicted safety factor. The plots are based on the effective stress strength parameters shown in Figure 4.8.

The results shown in Figure 4.11 compare the factors of safety obtained using average strength parameters from only direct shear tests, with those from both triaxial and direct shear tests. The cohesion values selected were tempered somewhat by values backfigured from the field observations at the existing nearby mines. For example, for the sandstone, friction angles of 33° to 45° were assumed as valid, and using stability charts, limiting ranges of cohesion were back-calculated. The observed cuts in sandstone were up to 24.4 meters (80 feet) in height, and had initial groundwater levels ranging from a depth of 9.1 meters (30 feet) to dry slopes. Considering a 60° highwall, a factor of safety of 1.0 along with the aforementioned conditions, estimated values of cohesion ranged from 57.4 to 119.9 kN/m^2(1.2 ksf to 2.5 ksf). Based on this data, a lower bound of 71.8 kN/m^2(1.5 ksf) was adopted for the average cohesion in the sandstone. Using a similar approach, the adopted lower bound of cohesion for the claystone was 100.6 kN/m^2 (2.1 ksf).

Figure 4.11 is useful in selecting the amount of dewatering or depressurization necessary at the site. The results show that for a 30.5-meter (100-foot) claystone highwall, the phreatic surface for the steady-state case should be lowered to a depth of at least 50 percent of the highwall height for a minimum factor of safety of 1.3. Depending on the average strength of the claystones in the highwall i.e., considering the two predominant claystone types and the applicable triaxial or direct shear lab test results, the safe highwall slopes lie in the range of 50° to 60°. This is a large range with great economic impact on the mining operation; one would prefer a more exact value of slope angle. However, the predicted range is realistic, considering the variability of the materials as represented by the laboratory test results. The results show that lowering the phreatic surface will be necessary in any event, which represents a clearly identified added cost to the mining operation. By comparison, the same plot indicates that dewatering is less important in the sandstone, regardless of the strength parameters. The results show that, for a 60° highwall, minimal dewatering is necessary in the sandstones. The amount that is required will probably occur by natural drainage during excavation.

Of interest is a comparison between the results for the claystone highwall shown in Figure 4.11*a* and the predicted factor of safety for the claystone immediately following excavation. A stability analysis was performed using the average undrained shear strength results from triaxial tests (Figure 4.9) and unconfined compression tests on the claystone. The unconfined compression tests gave shear strengths ranging from 95.8 kN/m² to 455 kN/m² (2 ksf to 9.5 ksf) with an average value of 335.3 kN/m² (7.0 ksf). Based on these results, an average undrained shear strength of 287.4 kN/m² (6.0 ksf) was adopted for the claystone

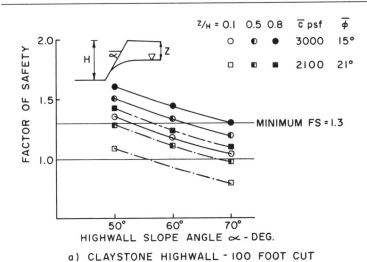

a) CLAYSTONE HIGHWALL - 100 FOOT CUT

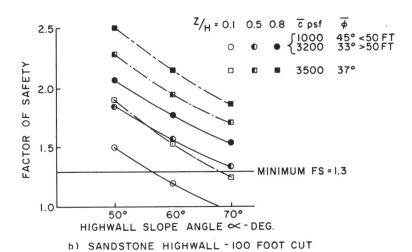

b) SANDSTONE HIGHWALL - 100 FOOT CUT

Figure 4.11. *Effect of dewatering and rock type on predicted highwall stability.*

highwall. Using this value, the predicted factor of safety was 2.1 for a 60° high-wall, which is significantly higher than the maximum factor of safety of 1.45 shown on Figure 4.11*a*, based on lowering the phreatic surface to 80 percent of the highwall height and the effective strength parameters.

Figure 4.12 shows the predicted variation in factor of safety with height for a 60° composite highwall. The strength parameters from triaxial tests were used for the claystone because of the high angle of the failure plane in this strata. The parameters selected for the sandstone were the lower and upper bounds for cohesion, together with the average friction angle from direct shear and triaxial tests. The results point out the importance of being able to establish accurate strength parameters. If the lower bound of cohesion is adopted, then essentially complete dewatering of the sandstone is necessary for highwalls greater than 30.5 meters (100 feet).

Attempts to predict the actual safety factor for a given slope more accurately than the range shown in Figure 4.12 may result in a high risk, particularly in the preplanning stage where the economic feasibility of the mine is being evaluated. It is recommended that, at this stage, detailed geotechnical studies be undertaken to at least define a probable range of strength parameters in these soft Tertiary

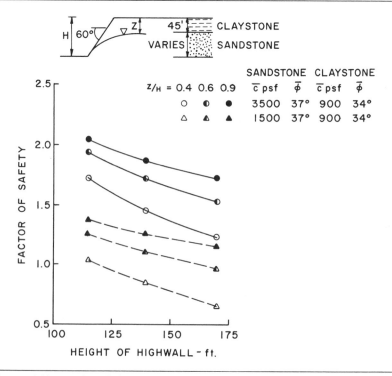

Figure 4.12. Predicted factor of safety versus highwall height and strength parameters.

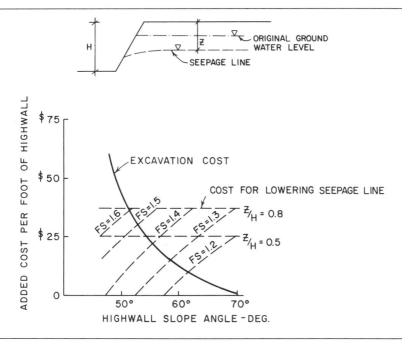

Figure 4.13. Excavation and dewatering costs for 100-foot highwall.

rocks. In addition, pumping tests should be undertaken to define the rate of groundwater drawdown among the various strata present.

Using the above data, a decision chart, such as that shown in Figure 4.13 may be prepared to help the mining engineer assess the economic potential of the mine. This figure was developed for a 30.5-meter (100-foot) highwall in claystone. The factor of safety is based on the average strength parameters from direct shear tests ($\bar{c} = 3.0$ ksf [143.7 kN/m²] $\bar{\phi} = 15°$). It assumes a 70° highwall as the steepest that will be cut, and all increased excavation costs are relative to this slope. This figure couples factor of safety with economics and helps identify the most cost-effective design of the mine highwall (a lowering of the slope angle versus lowering the phreatic surface or a combination of both). Using lower bounds for strength parameters, this type of information also allows management to assess the potential safety hazards and the dangers of failure, and to weigh them against the costs for mining the resource. An example of how one would use this data is to establish the minimum acceptable factor of safety, say 1.4. With no dewatering, the required highwall slope is approximately 50°, and increased excavation costs are $197 per linear meter ($60 per linear foot) of highwall. Alternatively, a 60° highwall is obtainable by lowering the phreatic surface to 80 percent of the slope height. The total increased costs for both lowering the phreatic surface plus excavation are in the order of $151 per linear meter ($46 per linear foot). Thus, the latter approach is more economical and does not sacrifice safety.

Limitations of Analysis

The results of the stability analysis shown in Figures 4.11 and 4.12 assumed the highwall to be composed of essentially intact rock. No consideration was given to the effect of either tension cracks or the surcharge from the dragline on the predicted factor of safety.

The effect of tension cracks on the factor of safety have been examined using a Morgenstern-Price, 1965, stability analysis. For slope angles of 60° or greater, and highwalls of 30.5 meters (100 feet), it was found that tension cracks up to 6.1 meters (20 feet) in depth had little effect on the predicted factor of safety unless they were water-filled. This seems logical when one considers the high angle failure surface near the top of the slope. The shear resistance in this area would be derived primarily from the cohesion which, in general, is relatively small when compared to the higher frictional resistance along the lower portion of the failure surface.

When assessing the impact of the dragline loading, the static load in combination with the dynamic loading from the dragline operation should be considered. In the authors' experience, there is no available data on the nature and magnitude of the dynamic loading; hence it was not considered in this analysis. When evaluating the effect of the static dragline loads, analytical methods are, at best, limited for evaluating the three-dimensional failure modes. Some methods (Azzous, 1978) have been developed for the undrained case where the highwall consists of claystone, but no generally accepted method is available for analyzing the long-term steady-state seepage case where effective stress parameters are applicable.

A conservative approach to analyzing the effect of dragline loadings is to assume that the dragline surcharge is to be a strip loading. In such a case a two-dimensional (Bishop, 1955, or Morgenstern-Price, 1965) analysis would apply.

Using this approach it was found that for 30.5-meter (100-foot) highwalls, the dragline lowers the factor of safety by some 15 to 20 percent. For highwalls in the order of 51.8 meters (170 feet) these values are reduced by up to 5 to 10 percent.

The predicted mode of failure for the dragline is of importance. The stability analysis indicates that for 30.5-meter (100-foot) highwalls at 60°, the failure surface will intersect the ground surface 15.2 to 18.3 meters (50 to 60 feet) behind the top edge of the slope. For draglines with capacities of 61.2 to 76.5 cubic meters (80 to 100 cubic yards), this failure surface will undermine about half the dragline pad, resulting in a toppling-type failure into the open pit. Such a failure would be catastrophic, and would no doubt result in significant damage to the equipment, serious injury to the operators, and reduced coal production.

Considering the range in strength values obtained from the laboratory tests, the uncertainty introduced into the factor of safety by ignoring tension cracks and dragline loadings is minimal when compared to the variability obtained by using the upper or lower bounds of the strength results. Therefore, the recommended approach is to ignore these two factors in the initial analysis of the highwall, but to account for their effects indirectly by stipulating that a minimum safety factor of 1.3 be used for purposes of establishing the design highwall. This value was adopted for the highwall design during the preplanning stage at the subject mine, and was the basis for considering the necessity for site dewatering.

Summary

The purpose in undertaking a highwall stability study is to establish the probable range of highwall slopes which may be safe at a mine site. More importantly, the studies identify, early in the mine planning stage, the potential hazards in mining a given site. This provides critical data necessary in developing the mining plan and assessing the economic feasibility of the mine.

This chapter attempts to convey the difficulties encountered and considerations involved in predicting the stability of open cuts in soft Tertiary rock. The overriding factors of concern are assessing the strength parameters for the various stratigraphic rock segments which will be encountered during mining, and evaluating the groundwater conditions within the highwall following excavation.

By far the greatest challenge lies with predicting the strength parameters and their changes with depth and/or lateral extent. In these predictions, allowances must be made to account for the variation of the strength parameters along the failure surface. Because of the range of strengths which may exist in a given mine site, it is recommended that *sensitivity plots* be developed for the typical highwall profiles. These consider several of the geotechnical parameters, and provide some of the guidance necessary to evaluate mine economics and develop mining plans. They also serve as a guide during mining for decisions about what action is necessary in the event that unstable slopes are encountered (dewatering, slope reductions, or both). To further aid in this decision it is possible to generate data demonstrating the economic impact of slope reduction versus dewatering.

Using all of this information, the safety risks involved and their associated costs may be identified before the mining operation commences. This gives both management and the mining engineer the necessary data to develop a safe, but not overly conservative, operation and provides a basis for considering potential problems in developing the economic feasibility of the mine. Should failures occur, the mining engineer has been forewarned and should be properly equipped to undertake corrective action.

During the early stages of mining, the accuracy of the *sensitivity charts* should be verified by instrumenting and observing the behavior of highwalls for each of the typical stratigraphic sections. As the observational data accumulates and experience is gained, the mining engineer can continually update and adjust slope angles, and the amount of dewatering to optimize the economics and safety of the mining operation. The observations of highwall behavior must be continuous, since the strength parameters and groundwater conditions can vary considerably over the mine site.

References

Azzous, A. S., 1977, "Three Dimensional Analysis of Slopes." Ph.D. Thesis, Massachusetts Institute of Technology.

Bishop, A. W., 1955, "The Use of the Slip Circle in the Stability Analysis of Earth Slopes." Geotechnique, Vol. 5.

Bishop, A. W. and Bjerrum, L., 1960, "The Relevance of the Triaxial Test to the Solution of Stability Problems." Proc. ASCE Research Conference on Shear Strength of Cohesive Soils, Boulder, Colorado.

Hoek, E., 1971, "Influence of Rock Structure on the Stability of Rock Slopes." Proceedings of the First International Conference on Stability in Open Pit Mining, The American Institute of Mining, Metallurgical and Petroleum Engineers, Inc., New York.

Kim, Y. C. and Cassun, W. C., 1977, "Economic Analysis Applied to Pit Slope Design—A Case Study." Transactions, Society of Mining Engineers, AIME, Vol. 262, December.

Moffitt, R. B.; Friese-Greene, T. W. and Lillico, R. W., 1972, "Pit Slopes—Their Influence on the Design and Economics of Open Pit Mines." Proceedings of the Second International Conference on Stability in Open Pit Mining, Metallurgical and Petroleum Engineer, Inc., New York.

Morgenstern, N. R. and Price, V. E., 1965, "The Analysis of the Stability of General Slip Surfaces." Geotechnique, Vol. 15, No. 1.

Stewart, R. M. and Seegmiller, B. L., 1972, "Requirements for Stability in Open Pit Mining." Proceedings of the Second International Conference on Stability in Open Pit Mining, The American Institute of Mining, Metallurgical and Petroleum Engineers, Inc., New York.

Slope Stability in Australian Brown Coal Open Cuts

5

R. Hutchings, Engineer, and R. J. Gaulton, Geologist,
Fuel Department, State Electricity Commission of Victoria,
Melbourne, Australia

Introduction

In common with other surface coal mining projects, the Latrobe Valley open cuts in Australia are susceptible to batter instability in three main areas. These are the overburden batters, the individual coal faces, and the overall batter system. In each of these areas, however, there are significant features of the Latrobe Valley open cuts which make aspects of the three described instability mechanisms quite different from many other surface open pit operations. Before outlining these instability problems with their unusual geotechnical and operational aspects, a brief description is given of the brown coalfields and the open cuts operated by the State Electricity Commission of Victoria.

Open Cut Operations

In the Latrobe Valley there are currently three operational brown coal open cuts with a fourth, at Loy Yang, planned to begin initial operation in the early 1980s. The location of these open cuts is shown in Figure 5.1. The extracted coal, which is a low-grade fuel having a net wet heating value ranging from 6.5 to 11.5 megajoules per kilogram (2,800 to 4,900 Btu per pound) and a gross dry heating value of approximately 25 megajoules per kilogram (10,700 Btu per pound), is mainly used for the generation of electricity in power stations sited close to the open cuts. Some concept of the expansion of these activities can be gathered from the statistics presented in Table 5.1.

Coal is mined by bucket wheel and bucket ladder excavators and is transported from the open cut to power station bunkers by train or conveyor. Excavation ratios of coal to overburden (volume to volume) are in the range of 3.1 to 1 and up to 5.9 to 1. In addition, coal reserves need to be kept at the operational faces because stockpiling may result in spontaneous combustion. The combination of the type of equipment used, the favorable ratios of coal to overburden,

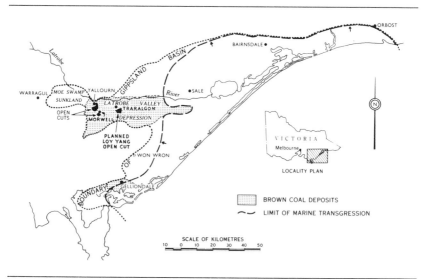

Figure 5.1. Map showing the extent of the Gippsland Basin in Victoria.

and the storage of coal at the operational faces, means that the open cuts do not resemble a normal strip mine operation.

Geology

The geology of the coalfields is relatively straightforward. As described by Gloe, 1960, the Latrobe Valley depression forms part of the coastal Gippsland Basin (Figure 5.1) in which offshore oil and gas deposits have been discovered. Within the depression, which is a structural sub-basin some 40 kilometers (25 miles) long and 16 to 24 kilometers (10 to 15 miles) wide, there are up to 670 meters (2,200 feet) of Tertiary sediments consisting of three main groups of thick brown coal seams interbedded with clays, sands, and silts. Known coal reserves total 108,000 million tons with 30,000 million tons currently considered to be economically extractable (Ministry of Fuel and Power, 1977).

Post depositioned deformation has occurred, and open cuts are located in areas where relatively thick single or multiple coal seams are overlain by minimal thicknesses of overburden.

Stability of the Overall Batters

Because of the relatively low unit weight of Latrobe Valley brown coal (saturated relative density of 1.12), water plays a very important role in the initiation of batter instability and in measures taken to control movement. The main stability mechanism is shown diagrammatically in Figure 5.2. Basically, complete blocks of coal may slide on an underlying clay layer which exhibits low residual strength properties. Direct shear tests indicate that representative residual effective

Table 5.1. Actual and Projected Coal Production Statistics from the Latrobe Valley Brown Coalfields

Year	Open cuts	Coal production (kilotons/year)	
		Individual	Total
1940	Yallourn	4,007	4,007
	Yallourn North	. . .	
1950	Yallourn	6,507	7,283
	Yallourn North	776	
1960	Yallourn	11,086	13,461
	Yallourn North	1,345	
	Morwell	1,030	
1970	Yallourn	10,267	23,155
	Yallourn North	403	
	Morwell	12,485	
1980	Yallourn	16,000	31,780
	Yallourn North	280	
	Morwell	15,500	
1990	Yallourn	20,000	61,780
	Yallourn North	280	
	Morwell	15,500	
	Loy Yang	26,000	

strength parameters for the clay are $C'_r = 7$ kilopascals (1.0 psi) and $\varphi'_r = 12°$ (Gloe, James, and McKenzie, 1973).

Regional jointing in the coal and stress relief crack patterns both provide ready access for surface water to percolate into the batter system. The subsequent rise in hydrostatic pressure, both along the coal-clay interface and in any

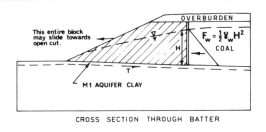

CROSS SECTION THROUGH BATTER

Figure 5.2. The block sliding mechanism.

perpendicular or near perpendicular joint, is then capable of initiating significant batter movement in a very short time.

Movement is dependent upon a supply of water and has occurred after high rainfall or following the failure of water mains in the open cut. Generally, the movement exhibits a slip-stick characteristic resulting in reduced hydrostatic pressures immediately after the movement has occurred, because of the resultant improvement in natural drainage and an increase in the volume of voids available to store the ingress of water. The repeated occurrence of such events could seriously disrupt operations and lead to the development of significant earth movements in urban areas adjacent to the open cuts (Hutchings, Raisbeck, and Fajdiga, 1977).

The slip-stick block sliding mechanism, with its potential to seriously disrupt urban and excavation activities, should be distinguished from the more catastrophic and dramatic circular arc, wedge, and planar failures often described (Brawner, 1977). Block sliding can be described as a serviceability failure mechanism requiring constant monitoring, maintenance, and preventive design in order to keep the frequency of occurrence, and the magnitude of movement, low.

Monitoring batter stability

The permanent batters of the 140-meter (450-foot) deep Morwell open cut are monitored for earth movement with regular surveys of pins installed up the batter system. In addition, bores tapping the unconfined groundwater table and the piezometric head of the coal-clay interface are drilled from benches down the batter along lines nominated for stability assessment. The arrangement of survey pin lines and stability assessment lines is shown in Figure 5.3. Water levels are monitored and compared with critical values determined from stability analyses. The occurrence of any unexpected rise in water level is investigated.

Recently, continuous monitoring of water level in critical bores has been instigated. As well as providing more data on day-to-day variations in water level, the continuous records from these bores are providing further insight into the reasons for periodic instability. Some of the more uncommon events which have caused dramatic rises in water pressure are the drilling of a bore from a bench using water circulation, fire protection spraying of the batters, and overflow from a water supply pressure relief valve.

Design methods to improve overall stability

With water being the most significant factor influencing batter stability, design methods for improved stability are aimed at either preventing the ingress of water into the batter system or aiding its quick release.

Examples of the former approach include the interception of runoff water in lined drains; the lining of drains with used conveyor belting, clay or half round pipe; increased monitoring of water mains for leakage; and the eventual removal of all but essential water mains from the batter system. Currently, clay covering of most of the exposed coal in the permanent batter system is being studied for both fire protection and batter stability reasons.

The second approach, of improving drainage from within the batters, has involved the extensive use of horizontal drains in the Morwell open cut. As shown

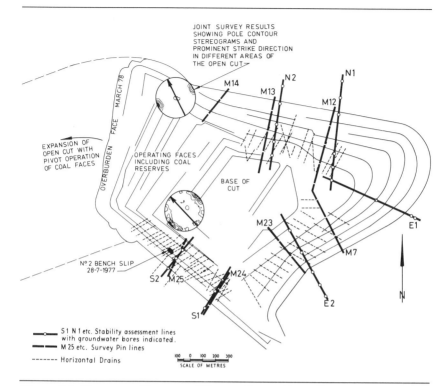

Figure 5.3. Features of batter stability studies at Morwell open cut.

in Figure 5.3, these drains are drilled from the lower level benches in order to lower the groundwater pressure in the critical toe region of the overall batter. They are between 200 and 300 meters (650 to 1,000 feet) long and have been installed 50 to 70 meters (160 to 230 feet) apart. They are lined with 100-millimeter (4-inch) slotted pvc pipe, and for most of the year they only dribble water; but, in periods of very heavy rainfall or when water is accidentally discharged onto the batters, the drains flow full.

It is a paradox of the block sliding mechanism that another way of improving stability is to design steeper overall batters. With all other things equal, this increases the weight of the toe region of the batter and thus the friction forces. Where access roads, excavator transfer ramps, etc., are not the dominating criteria, plans are prepared with overall batter slopes of between 2.0 and 2.5 horizontal to 1 vertical.

Recent failure of southern permanent batter

On July 18, 1977, a significant proportion of the 2-kilometer (1.2-mile) long overall southern batter system of Morwell open cut moved by up to 250 millimeters (10 inches). A fall of approximately 1,800 tons of coal from No. 2 level bench was the first indication that an instability condition existed. The location of this fall

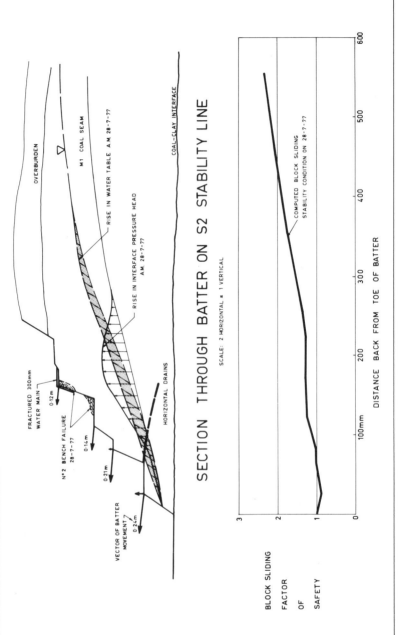

Figure 5.4. Block sliding stability of southern batter July 1977, Morwell open cut.

is shown in Figure 5.3. Initially, operations were concerned for the safety of a conveyor situated on the next lower bench. However, survey movements along the M25 pinline (see Figure 5.3) and water level measurements from bores on the S2 stability assessment line indicated that stability of the overall batter was the most serious problem, and that the No. 2 bench, while associated with the initiation of the overall failure, was in no immediate danger of collapsing further.

The sequence of events prior to the failure was as follows. Rainfall for the previous three days had totaled 78 millimeters (3 inches). In addition, a 300-millimeter- (12-inch-) diameter water main immediately above the No. 2 bench slip had shown signs of leakage during the previous day. Attempts had been made to repair the pipe, but they were not successful and leakage continued throughout the night. From an examination of continuous flow charts for the complete water supply and dewatering system, it is estimated that the No. 2 bench slip occurred at approximately 5:30 a.m. This movement fractured the water main and carried the dropper section of the pipe down the batter. The coal fall also demolished an unoccupied crib hut which was situated adjacent to the conveyor line on the next level down.

By the time the water main was isolated at 8 a.m., some 1,500 tons of water had been released over and into the batter system. The resultant rise in water table and interface pressure head is shown plotted in Figure 5.4.

Block stability analyses indicated that the batter was unstable, and this was confirmed by survey data which showed that the toe of the batter had moved 250 millimeters (10 inches) with progressively less movement at higher levels. These movements and the results of the stability analyses are also shown plotted in Figure 5.4. Horizontal bores on levels 5 and 7 below the site of the bench failure were flowing full at up to 5 liters per second (78 U.S. gallons per minute) each. Movement surveys and water level measurements during the next hours and days indicated that the batter had restabilized within 24 hours of the bench slip occurring.

Stability of Individual Benches

In the Latrobe Valley open cuts, individual benches are either of coal, clay, or silty clay material. The most common failure mechanism with these latter overburden materials is a slip circle. With the coalfaces, however, jointing patterns make planar or wedge failures likely.

Overburden fire holes

A geologic feature which has a significant influence on the occurrence of overburden slips is the existence of clay-filled fire craters in the upper coal surface.

These Tertiary burn craters are saucer-shaped depressions up to 300 meters (1,000 feet) in diameter and 45 meters (150 feet) deep, extending at the extreme, over two normal coal winning faces.

Both the character and the strength of the infill material in these burn holes—C_u of 12 to 100 kilopascals (1.7 to 14 psi), φ_u of 3° to 12°, and a liquid limit of 90 percent—and its structure, mean that normal overburden excavation at a batter angle of 45° cannot be safely undertaken.

Slip circle failures involving up to 150,000 tons of material have occurred, and auxiliary excavation of the clay-filled fireholes using dragline, shovels, and trucks has been adopted to minimize risk to major plant.

Jointed coalfaces

The brown coal in the Latrobe Valley open cuts behaves as a soft rock with a friction angle of approximately 35° (Rosengren, 1961). The Morwell 1 seam is extensively jointed, with the joints having a remarkably consistent strike direction from the north-northwest to the south-southeast, as shown in Figure 5.3.

In order to provide detailed advice on the stability of the coalfaces, the joint pattern is surveyed annually by geology students working in their long vacation. Each survey covers at least three full-length operating faces and involves upwards of 3,600 meters (12,000 feet) of traversing. Coal joints have been grouped into three categories on the basis of their true dip:

1. High angle joints   >80° dip
2. Intermediate angle joints   ≥40° ≤80° dip
3. Low angle joints   <40° dip

The high angle joints comprise over 90 percent of all jointing found within the Morwell open cut. Most high angle joints are smooth-walled, quite regular, and very extensive, individual joints passing through the full thickness of the seam being quite common.

Although intermediate angle joints are rare and constitute less than 3 percent of the total joint population, their effect on the stability of the faces is very significant. Planar failures such as that shown in Figure 5.5 can readily occur when orientation of the working faces approaches the strike of these joints. The batter shown in this photograph is 20 meters (65 feet) high with an orientation of 40°,

Figure 5.5. Planar failure of 20-meter-high operating face at Morwell open cut.

while the joint has a strike of 44° and a dip of 48°. Several hundred tons of coal were involved in the slip, and this would be sufficient to cause damage to a bucket wheel excavator operating at the face. Such an accident occurred in October 1962 when No. 3 bucket wheel excavator was extensively damaged by a coal fall. The operating faces then had an orientation of 330°, which is close to the strike of the major jointing system.

Statistical analysis
In order to advise on batter stability problems which could arise with further expansion of Morwell open cut, the joint data is statistically analyzed to give an assessment of the likelihood of wedges or blocks of coal sliding from the coalface. The method used assesses the probability of occurrence of a critical wedge or plane failure given a specific batter and a known joint pattern. "Daylighting" wedges are detected by mathematically combining each joint with every other joint. For n joints surveyed, a total number of $n(n-1)/2$ wedges are generated by this procedure. Joints which form true planar failures are specifically noted due to the inherent instability of such situations.

The combining of all joints with every other joint is statistically justifiable only if the joint data is a true sample of the overall joint population. This implies that the physical position of any joint should have no effect on joint characteristics. Limitations are therefore placed on combining joint information from different areas of the open cut as there are variations in joint pattern with depth and, to some lesser extent, a variation along any one face.

Some results of the statistical analyses are shown in Figure 5.6 where varia-

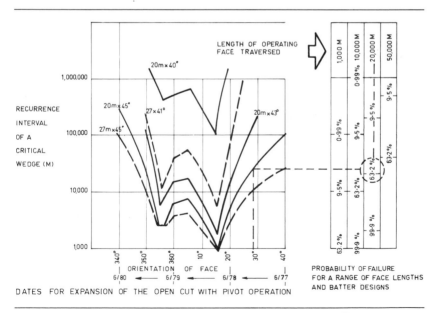

Figure 5.6. Statistical analysis of coal face stability in Morwell open cut.

tions in recurrence interval for wedge failures are plotted with respect to orientation of the coalface. The analyses showed that the most significant single parameter influencing the probability of failure was the orientation of the batter being examined. Other factors, in their relative degree of significance, were batter slope and batter height.

To maintain a constant risk of failure during pivot operation of the coalfaces over the next few years, batter angles for the upper coal cuts have been reduced from 45° to 40°. As indicated in Figure 5.6, the likely failure recurrence interval for a 20-meter (66-foot) × 45° batter at an orientation of 40°North is 100 kilometers (60 miles), and this is equal to the recurrence interval for a 20-meter × 40° batter pivoted at a critical orientation of 15°North.

It can be seen that coal operating face stability will be a critical problem during pivot operation of the Morwell open cut coalfaces over the next few years, and studies of the type described are currently being extended to cover all levels of the open cut.

Conclusions

Both the overall slopes and operating faces of the Latrobe Valley open cut excavations have some unusual stability problems. The application of relatively simple analytical models has, however, given sufficient information for the design of safe batter systems. Reliable and orderly operation of the open cuts adjacent to urban areas has thus been possible.

Acknowledgments

This information is published with the permission of the State Electricity Commission of Victoria. Acknowledgment is made to the many officers who have contributed, over a period of many years, to the work described in this paper.

References

Brawner, C. O., 1977, "Open Pit Slope Stability Around the World." Canadian Mining and Metallurgical Bulletin, July.

Gloe, C. S., 1960, "The Geology of the Latrobe Valley Coalfield." Proceedings, The Australasian Institute of Mining and Metallurgy, No 194.

Gloe, C. S., James, J. P. and McKenzie, R. J., 1973, "Earth Movements Resulting from Brown Coal Open Cut Mining, Latrobe Valley, Victoria." Subsidence in Mines, 4th Annual Symposium, Illawarra Branch, The Australasian Institute of Mining and Metallurgy.

Hutchings, R., Fajdiga, M. and Raisbeck, D., 1977, "The Effects of Large Ground Movements Resulting from Brown Coal Open Cut Excavations in the Latrobe Valley, Victoria." Conference on Large Ground Movements and Structures, University of Wales, Institute of Science and Technology, Cardiff.

Ministry of Fuel and Power, 1977, "Brown Coal in Victoria, The Resource and its Development."

Rosengren, K. J., 1961, "The Structure and Strength of Victorian Brown Coals." Master of Engineering Science Thesis, University of Melbourne.

The Use of Movement Monitoring to Minimize Production Losses Due to Pit Slope Failures

6

D. C. Wyllie, Senior Engineer, Golder Associates Ltd.,
Vancouver, British Columbia, Canada,
and F. J. Munn, Senior Mining Engineer, Cardinal River Coals,
Hinton, Alberta, Canada

Introduction

Pit slope failures are not uncommon events and many mines operate with unstable slopes. In fact, the existence of unstable slopes shows that the slope angle is near to its limiting value and that the stripping ratio is being minimized. However, slope failures are not acceptable when men working in the pit bottom are injured, equipment is damaged, or when there is severe disruption to operations.

This chapter describes techniques that have been used successfully at many mines to measure slope movement and predict when stability conditions are deteriorating and operations should cease. The method is based on the fact that a moving slope will accelerate before failure occurs. By measuring the rate of movement and identifying acceleration, a warning of hazardous conditions can be obtained (Brawner et al, 1976; Kennedy, 1971). In this way, operations can safely continue beneath moving slopes so that production losses are minimized. Methods of measuring slope movements and interpreting the results are discussed and a case study is described where mining continued for 17 months beneath a continuously moving slope.

The techniques described are most applicable to the failure of slopes where movement of several tens of meters (or feet) occurs before failure. However, with slight modifications, the techniques could successfully be applied to the study of the stability of foundations for structures such as conveyors and concentrators which can only tolerate movements of a few millimeters (or inches) (Watt, 1970).

The successful application of movement monitoring in the study of slope stability depends upon carrying out three important steps:

1. Recognizing the type of failure and the probable causes of failure.
2. Making accurate and unambiguous movement measurements.
3. Correctly interpreting the results to determine when acceleration occurs.

Types of Slope Failure

Slope failures can be of several different types and each has certain characteristics, which include the direction in which the movement occurs. The correct identification of the type of failure will often be helpful in designing monitoring systems and in interpreting movement results. Three of the most common types of failures are discussed below.

Circular failures

Circular failures (Hoek and Bray, 1977) are the most common type where materials are reasonably homogeneous. They derive their name from the fact that the failure surface has the shape of an arc of a circle, shown in Figure 6.1. They occur in soil, weathered and soft rock, and in highly fractured rock and in waste dumps. They can vary in size from a few meters (or feet) in height to several kilometers (or miles) across, and movement can occur on slopes as flat as 5°. Where there is no distinct failure surface, movement is often slow, continues for some time, and considerable displacement takes place before failure occurs. The first sign of instability is usually the opening of tension cracks along the crest of the slopes, followed by slumping of the crest, and lateral movement of the toe. The final collapse of the slope can be rapid.

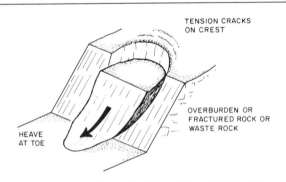

Figure 6.1. Circular type slope failure.

Toppling failures

Toppling failures are less common than circular failures, although recent studies (Goodman and Bray, 1976) have shown that they have a widespread occurrence. These failures occur in competent rock containing well-defined bedding planes or joints which dip into the slope. This structure forms tall, thin slabs, and when the slope is excavated these slabs topple because the line of action of the center of gravity lies outside the base. This horizontal movement opens up tension cracks along the crest, although there is little movement at the toe of the slope. In hard rock, toppling may continue for some time until a slab suddenly falls from the crest (Figure 6.2). In softer rock the slabs may bend which may eventually lead to a circular failure as the rock becomes highly fractured. Toppling may also occur where discontinuities dip steeply out of the slope.

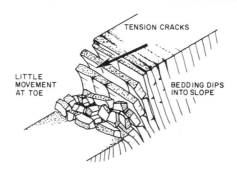

Figure 6.2. Toppling type slope failure.

Planar failures

Planar failures (Hoek and Bray, 1977) also occur in hard rock, but in this case continuous bedding or joint planes dip out of the slope as shown in Figure 6.3. If the angle of the slope is steeper than the dip of the fractures, i.e., the slope undercuts the fractures, blocks of rock may slide.

In a dry slope in which the fractures are unhealed, sliding will only occur if the dip angle of the fracture exceeds the friction angle of the surface, which may be as high as 45° in granite and as low as 5° if it contains a clay infilling. Cohesion on the failure plane will improve stability conditions, but if the slope is saturated, sliding may occur at dip angles flatter than the friction angle. Because failure takes place on a distinct plane the block will move parallel to the plane, and failure is often sudden with little warning of instability. Planar-type failures of considerable size could occur in such rocks as well-bedded sandstone and granite with sheet joints which could have continuous lengths of tens of meters (or feet). Failures may also occur where fractures of different orientations combine to form wedges of unstable rock.

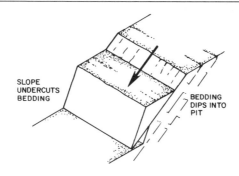

Figure 6.3. Planar type slope failure.

This chapter will not discuss the causes of instability at length since they are well documented elsewhere (Hoek and Bray, 1977). However, the most usual causes of instability can be summarized as follows:

1. *Geological conditions.* The strength of different rock types and their distribution within the slope, the degree to which the rock is fractured, the strength along these fractures and their orientation relative to the slope will all have an effect on the maximum stable slope angle.
2. *Groundwater.* Groundwater pressure within a slope can often cause instability because the pressure reduces the force that resists failure and increases the force that tends to displace the potentially unstable rock. Groundwater pressure may increase after periods of heavy rainfall, in the spring in heavy snowfall areas, and also in the winter when the face freezes and seepage is inhibited.
3. *Blasting.* The forces produced by blasting in open pit mines are often sufficient to break the rock for some distance behind the face. This reduces the rock strength and can lead to slope instability. The distance to which damage occurs is directly related to the peak particle velocity produced in the rock by the detonation of the explosive, and the particle velocity is in turn related to the weight of explosive detonated per delay. Modification of delay sequences to reduce explosive loads per delay can often have a beneficial effect on stability.

Methods of Measuring Slope Movements

When setting up a movement monitoring system there are a number of factors which should be considered in order to produce reliable and unambiguous results. The accuracy of the measurements should be compatible with the magnitude of the movements expected to be critical; for example, the critical movement for a crusher foundation is much less than that of a pit slope. The size and type of failure, and the rapidity with which movement results must be available, will also determine the measurement method. Furthermore, the method should be free of operator bias, independent of weather conditions, and be operable at night if operations must continue around the clock. This chapter, and the case study, describe the methods used to rapidly measure substantial movements of a large slide.

Crack width measurements

In almost all slides the opening of tension cracks on the crest of the slope is the first sign of instability. Measurement of the width of these cracks will often be representative of the movement of the slide itself.

The simplest means of measuring crack widths is to set pairs of steel pins, one on either side of the crack, and measure the distance between the pins with a steel tape or rule as shown in Figure 6.4. The advantages of this system are that the equipment is readily available and can be set up quickly, and that readings can be made and results analyzed immediately. The disadvantages are that the movement measured is not absolute, especially if both pins are on unstable ground; so this limits the system to small failures. Furthermore, vertical movement is not

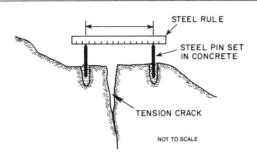

Figure 6.4. Crack width measurements show movement of crest of failure.

readily measured. Also, safe access to the crest of the slope must be possible in order to make measurements, and this will probably become dangerous when the width of the crack is several meters (or feet).

As the slide becomes larger and it is no longer possible to use pairs of pins across the cracks, tensioned wire extensometers can be used to measure movement over a greater distance as can be seen in Figure 6.5. In this method, a wire is stretched across the tension cracks between a measurement station established on stable ground and an anchor is secured on the crest. Relative movement between the station and the anchor is indicated by an adjustable block threaded on the wire which moves along a steel rule. This type of equipment can be easily manufactured in the mine shop, and need not take the exact form shown in Figure 6.5.

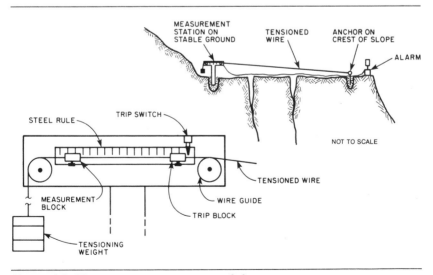

Figure 6.5. Wire extensometer on crest of slope.

Wire extensometers have much the same advantages and disadvantages as crack width measurements. The main advantage is that it is easier to measure movement over longer distances so that the measurement station can be established some distance from the unstable crest. In addition, the wire can be extended over the crest so that movement of the toe of the slope can also be measured. However, if this is done, corrections for thermal expansion and contraction of the wire may have to be made.

Another feature of the wire extensometer is that it can incorporate an alarm device to signal operators in the pit bottom when the pit should be cleared because of sudden movement of the slide. In Figure 6.5 the alarm device shown consists of a switch, which is mounted on the measurement station that is tripped by a second block threaded on the tensioned wire.

Surveying

As the slide becomes larger, crack width measurements will probably not be possible because a stable reference point will be difficult to find. Therefore, remote measurement using standard surveying techniques will be required. The selection of the most appropriate method will depend upon the degree of accuracy required and the physical restraints at the site.

The general principles of any surveying technique are shown in Figure 6.6. Instrument stations are established opposite the slide, and their positions are determined from a reference station on stable ground some distance from the pit. It is essential that the position of the instrument stations be checked against the reference, because the slope beneath the instrument stations may also be moving. Monitoring points are established on the slide, and by regularly determining their positions the movement of the whole slide can be found. These points should also be established behind, and to either side of, the expected extent of failure, so that the limits of instability as well as any increase in its size can be determined.

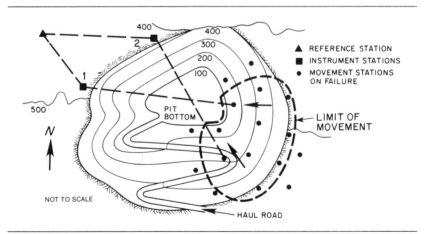

Figure 6.6. Remote measurement of slope movement.

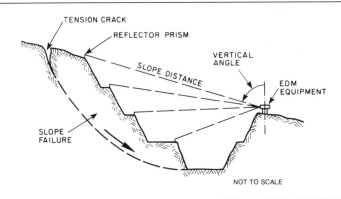

Figure 6.7. Slope distance measurement with EDM equipment.

Electronic distance measuring

Although standard triangulation methods can be employed, the widespread availability of electronic distance measuring (EDM) equipment has made it possible to measure displacement to an accuracy of less than 15 millimeters (0.6 inches) over sight distances of 2 kilometers (1.2 miles). These readings are made in a few seconds so that an almost continuous record of movement can be maintained. The instruments also have built-in corrections for variations in temperature and barometric pressure, and can be used at night if targets are illuminated. The targets themselves consist of reflector prisms costing as little as $75 each, which can be readily mounted on pieces of reinforcing bar driven into the ground, or grouted into drill holes. Most instruments employed in surveying work use an infrared beam which is adequate for most applications. Over extreme distances a laser instrument may be required. One disadvantage of the surveying technique is that it is not possible to make readings during heavy rainstorms or snowstorms, or when clouds obscure the targets, and thus a backup system of extensometers may be useful during extended periods of poor weather. Access to the slope to inspect prisms is also desirable.

The simplest method of surveying involves measuring the distance between the instrument station and prisms on the slope, as can be seen in Figure 6.7. For this method to be accurate it is essential that measurements be made parallel to the expected direction of movement; otherwise only a component of the movement will be measured. This is illustrated in Figure 6.6 where the northern half of the slide, which is moving west, is monitored from station 1 and the southern part, which is moving northwest, is measured from station 2. Information on the approximate vertical movement can be obtained by measuring the vertical angle to each station as well as the slope distance. This will give an indication of the mode of failure, since a toppling failure will tend to move horizontally, while in a circular failure the prisms will tend to move parallel to the failure surface.

Much additional information on the mechanism of slope failure can be obtained by finding the coordinates and elevation of each station, from which vectors of movement between successive readings can be calculated. A number

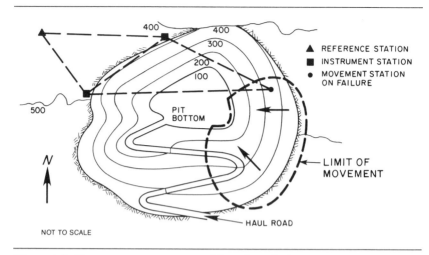

Figure 6.8. Slope movement measured by triangulation.

of ways in which this can be done are illustrated in Figure 6.8. If there is only one instrument station, angles can be turned from the reference station to each prism, and the distance measured with EDM equipment. If there are two instrument stations, the position of each prism can be determined either by triangulation, or by trilateration using EDM equipment. Best results are obtained if the three points form an equilateral triangle, and this should be taken into account when setting out the baseline between the instrument stations.

Another alternative, which does not require the measurement of any angles, is to determine the distance of the prisms from three stations forming a tetrahedron (Hedley).

EDM measurements are rapid and accurate, and surveying is useful in that it gives the three-dimensional position of each prism. Surveying does have the disadvantage though that the measurements and the calculations are time-consuming and results are not immediately available. Triangulation, under ideal conditions, using a 1-second theodolite with all angles doubled, and an EDM measuring to ±1 millimeter (±0.04 inch) over sight distances of 305 meters (1,000 feet), can give errors in coordinate positions of as little as 3 millimeters (0.12 inch) (Yu and Hedley). However, it is likely that mine surveyors doing routine measurements in all-weather conditions using equipment in less than perfect adjustment will obtain average errors of ±102 millimeters (4 inches) to ±152 millimeters (6 inches). For this reason, coordinate determinations should only be carried out when the expected movement distance between readings is greater than the magnitude of error.

Photogrammetry

On some large slides where it is not possible to survey the whole moving area, the use of photogrammetry may be considered. It is likely that the minimum error in coordinate position that can be obtained with this method is ±152 millimeters (6 inches); and while wide coverage will be obtained, results will not be available for several days or even weeks, and photographs can only be taken on cloud-free days.

Borehole methods

In many cases, it is useful to know the position of the failure surface so that the volume of the sliding mass can be calculated, the type of failure identified, and stability analyses carried out. Two methods can be used:

1. The "sond" method consists of drilling a hole to below the expected depth of the surface, casing the hole, and then lowering a length of steel on a piece of rope down the hole. As the slope moves, the casing will be bent, and eventually it will not be possible to pull the steel past the distortion. This will indicate the base of the failure plane. In a similar manner, a sond lowered from the surface will show the top of the failure.
2. If the positions of several failure surfaces and the rate of movement is required, an inclinometer instrument can be used. This instrument precisely measures small movements over the length of the hole and also gives the plan direction of movement (Soiltest, Inc.). However, if the movement rate is great, the casing will bend at the failure surface, and this expensive instrument may be lost in the hole.

Interpretation of Movement Results

In order to use monitoring to successfully decide when operations may, or may not, continue below a moving slope, the movement data must be correctly and rapidly interpreted. It is also imperative that everyone working in the pit understands the purpose and value of the monitoring system. It has been found that greater acceptance will be achieved if the system is explained to the union executive and equipment operators.

The most useful method of displaying movement data is to plot cumulative slope movement against time illustrated in Figure 6.9. This graph will readily show any increase in the rate of movement that is indicative of deteriorating

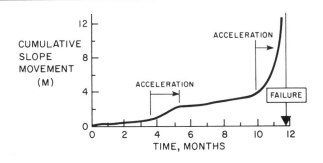

Figure 6.9. Typical shape of movement/time plot preceding failure.

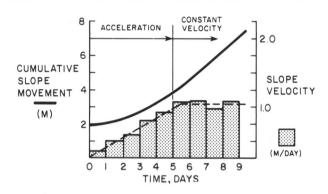

Figure 6.10. Velocity/time plot highlights changes in movement rate.

stability conditions. Since the appearance of the graph is dependent on the scales chosen for the axes, these dimensions should be carefully selected to insure that acceleration is clearly identifiable. This means that monitoring should start when instability first becomes apparent so that the steady rate of movement can be established. The frequency of measurements will depend upon the rate of movement and whether equipment is operating in a dangerous location. In non-critical areas monthly readings may be sufficient, but in critical areas hourly readings may be necessary. Figure 6.9 shows that sometimes several cycles of acceleration may occur before failure and that the total displacement is usually substantial. Also, the acceleration period will often have a duration of several days or weeks, thus producing an adequate warning of failure. However, it should be noted that planar type failures may occur with much less warning.

Further information on stability can be obtained by plotting movement velocity against time, where the gradient of the graph indicates the acceleration of the slope, seen in Figure 6.10. In this figure, the slope accelerated for the first five days and then moved at a constant velocity. If it were necessary to halt operations during the acceleration period, it may be possible to start mining if the slope continues to move with zero acceleration. Frequent monitoring would be required under these circumstances.

Monitoring data can also provide information on aerial extent, mechanism, and depth of failure. In Figure 6.11, contours of slope velocity plotted on a pit plan show the size of the slide and the fastest moving areas. These contours can be used to decide how mining should be scheduled to minimize production losses. For example, for the slope failure shown in Figure 6.11, all mining should be completed as soon as possible in the southeast corner of the pit before failure of the fastest moving area of the slide takes place and buries the coal. Although movement of the south wall may eventually cut the haul road, the present rate of movement is slow, and this road may continue to be used while a new access is put in.

Plan plots of displacement vectors obtained from triangulation can show the direction of movement, and vectors plotted on section often have dip angles

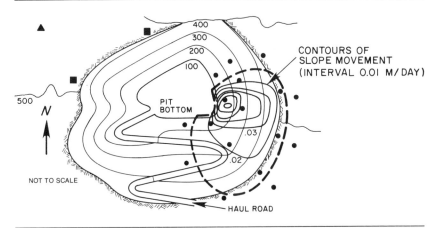

Figure 6.11. Contours of slope movement show extent and relative movement rates of slide.

parallel to the failure surface beneath them, which may indicate the geometry and mechanism of failure, illustrated in Figure 6.12. Thus, in a circular failure, prisms near the crest will tend to have movement vectors with dip angles as great as 45°, while prisms at the toe will move approximately horizontally or even slightly upwards. If the toe of the failure decreases in elevation as the pit is deepened, the dip angle of the movement vector of the lower prism in Figure 6.12 would increase, showing that the slide was becoming deeper.

If monitoring of a large slide continues for some time, a considerable amount of data will soon be accumulated and the plotting of movements and vector will become most time-consuming. In fact, it may not be possible to make interpretations fast enough to make critical production decisions. Fortunately, storage of

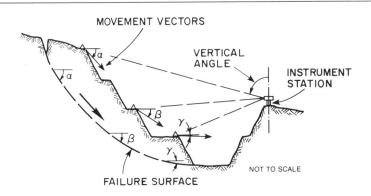

Figure 6.12. Dip angle of movement vectors show approximate depth of failure.

survey data, calculation of vectors, and plotting of movement graphs is an ideal application for computers. In this way, movement plots of any stations over any time span can be prepared in a few minutes.

A Case Study of the Application of Movement Monitoring

The failure discussed in this case study occurred at the Cardinal River Coals Ltd. mining operation located in the Alberta foothills approximately 300 kilometers (186 miles) west of Edmonton. Mining in the area, which is known locally as the Coal Branch, was continuous between 1900 and 1950 from a variety of sites, but between 1955 and 1968 mining activity was abandoned.

The present mine commenced production from the old Luscar mine in 1968. Production, which comes from a number of open pits, has increased in several increments from one million tons annually to just under two million tons in 1977. The coal being mined is of high-grade, metallurgical quality.

Mine Geology

The coal occurs in the Luscar Formation which is of Lower Cretaceous age. Two major faults, the Nikanassin Thrust to the southwest and the Folding Mountain Thrust to the northeast, form the boundaries of the Luscar Formation in the area. Intense deformation has folded these rocks into a series of synclines and anticlines which trend approximately north 60° west. Minor faulting and widespread jointing has accompanied the folding. Figure 6.13 is a typical cross section of the southern end of the property.

The rocks of the Luscar Formation consist of an interbedded sequence of shales, sandstones and coal. There is one minable coal seam which is known locally as the Jewel Seam. This seam varies in thickness from 9.1 to 12.2 meters (30 to 40 feet) along the limbs of the folds to 61 meters (200 feet) at the crests and troughs.

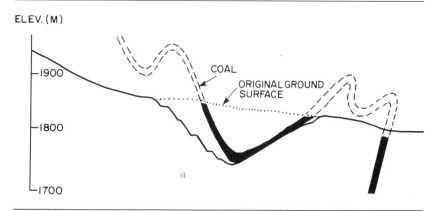

Figure 6.13. Cross section through southern end of coal property.

Mine Operations

Mining is carried out along four baselines, which generally follow the outcrop of the coal and are parallel to the trend of the synclines and anticlines. A series of pits is designed along each of the baselines. These pits vary in size from 3.8 million to 34.4 million cubic meters (5 million to 45 million cubic yards), and in stripping ratio from 5 to 1 to 15 to 1 clean coal.

Currently there are three active pits. These were designed for bench heights of 10 meters (33 feet), a safety berm every second bench, 24.4-meter- (80-foot-) wide roads, and a maximum road grade of 8 percent. Wall slopes were based on geotechnical evaluations made by consultants, and vary in angle from 35° to 55°.

The complexity of the geology requires exploration drilling at centers of 30.5 meters (100 feet) for final designs. Coal interpretation is done from cross sections, and pit designs are made on plan from coal contour maps made from the cross sections.

Mining Plan

Cardinal River Coals Ltd. uses a conventional truck and shovel open pit mining method. The annual mining rate is 1.94 million tons of clean coal, and 13.6 million bank cubic meters (17.8 million bank cubic yards) of rock, which gives an average stripping ratio of 9.2 to 1. The mine operates 24 hours per day, seven days per week, with shutdowns only for statutory holidays.

The mining operation follows the following sequence:

1. *Rock Removal.* All rock is drilled using four Bucyrus-Erie 45-R drills, and then blasted with a powder factor of about 1.7 kilograms per cubic meter (1 pound per cubic yard). Four P & H 2100 shovels are used, with 26 Unit Rig M-100 and Wabco 120B rock trucks. Shovel production averages 9,940 bank cubic meters (13,000 bank cubic yards) per scheduled day.
2. *Coal Loading.* Coal that has been exposed by the shovels is removed using front-end loaders, and rock trucks are used to haul the coal either to the pit raw coal stockpiles or to the preparation plant. The final pit bottom coal is removed by dragline, which can amount to 10 percent of the coal in the pit when mining a syncline.
3. *Back Filling and Reclamation.* The last step in pit development is backfilling and reclamation. Backfilling is not only part of the pit reclamation program, but is instrumental in stabilizing the pit walls. Following backfilling, the final surface is shaped approximately to the original contours, then covered with soil material and seeded.

Movement Monitoring Program

The 50-A-2 Phase II pit was designed to expose 1.33 million tons of raw coal by moving 5.7 million bank cubic meters (7.45 million bank cubic yards) of rock. It extended from the 1,860-meter (6,100-foot) elevation to pit bottom at the 1,710-meter (5,600 foot) elevation. The mining operation in this pit was carried out almost entirely under a moving wall.

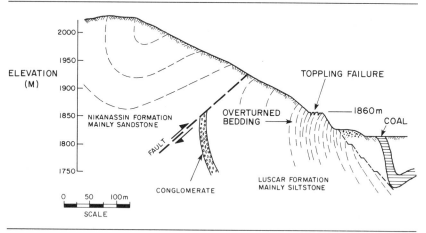

Figure 6.14. Slope failure, March 1975.

The slope failure occurred on the south side of the pit where the coal seam was folded into an asymmetrical syncline, the southern limb of which is near vertical, as shown in Figure 6.14. The Luscar Formation forming the footwall was comprised mainly of a sandy siltstone which was highly to moderately broken (average RQD = 25 percent), and contained several carbonaceous and clay seams. A thrust fault, which at this point lies about 305 meters (1,000 feet) from the coal seam, had caused the Nikannisan Formation, which is comprised mainly of a more competent sandstone (average RQD = 50 percent), to overlie the Luscar Formation.

On the pit crest at an elevation of 1,860 meters (6,100 feet), the bedding in the footwall was overturned and dipped into the slope at an angle of about 70°. Since the bedding spacing at this location was about 2.1 to 3.05 meters (7 to 10 feet), the beds formed tall, narrow columns that underwent a toppling failure, seen in Figure 6.2.

Mining commenced in March 1974, and by October 1974, when mining on the 1,840-meter (6,033-foot) bench, cracks were observed on the 1,860-meter (6,100-foot) bench. In February 1975, when the pit bottom was at 1,810 meters (5,933 feet) deep, new cracks due to horizontal toppling movement were observed on the pit crest. At this time, it was essential that the next bench be mined back to the design line at the toe of the slide to insure that all the coal could eventually be mined in the pit bottom. Therefore, a monitoring system was set up on the 1,860-meter (6,100-foot) bench, which consisted of wire extensometers, seen in Figure 6.5, and slope distance measurements with a Kern DM1000 infrared distomat, and 70-millimeter- (2.75-inch-) diameter reflector prisms. It was found that the rates of movement measured by the two systems were comparable; so it was decided that the extensometers could safely be used to measure movement during night operations. The extensometer incorporated a trip switch that activated a bank of flashing lights when a movement exceeding 25.4 millimeters (1 inch) occurred.

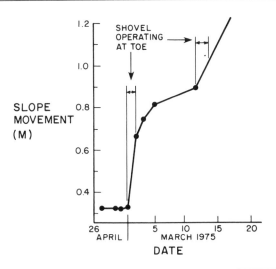

Figure 6.15. Slope movement, March 1975.

When readings were first made in February 1975, the average rate of movement of the crest of the pit slope was 15 millimeters (0.6 inches) per day, but several increases in the rate of movement occurred when the shovel was digging at the toe of the slide, illustrated in Figure 6.15. At this time, readings were taken as frequently as once every hour, and operations would continue until the movement rate reached 25 millimeters (0.96 inches) per hour, at which time operations would be halted and the equipment moved out. However, the movement rate soon decreased, and after about 10 days when the rate was again 15 millimeters (0.6 inches) per day, operations continued. In this way the slope was mined back to the design toe with only a slight delay to the mining schedule. Despite a total movement of about 7 meters (23 feet) no significant failure took place and little movement occurred at the toe of the 1,860-meter (6,100-foot) bench.

By April 1975, the rate of movement had decreased to 6 millimeters (0.24 inches) per day, and this steady deceleration continued throughout the spring. At this time, the wire extensometers were abandoned, and monitoring was continued using slope distance measurements only. Prisms were established on the 1,860-meter (6,100-foot) bench at an elevation of 1,900 meters (6,230 feet) on the hillside above. Over the next year to March 1976, movement continued at an average rate of 6 millimeters (0.24 inches) per day, and although a total movement of about 1.8 meters (6 feet) occurred (see Figure 6.16), mining continued uninterrupted in the pit below.

In April 1976, the slope started to accelerate, and over the next two months a total movement of about 30.5 meters (100 feet) occurred on the hillside above the pit, and the maximum movement rate reached several tens of meters/feet per day. The plot of slope movement against time gave adequate warning of deteriorating stability conditions by demonstrating the acceleration of the slope, and a

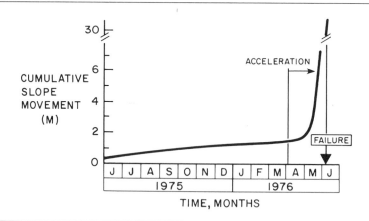

Figure 6.16. Typical movement plot during year preceding failure.

plot of acceleration clearly showed that failure was imminent (Figure 6.17), and mining was abandoned. The width of movement was about 244 meters (800 feet), and it extended from the hilltop at an elevation of 2,030 meters (6,650 feet), where tension cracks several meters wide and up to 9.1 meters (30 feet) deep (see Figure 6.18) opened to an elevation of about 1,800 meters (5,900 feet). This was shown clearly by contour plots of movement. The toe of the slide was in the original toppling failure and did not extend to the pit bottom. This was confirmed by putting in a sond from the 1,860-meter (6,100-foot) bench. This extensive and rapid movement had the effect of causing further deformation of the overturned beds below the 1,860-meter (6,100-foot) bench until the dip angle had decreased

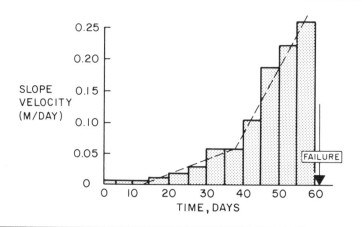

Figure 6.17. Slope velocity in the two-month period preceding failure.

Figure 6.18. Tension cracks on crest of slope, June 1976.

from the original 70° to about 30° (Figure 6.19). At this time, the most severely fractured beds on the crest of the pit failed, and two rockfalls totaling 573,450 bank cubic meters (750,000 bank cubic yards) occurred in early June 1976, which partially covered the exposed dragline coal in the pit bottom.

After the failures occurred the movement rate gradually decreased, and within about one month it was decided that there was no immediate risk of another failure occurring; so mining was restarted in the pit bottom to recover the exposed coal. In order to ensure the safety of the equipment operators, new prisms were established on the remnants of the pit crest and on the hillside above, and monitoring was carried out continuously. The field criteria for evacuating the pit was a movement exceeding 30 millimeters (1.2 inches) per hour. Furthermore, operations were not carried out at night, because adequate monitoring would have been difficult. Using this system, mining continued although the slope above was still moving as fast as 30 millimeters per day (1.2 inches). When

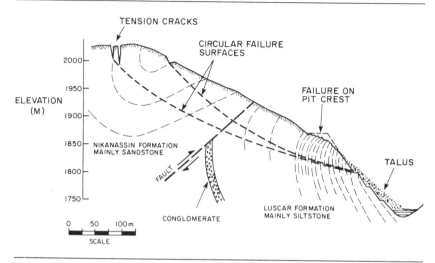

Figure 6.19. Slope failure, June 1976.

all the remaining coal that was exposed had been mined out, the pit was abandoned and backfilling was started. By March 1977, the backfill had reached the approximate elevation of the toe of the slide, and soon thereafter the movement rate decreased rapidly from about 6 to 0.6 millimeters (0.24 to 0.02 inch) per day, which is indicative of the support provided by the fill. This is illustrated in Figure 6.20.

Impact of Slope Failure

The immediate impact of this failure on costs was the expense of monitoring. The team for monitoring and evaluating slope movement consisted of:

Figure 6.20. Pit slope at completion of backfilling.

1. Two contract surveyors, working a 12-hour day, 7 days per week.
2. A geotechnical consultant.
3. A staff mine engineer to coordinate the monitoring activity.

During the four months the costs incurred (which do not include those for in-house personnel) were:

Contract surveyors	$ 76,000.00
Consultant fees	39,500.00
Monitor shelters (3 at $1,000)	3,000.00
Prisms (43 at $125)	5,375.00
Aerial photogrammetry and mapping	7,500.00
Total	**$131,375.00**

This was the first extensive monitoring program undertaken by Cardinal River Coals Ltd (CRC). It was successful in permitting an entire pit to be mined under a wall which was known to be moving. This experience proved monitoring to be an indispensable tool in the evaluation of wall stability. Monitoring now forms part of the ongoing program of geotechnical evaluation of all pits. This program consists of:

1. Monitoring all walls.
2. Piezometer nests in all walls.
3. Monthly geotechnical inspections and a monthly geotechnical report consolidating mining activities, weather data, piezometer readings, monitoring results, and geotechnical inspection reports.

The 50-A-2 failure has underlined the importance of wall control. In addition to the above geotechnical evaluation program CRC has also introduced a program for controlled blasting in all final walls, and modified pit designs to reduce the length of the pits along strike. Through these procedures it is hoped that there will be no reoccurrence of the 50-A-2 Phase II slide and subsequent loss of coal.

Acknowledgment

The authors would like to express their appreciation to Cardinal River Coals for making available data used in the case history described in this chapter.

References

Brawner, C. O., Stacey, P. F., Stark, R., 1976, "A Successful Application of Mining with Pit-wall Movement," Canadian Mining and Metallurgical Bulletin.

Goodman, R. E., Bray, J. W., 1976, "Toppling of Rock Slopes." Speciality Conference on Rock Engineering for Foundations and Slopes, Boulder, Colorado.

Hedley, D. G. F., "Triangulation and Trilateration Methods of Measuring Slope Movement." Canadian Department of Energy, Mines and Resources, Mines Branch, Mining Research Center, Internal Report 72/69.

Hoek, E., Bray, J., 1977, "Rock Slope Engineering." Second edition, IMM, London.

Kennedy, B. A., 1971, "Some Methods of Monitoring Open Pit Slopes." 13th Symposium on Rock Mechanics, Urbana, Illinois.

Oriard, L. L., 1970, "Dynamic Effects on Rock Masses from Blasting Operations." Slopes Stability seminar, University of Nevada.

Soiltest, Inc., Illinois, "Slope Meter Detects Angular Deviations from Vertical."

Watt, I. B., 1970, "Control of Early Warning of Potential Danger in Open Pits." Symposium on Planning Open Pit Mines, Johannesburg, South Africa.

Yu, Y. S., Hedley, D. G. F., "A Trial of Monitoring Slope Wall Movement at Hilton Mines Using a High Precision Theodolite." Canadian Department of Energy, Mines and Resources, Mines Branch, Mining Research Center, Internal Report 73/18.

The Influence of Blasting in Mine Stability

7

T. N. Hagan, Senior Mining Engineer,
ICI Australia Ltd., Melbourne, Australia,
J. S. McIntyre, Senior Mining Engineer,
Utah Development Company, Goonyella, Australia,
and G. L. Boyd, Senior Engineering Geologist,
Utah Development Company, Brisbane, Australia

Introduction

In a large overburden blast at any one of several Australian coal strip mines, energy is liberated at a rate of approximately 7.5×10^8 kilogram calories per second (2.32×10^{12} foot-pounds per second). This is almost an order of magnitude greater than that generated by North America's combined electrical power plants during peak load periods. But this colossal explosive's energy concentration is heightened by geometrical as well as temporal limits of the blast. Initially, at least, this 750 million kilogram calories per second of energy is liberated over an area less than half that of a regular ice hockey rink.

Where such explosive's energy concentrations are unleashed, therefore, it is not surprising that the structural strength of the rock alongside the blasted volume can be reduced to a very small percentage of its original (in situ) value. New fractures and planes of weakness are created and (naturally-occurring) discontinuities which may well have been tight and/or weakly cemented initially are opened up and loosened. As a result, there is an overall increase in rock mass discontinuity, and the fractured wall is left with a greater instability and rock fall potential.

The stabilities of newly exposed highwalls in surface coal mining operations are important to safety, productivity, and profitability. Large failures can result in appreciable delays to operating schedules.

Factors contributing to instability in the highwall are:

1. The textural and, more particularly, the structural characteristics of the overburden.
2. Groundwater.
3. Blasting.

The relative effects of these factors on stability can vary considerably from one mine to another and even between areas in any given mine. But even where blasting might prove the least influential, it must be appreciated that within the existing operational constraints, this factor is the most amenable to modification. Because the nature and effects of overburden blasting can be controlled within wide limits, the operator's efforts to promote stability can be most profitably directed by minimizing blast-induced damage.

In order to fully appreciate the influence of blasting on pitwall stability, of course, it is first necessary to understand:

1. How blasts create overbreak.
2. The relationship between overbreak and instability.

These are prerequisites of effective blast design.

Overbreak Mechanisms

There are six identifiable breakage mechanisms which are responsible for blast-induced overbreak:

1. Crushing.
2. Radial cracking.
3. Internal spalling.
4. Fracturing along boundaries of modulus contrast.
5. Gas extension of fractures.
6. Release-of-load fracturing.

The above mechanisms are listed in their approximate chronological order of occurrence but by no means in their order of importance. In most sedimentary strata, mechanisms 5 and 6 (see list above) have by far the greatest ability to create excessive overbreak.

The intensity and extent of overbreak resulting from mechanisms 1 to 4 are influenced almost entirely by the characteristics of the rock and the back-row charges. The presence of an effective traveling free face (in front of back-row blastholes) does not prevent fracturing by these mechanisms, the amount of breakage produced in the solid stratum beyond the blast boundary being equal to that in the burden rock. The nature of overbreak created by mechanisms 1 to 3 is shown in Figure 7.1. Mechanisms 5 and 6 (the potentially dominant modes of overbreak) are affected by these two factors and, to a greater degree, by several design parameters of the entire blast (for example, burden, initiation sequence, and others).

Crushing

An annular zone of crushed or permanently compressed material is often formed immediately around the blasthole wall. This occurs where the peak of the cylindrically expanding radial compressive wave (generated by the charge) exceeds the dynamic compressive breaking strain or the plastic yield of the material. Fracture/deformation occurs during a period of volume compression through the collapse of intercrystalline or intergrain structures. At points beyond the

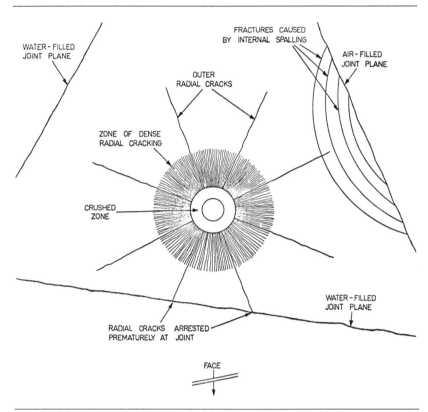

Figure 7.1. Crushing, radial cracking around a back-row blasthole and internal spalling are shown in this sketch.

crushed zone boundary, the peak compression of the primary wave falls to a value below that required to propagate this mechanism (Hagan, 1977a).

The thickness of this annular zone is usually small, typically less than twice the blasthole diameter d. But where a very powerful explosive is used in a highly porous material, this thickness can be as great as about $7.5d$. Should the rock be well-bedded and/or well-jointed, as is usual in formations overlying coal, the crushed zone thickness will be appreciably less than $7.5d$.

Radial cracking

When the strain wave front passes, a cylindrical shell of rock immediately around the blasthole (or, if crushing occurs, around the crushed zone boundary) is subjected to intense radial compression; and tangential tensile strains develop. If these strains exceed the dynamic tensile breaking strain of the rock, a zone of dense radial fractures is formed immediately around the blasthole (or crushed zone boundary). This intense radial fracture zone terminates quite abruptly at that radial distance where the wave's tangential tensile strain attenuates to a value which is incapable of generating new cracks. The width of this zone decreases with:

1. A decrease in the peak strain at the blasthole wall.
2. An increase in the rock's dynamic tensile breaking strain.
3. An increase in the rate of (strain wave) energy absorption in the rock.

Beyond this inner zone, some radial cracks, evenly distributed around the blasthole, are propagated into the surrounding rock mass, provided wave-induced tension is applied normal to the crack tip (see Figure 7.1).

Where the peak strain at the blasthole wall cannot cause crushing, crack lengths in both the inner and outer radial fracture zones increase linearly with peak strain. Peak strains exceeding those that are required for crushing do not increase the length of the crack, but result only in additional crushing around the blasthole.

An approximately linear relationship exists between crack extension and length of preexisting cracks; longer cracks are preferentially extended. Where subvertical joints intersect the blasthole alongside the charge, such joints will open under the action of the strain wave and limit the development of radial tensile fracturing in other directions. Joints which are parallel to but some distance behind the blasthole, interrupt and arrest radial crack propagation (see Figure 7.1).

In the coarser-grained overburden strata (e.g., sandstone), joints are usually broadly spaced and extensive; the probability of blastholes being remote from a joint, therefore, is relatively high, and the full development of radial cracking is promoted.

In finer-grained strata (e.g., siltstone, claystone), joints are more closely spaced and of more limited extent and, therefore, are more capable of localizing radial cracking through preferential extension of the joint(s) and radial crack termination.

Backward-facing radial cracks from a given blasthole tend to create interlocking segments of rock beyond the blast perimeter. Where radial cracks from adjacent blastholes overlap and/or intersect other discontinuities, the newly formed pit-wall may suffer local wedge or block failures.

Internal spalling

Where the compressive strain wave strikes an effective free face (usually a pronounced rock/air interface), a reflected tensile wave is created. If this reflected wave is sufficiently strong (the strength of rock in tension usually being only about 5 to 15 percent of that in compression or shear), spalling or slabbing occurs progressively from the effective free face back towards the blasthole.

Insofar as they affect reflection of the compressive wave, wide air-filled cracks, joints, or bedding planes beyond the peripheral blastholes can be regarded as effective free faces. Internal spalling produces more intense overbreak between the blasthole and such discontinuities but, because of reflection and dispersion of the outward-propagating strain wave, overbreak beyond the joint is reduced (see Figure 7.1).

Because spalling of the (external) free face does not normally occur for the burden distances commonly used in rock blasting, it can be assumed that internal spalling can occur only within about 25d of the blasthole. Internal spalling is expected to become appreciable, however, only within about 10d of the blasthole.

At water-filled cracks, fine air-filled cracks, moderate density changes in the strata, etc., the fraction of the strain wave energy reflected is usually insufficient to cause internal spalling, most of the strain energy being reflected (see Figure 7.1).

In highly jointed strata, internal spalling together with radial fracturing (and perhaps crushing) encourages intense overbreak between the blasthole and the first set of joints which surrounds the blasthole. Overbreak beyond this first set of joints is not caused to any significant degree through these three mechanisms. Where the mean distance between consecutive joint planes is small, therefore, the extent of fracture resulting from these modes of overbreak is also small. But, as a result of the dominant effects of mechanisms 5 and 6 (see discussion under "Gas Extension of Fractures" and "Release-of-Load Fracturing"), well-bedded and well-jointed strata usually exhibit a greater overall degree of overbreak.

Fracturing along boundaries of modulus contrast

Propagation of the compressive strain wave along adjacent beds which possess a significant contrast in deformation moduli will lead to differential strains being generated at the boundary. The magnitude of this strain differential may well cause shear fracturing by relative displacement of the beds.

This mechanism of fracture varies with distance from the blasthole as described in the following paragraphs.

1. In the rock immediately outside the crushed zone, the outward-propagating strain waves in the contrasting beds are approximately in phase, i.e., rock elements lying opposite each other across the boundary experience differing degrees of compression. In this instance, breakage will ensue only if the modulus contrast between the beds is sufficiently large for the resulting strain differentials to exceed the shear yield strain of the boundary.
2. At greater distances from the blasthole, the strain waves become increasingly out of phase (since wave velocity is a function of material density and deformation modulus), the wave front being ahead in the bed with the higher deformation modulus. Following the compressive wave is a zone of tensile relaxation. Beyond a certain distance from the blasthole, therefore, compressive and tensile phases of the wave are in opposition, and rupture along the boundary results from strain differentials exceeding the shear yield strain of the boundary.

Beyond this outer zone, the combined effects of wave attenuation and wave front separation are insufficient to cause extension of the bedding plane rupture by this mechanism.

This mode of rupture may well explain the presence of first-time fractures as well as fresh slickensided bedding surfaces observed in the base of pitwalls (high-walls), the toe of the wall lying coincident with or immediately behind the back-row blastholes of the previous strip.

Gas extension of fractures

During or after the formation of radial cracks and internal spalling, the gases start to expand and penetrate into strain wave-induced fractures and natural discontinuities. The high-pressure gases jet into, wedge open and, hence, extend the radial cracks. Without subvertical joints, this mechanism can increase the lengths of strain wave-generated radial cracks by a factor of about 5 (Dally et al., 1975).

Because stress concentration at the crack tip increases with crack length, the longest crack (whether this be a joint or a strain wave-generated crack) is the least stable and requires the lowest critical pressure. Therefore, longer fractures always extend first and propagate at a higher velocity than shorter adjacent fractures. The further they get ahead, the greater is the velocity difference until the shorter ones stop altogether. Where a well-defined discontinuity intersects the blasthole alongside the charge, therefore, high gas flows cause this joint to be preferentially expanded by a wedging action.

In well-bedded and well-jointed strata, the extensions of blast-generated cracks are usually completely masked by those of natural discontinuities. Indeed, pre-existing cracks such as joints and bedding planes frequently dominate both the nature and extent of fracturing beyond the design boundary of the blast. Because major crack development is along such joints and bedding planes, uniform breakage in the overbreak zone occurs only where such discontinuities are closely spaced and quite evenly distributed throughout the rock mass. Where joints are on wide centers and/or exhibit a very uneven distribution, fragmentation in the overbreak zone will be coarse and the profile of the newly formed highwall will be highly irregular.

Where the longitudinal axis of the blasthole is normal to discontinuities (usually bedding planes), the widening and extension of these can be assisted only to a limited extent by the strain wave (see section under "Fracturing Along Boundaries of Modulus Contrast"). The invasion of high-pressure explosion gases is almost totally responsible for the wedging open and extension of bedding fractures well beyond the blast perimeter.

The high-pressure explosion gases use the radial cracks and any natural discontinuities which intersect the charged section of the blasthole as access routes to fractures created by internal spalling and, more particularly, the network of joints and bedding planes beyond the immediate vicinity of the blasthole.

Where bedding planes (perhaps already weakened by the mechanism described in "Fracturing Along Boundaries of Modulus Contrast") are horizontal or near horizontal, the upward push exerted (by the invading gases) on the walls of this opening tends to cause rotational uplift and, hence tensile fracture across the bedding. The amount of breakage by this mechanism will increase towards the top of the bench, with the brittleness of the beds, and with decreases in the thickness of the beds.

Quite unlike weak, close-jointed and close-bedded strata, strong massive rocks rarely exhibit visible cracking behind back-row blastholes, especially where the burden distance, initiation sequence, and delay timing facets of blast design have received sufficient attention. In these circumstances, therefore, the penetration of backward-facing radial cracks by high-pressure gases is considered to be relatively limited.

Release-of-load fracturing

The initial strain wave never carries more than 20 percent of the blast energy into the rock (Harries, 1973). Before the strain wave reaches the (effective) free face, however, the total energy transferred to the strata by the initial compression of the rock is claimed to be as much as 60 to 70 percent of the blast energy (Cook et al., 1966). After the compressive strain wave has passed, a state of quasi-static equilibrium exists, the pressure of blasthole gases being balanced by the strain at the blasthole wall (or at the crushed zone boundary). When the pressure in the blasthole subsequently falls (as gases push the burden rock forward and escape via the stemming column), this strain energy is very quickly relieved rather like a compressed coil spring being suddenly released. The rock mass beyond the blast boundary responds by failing in tension in planes parallel to the blast. The tensile forces so generated will preferentially seek out subvertical joints.

When large multirow blasts with an in-line initiation sequence are fired (see Figure 7.2), therefore, all blastholes in any given row act in unison to create tensile fractures in the rock mass parallel to and over the length of the back row. As one would expect, there is also some reinforcement of the effects (beyond the

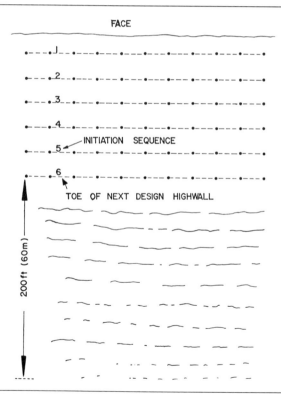

Figure 7.2. Subvertical fractures created by release-of-load behind a square-in-line overburden blast.

back row) produced by the individual rows. In coal measure sediments, large blasts of this type can cause vertical release-of-load cracks parallel to and up to 60 meters (200 feet) behind the next highwall face (see Figure 7.2).

Instability and Controlling Factors

The association between overbreak mechanisms and pitwall instability, considered here, draws from investigations and observations at a number of open pit coal strip mines in Central Queensland, Australia. At each operation, blasting of overburden material has been carried out predominantly by multirow in-line firings, with some experimentation made into the staggered "V1" and square "V" patterns.

Overbreak from square in-line blasting is observed to take the following form:

1. Release-of-load tensile fracturing up to 60 meters (200 feet) behind back-row blastholes.
2. Deformation of bedding planes exposed in the highwalls, probably resulting from mechanism 4 and/or mechanism 5.

Overbreak mechanisms 2 and 3 have not been observed at these operations. Mechanism 1 is observed as the *bulling* (i.e., increase in diameter) of the near-surface section of the blasthole by compression through nonelastic deformation of weathered clay sediments.

Geological setting

The geological setting in which blast-induced overbreak is developed is represented in Figure 7.3, with dominant beds described in Table 7.1. Pertinent features of the geology include the following:

Weathering. Weathering extends between 15 and 40 meters (50 and 130 feet) below ground surface, with greatest penetration achieved in sandy facies sedi-

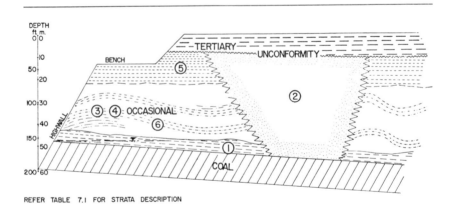

REFER TABLE 7.1 FOR STRATA DESCRIPTION

Figure 7.3. Representation of overburden geology.

ments via extensive jointing and higher relative porosity. Products of weathering include kaolinized feldspar, montmorillonite clay, and mixed layered montmorillonite-illite-chlorite clays.

Bedding. Bedding, either as thin laminae or thick beds, is sheetlike and can persist in the pitwall over distances up to 1,370 meters (4,500 feet), commonly 460 meters (1,500 feet). Similar persistence down dip can be demonstrated.

Although the regional dip of beds including coal and Unit 1 sediments (see Figure 7.3) is 3° to 7° east, postdepositional warping of Units 3, 4, 5, and 6 (see Figure 7.3) locally cause westerly dip directions (i.e., towards the pit).

Some bedding planes (recovered in core taken at sites remote from the operation) exhibit natural polished and faintly slickensided clay surfaces which appear weakly cemented. Secondary, or conjugate surfaces to bedding, which are curviplanar in shape, generate from polished bedding surfaces but attenuate usually within 3 to 5 meters (10 to 15 feet).

Jointing. The relative orientation of regional joint sets to the dominant orientation of pits results in one joint set usually at a shallow angle to, and often exposed in the pitwalls. Joint sets are subvertical and widely spaced up to 19 meters (60 feet) in massive sandstone (see Unit 2, Figure 7.3), reducing to between 1.5 and 6 meters (5 and 20 feet) in fine-grained, layered strata (see Units 3, 4, 5, and 6, Figure 7.3).

Groundwater

Groundwater appears to be confined within the coal seam. However, a piezometric level is occasionally found where water-filled subvertical joints are intersected in drill holes.

Regional groundwater levels measured over the lease area vary between 35 and 45 meters (115 and 150 feet) below ground surface.

Overburden properties

Relevant mechanical and chemical properties of overburden material representative of the principal strata (refer to Figure 7.3 and Table 7.1) are summarized in Table 7.2.

The principal effects of weathering on geomechanical properties are those associated with alteration of soft sedimentary rocks to fissured clay equivalents. The majority of material types lining or forming the surfaces to bedding and jointing deform on shearing according to the Terzaghi shear strength equation for clay materials.

$$\text{Shear strength } (\tau) \ = \ C + \sigma_n \tan\phi$$

where C is the cohesion

         ϕ is the angle of internal friction, and

         σ_n is the normal stress across the plane of deformation.

In addition, undisturbed (strain softening) clay materials subject to shearing stresses will yield (or displace) until their shear strength attains a peak value. With further displacement, shear strength decreases to a residual value, which

Table 7.1. Description of Dominant Stratigraphic Units

Unit	Texture	Clay composition		Structure
		(Minerals)	(%)	
1. Lacustrine muds	Interlaminated (20mm spacing) to thinly interbedded (60-200mm spacing) siltstone-claystone with minor sandstone and coaly beds; dips conformably with coal seam at 3°-7°	Illite-Montmorillonite	40-75	Sheet bedded, conformable with coal, with polished and slickensided bedding and off bedding shear planes
		Quartz	14-33	
		Kaolinite	7-12	
		Illite	5- 9	
		Chlorite	2- 5	
2. Channel sandstone	Fine-to medium grained trough and cross-bedded sandstone beds— thick (>600mm) to massive (>2000mm); dips are random, cross-beds dip up to 25° locally	Montmorillonite	28	Local channel filling deposits of limited strike length in pit, or as point bar deposits associated with channel accretion
		Illite	20	
		Kaolinite	21	
		Quartz	12	
		Feldspar	7	
		Dolomite	6	
		Chlorite	trace	
3. Overbank deposits	Interlaminated to medium interbedded (200-600mm) claystone/siltstone with minor sandstone (clayey), beds dip with coal seam or as modified by domal warps where dip angles vary between 5° and 15°	Illite-Montmorillonite	55-71	Rhythmically bedded (varved appearance) laterally persistent, often truncated, pinched out, subject to lateral facies change with (4) below
		Illite	12-22	
		Quartz	2-25	
		Kaolinite	3-17	
		Montmorillonite	0- 7	
		Feldspar	1- 5	
		Chlorite	trace	

	Description	Clay composition		Remarks
4. Sheet sandstone-siltstone	Fine-grained clayey sandstone and siltstone thickly bedded with light and dark grey clay rich partings/laminations, beds dip with seam or as modified by domal warps; dip angles vary between 5° and 15°	Illite-Montmorillonite Montmorillonite Illite Kaolinite Quartz Feldspar Chlorite	41-75 7-29 3-16 5-54 1-20 1 trace	Rhythmically banded, graded coarse upward, laterally persistent, subject to lateral facies change with (3) above
5. Drape deposits	Thinly laminated (<6mm), grey to dark brown claystone; beds dip with underlying structures, dip angles are between 3° and 6°	Illite-Montmorillonite Illite Kaolinite Quartz	70 14 7 9	Base assumes topography of surface to underlying material. Laminae are sheetlike, laterally persistent.
6. Coaly beds	Seam splits with associated over and under clays; dip with seam or modified by domal warps; dip angles are between 5° and 15°	Clay composition as for drape deposits		Laterally persistent, delineates domal warp structures; occasionally faulted.

Table 7.2. Summary of Mechanical and Chemical Properties of Overburden Materials

Unit (See Figure 7.3)	Condition	Density (kg/m³)	Tensile strength		Unconfined compressive strength (psi [MPa])	Young's Modulus (psi [MPa])	Poisson's Ratio	Anisotropy index (Ia)	Slaking test (ISRM) durability	Moisture (%)
			(Brazilian) (psi [kPa])	(Point load) (psi [kPa])						
1	Unweathered	2,430	450 (3,100)	6-180 (40-1,200)	3,400 (23.43)	48×10^4 (3,300)	0.26	3 to 6	med. (high plast.)	8
2	Unweathered	2,230-2,520	450-910 (3,100-6,270)	160-460 (1,100-3,200)	8,600-8,900 (59.25-61.32)	155×10^4 (10,700)	0.14-0.29	1 to 2	med.-high (high plast.)	11-14
∥	Weathered	1,930-2,250				$20\text{-}80 \times 10^4$ (1,380-5,500)		1 to 2		8-20
3	Unweathered	2,170-2,380	238 (1,640)	6-95 (40-650)	5,250-9,080 (36.17-62.6)	$3\text{-}60 \times 10^4$ (200-4,130)	0.13	1 to 4	med. (high plast.)	11
4	Unweathered	2,090-2,450	500 (3,445)	6-460 (40-3,200)	7,700-9,000 (53.05-62.01)	$22\text{-}160 \times 10^4$ (1,520-11,000)	0.18-0.23	1 to 3	low-med. (high plast.)	9-14
					No data for units 5 & 6					
Carbonate cemented sediments		2,610-2,720	538 (3,700)		13,232 (91.2)	302×10^4 (20,800)	0.37	≈1	high (med. plast.)	7

remains unchanged with continued displacement. Residual shear strength values are about 50 percent of peak shear strengths.

Increasing degrees of disturbance alter the clay fabric so that polished bedding surfaces develop reduced peak strength values and slickensided surfaces develop low to zero peak strength values. The shear strengths available (along such surfaces), therefore, are dictated by residual shear strength values.

Nature of instability

The results of field observations and investigations of pitwall instability at a number of open pit coal mines has defined the character of highwall failure and the controlling mechanisms. The typical development of highwall instability (summarized in Figure 7.4) includes:

1. Appearance of subvertical separation cracks at the surface behind and parallel to the wall in association with move out at the toe of the wall in Unit-3 or Unit-4 type sediments (see Figure 7.4a).
2. Mobilization of the mass over discrete bedding planes with lateral displacement associated with breakup of the mass into a network of upright blocks and columns; at this stage some collapse occurs at the toe (see Figure 7.4b).
3. Lateral and vertical displacement of columns at the rear of developing failure by sliding of these columns over an inclined failure surface which usually remains buried (see Figure 7.4c).

Elements of failure

The important elements of the mechanism promoting highwall instability (described above) were found to be:

1. Bedding surfaces forming the base plane of failure, invariably dipping toward the pit, and in most instances observed as possessing highly polished and slickensided clay surfaces (prior to failure).
2. Subvertical tension cracks or possibly surfaces of separation observed to reach depths equivalent to half the wall height, to form the head or scarp of the failure and to contribute to the breakup of the mobilized mass into blocks and upright columns (see Figure 7.4).
3. An inclined shear surface (that is not always observed) estimated from surface block movements to dip at about 50° to 55° to the horizontal, thus cutting across bedding planes.
4. The ingress of ground or surface water into the above failure surfaces.

These four elements controlling pitwall instability are summarized in Figure 7.5; their association with blast-induced overbreak is outlined in the following section.

Association Between Overbreak and Instability

The association between blast-induced overbreak and instability has been the subject of investigation at Utah Development Company's Goonyella mine, in

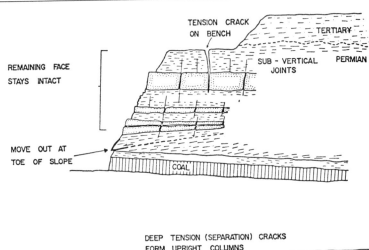

(a)

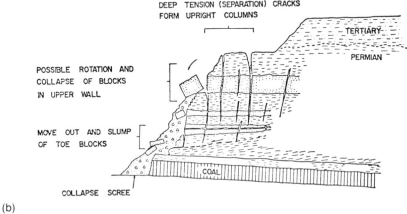

(b)

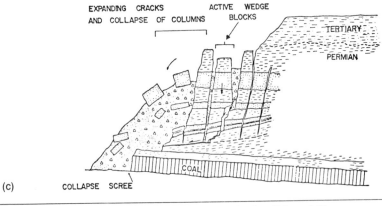

(c)

Figure 7.4. Sequence of deformations during highwall instability.

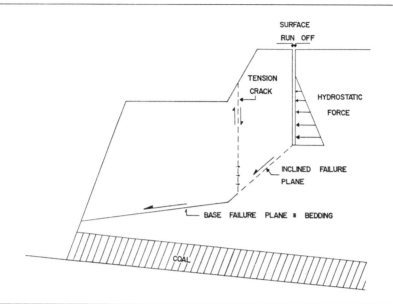

Figure 7.5. Dominant elements controlling highwall instability.

Central Queensland. Observations of a number of failures at this site have strongly suggested the following associations.

Base failure planes

As previously described under "Nature of Instability," each failure is associated with at least one discrete plane of rupture in pitwards-dipping sediments. This feature allows the mobilization of large sections of the highwall, sliding on bedding surfaces up to 2 millimeters (0.08 inch) thick. Particular features of these base failure planes observed prior to failure are as follows:

1. The surfaces are slickensided (grooved) and orientated normal to the strike of the pit, their degree of development and fresh appearance suggesting that their origins are due to recent (mining) displacements.
2. Measured displacements of recent slickensides vary between 20 and 50 millimeters (0.75 inch and 2 inches).
3. Extensive rupturing of bedding surfaces found developed between beds of marked competency contrast (e.g., see Table 7.2, carbonate cemented materials cf. Unit 3 sediments).
4. Displacement of undulating (wavelike) bedding planes leading to half phasing of these surfaces and, as a result, numerous elongate cavities develop along the bedding plane.

The displacements recorded above are more than sufficient to develop residual shear strengths along these surfaces, which is supported by similar strength being obtained from back analysis of monitored highwall failures. The extent to which blast-induced overbreak is responsible for these disturbances is examined below.

Slickensided bedding planes. From the above observations and measurements, it was concluded that the deformation of bedding planes to form failure surfaces can occur through:

1. Geological movement.
2. Movement due to pitwall relaxation.
3. Movement promoted by recent major energy release (blasting).

In order to establish the origin of these slickensided bedding surfaces, a survey of their frequency with increasing distance from the pit was conducted. A series of boreholes drilled along traverses normal to the strike of the pit was logged in detail, and the frequency of slickensided surfaces recorded. The results of this survey are shown in Figure 7.6 for:

1. Slickensided bedding surfaces (see Figure 7.6*a*).
2. Polished and occasionally slickensided conjugate planes to bedding (see Figure 7.6*b*).

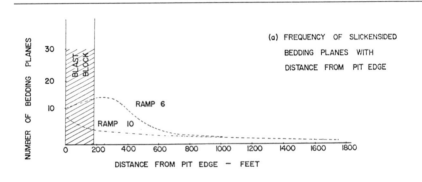

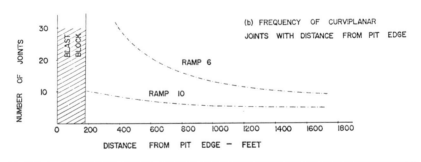

Figure 7.6. Frequency of disturbed bedding and joint planes with distance from blasting recorded in survey.

The resultant plots show the following:

1. Sheared or polished bedding plane frequency increases with proximity to the blast block.
2. Both graphs exhibit a drop-off in frequency with increasing distance from the blast perimeter, finally maintaining a residual value likely to reflect the frequency of naturally-occurring disturbed surfaces.
3. Peak disturbance extends to between 43 and 60 meters (140 and 200 feet) beyond the blast perimeter, and between 90 and 110 meters (300 and 360 feet) beyond the previously excavated highwall. This maximum distance between wall edge and the limit of peak disturbance represents 3.5 times the wall height and, therefore, is likely to be outside the zone of significant movement due to wall destressing and relaxation.
4. The principal difference in the graphs for the two areas (Ramp 6 and Ramp 10) is the predominance of thick-bedded to massive sandstone at Ramp 10 (see Unit 2, Figure 7.3) compared with laminated and thinly bedded argillites at Ramp 6 (see Units 3, 4, 5, and 6, Figure 7.3). At Ramp 6, the failure frequency is three times that experienced at Ramp 10.

It would appear that the low frequencies of disturbed bedding planes remote from the pit reflect geological background values, while the distance over which peak disturbance is recorded exceeds that likely for wall relaxation effects. Thus blasting would appear to be the only remaining activity with an energy input sufficient to affect disturbances of such magnitude and extent.

Strain-induced fracture surfaces. As discussed earlier in "Fracturing Along Boundaries of Modulus Contrast," the passage of the compressive strain wave (due to blasting) develops differential strains at boundaries of beds of differing moduli. Where strains exceed the yield limits of the boundary, rupture (and possibly displacement) follows, and the plane is then exposed to the effects of invading gases. Where these beds are observed in the newly exposed highwall (and adjacent to the previous blast), partings along bedding planes are common. Highwall instability has been observed along base planes comprising such beds.

Displaced bedding surfaces. Figure 7.7 shows the open texture of a black clay bedding plane which commonly forms the base failure plane at a number of localities throughout the mine. The undulating surface, as shown, has an estimated wavelength of 100 to 400 millimeters (4 to 16 inches).

As a result of disturbance to the plane, the upper and lower bedding surfaces exhibit a relative lateral displacement of up to 50 millimeters (2 inches). Undulations cause the plane to open as out-of-phase sections of the undulation oppose each other. Polishing and slickensiding are also present on the displaced surfaces.

The distance over which such disturbance extends behind the wall (and hence the previous blast perimeter) is unknown, but it is evident that very considerable forces are necessary to achieve the displacement measured along bedding overlain by up to 50 meters (165 feet) of overburden. The strong but inconclusive evidence indicates that explosion gas invasion and release-of-load displacements are the overbreak mechanisms most capable of generating such forces.

Figure 7.7. Displacement (half phasing) along undulose bedding plane causing plane dilation and openings.

Tension cracks

With the square in-line pattern of overburden blasting, cracking at ground surface can be detected at distances up to 60 meters (200 feet) behind the last row of blastholes (see Figure 7.2). Other observed features of tension cracks caused by blasting include the following:

1. Cracks parallel to the long axis of the blast block become tangential and indistinct at either end of the block.
3. Tension cracks due to blasting but which form the upright scarps of failures extend to depths exceeding half the wall height. This exceeds that which would result from wall destressing alone (Hoek and Bray, 1974).
3. In most instances, tension cracks represent planes of separation along extensive joint sets; this would be expected from the release-of-load fracture mechanism.
4. A survey of the extent and separation widths of tension cracks behind a blast block indicated that, over a distance of 60 meters (200 feet) behind the blast perimeter, the total width of cracks was approximately 1 meter (3.3 feet). This figure reflects total tension crack development immediately after blasting and also the net displacement due to the release-of-load mechanism associated with square in-line blast patterns.
5. The variation of crack separation with distance behind a blast is summarized in Figure 7.8. Greatest crack separation occurred within 20 meters (65 feet) of the blast perimeter, and separation widths of up to 38 centimeters (15 inches) were recorded. Crack widths decreased rapidly between 20 and 60 meters (65 and 200 feet).

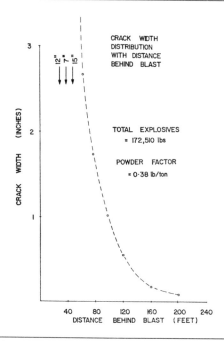

Figure 7.8. Distribution of crack widths with distance behind blast perimeter.

Inclined failure surfaces

Although directly observed on few occasions, this feature can be deduced from the movement of blocks and columns in the mobilized mass.

A back analysis of a pitwall failure using limit equilibrium techniques for a two-wedge failure model (see Figure 7.5) did not produce computed factors of safety approaching unity along the base plane. This initial case examined pure sliding failure along the inclined surface as a first time rupture through the material. However, by altering the mechanism of rupture to one involving a steplike failure of bedding planes and inclined preformed fractures, factors of safety approaching 1.0 could be computed.

The inclined fractures necessary to simulate failure in this stability model probably originate as a result of blast-induced overbreak. Conjugate shear planes, also disturbed by blasting, may interconnect bedding slip planes.

Groundwater

Water influences pitwall instability by:

1. Slaking and softening the clays along bedding surfaces, thereby causing a significant reduction in the shear strength of the bedding features.
2. Infilling tension cracks and exerting hydrostatic forces along the crack surfaces behind the highwall.

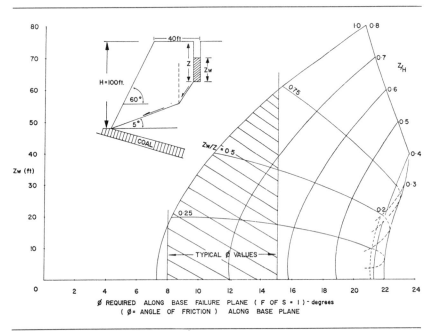

Figure 7.9. Sensitivity of highwall stability to water-filled tension cracks.

Figure 7.9 indicates the sensitivity to stability of water-filled tension cracks.

Surface runoff is drained by bulling of blastholes resulting from blast-induced compression of the material surrounding the borehole and by open tension cracks (created by release-of-load fracturing) developed beyond the blast perimeter.

The effect of the penetration of runoff down into the rock mass behind the highwall is indicated in Figure 7.10, showing eroded Tertiary clays deposited over (blast) displaced bedding planes in the base of the wall. Transport of clays in this manner indicates the degree to which the ground beyond the blast perimeter must be broken.

Experience at Goonyella Mine

Historically, overburden blasting at Goonyella Mine has been carried out using vertical blastholes, usually in a 9 × 9-meters (30 × 30-feet) configuration (either square or staggered), fired in rows parallel to the highwall. In 1976, experiments were carried out using blasthole patterns with the following burdens and spacings:

9 × 9 meters (30 × 30 feet)
8 × 11 meters (26 × 36 feet)
10 × 10 meters (33 × 33 feet)
9 × 12 meters (30 × 39 feet)

Figure 7.10. Bedding (failure) surfaces coated with eroded and transported Tertiary clays.

The initiation sequence was changed to "V1" and then to "V" (see Figure 7.11) and the inter-row delay was increased from 17 to 25 milliseconds. Since these trials were started, 38 major blasts have been fired, using various combinations of the patterns and initiation sequences listed above.

The most immediately noticeable effect of both the "V1" and "V" sequences was the very considerable reduction in the visible overbreak damage behind the blast area. As already reported (McIntyre and Hagan, 1976) this was attributed to:

1. Delayed rather than simultaneous detonation of back-row blastholes.
2. Improved effective spacing:effective burden (S_e to B_e) ratios.
3. Greater inter-row delays and, because of the greater numbers of effective rows, much longer blast durations.
4. Smaller numbers of blastholes detonating simultaneously.

Apart from four very localized minor block failures at fault boundaries, there have been no highwall failures recorded in areas where "V1" or "V"-type firings have been used. Overbreak beyond the design highwall has been reduced from approximately 60 meters (200 feet) to less than 4.5 meters (15 feet), although

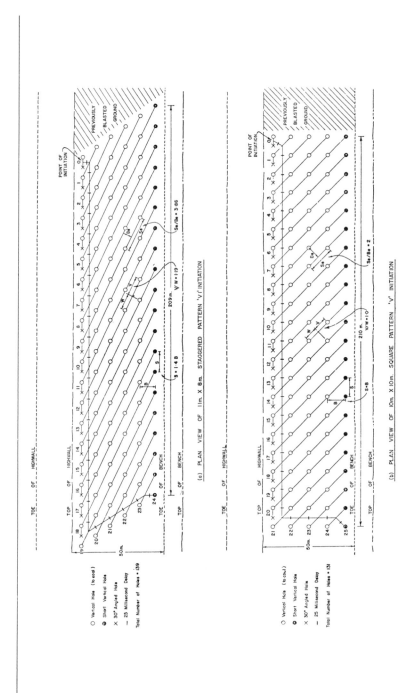

Figure 7.11. Layout of various angled initiation patterns.

some surface cracking adjacent to the blast and along the strike is still experienced. The latter does not cause any operational problems, however, as it is usually confined to within 9 meters (30 feet) of the end of the blast, and is subsequently broken by firing of the adjacent block.

To date, the most acceptable practicable pattern appears to be the 10 × 10-meter (33 × 33-foot) square pattern, fired in the "V" initiation sequence. This produces good fragmentation, suitable but not excessive throw and negligible overbreak; it is straightforward from both the drilling and tying up viewpoints. The S_e to B_e ratio for this pattern is 2.0, which, while suboptimum, gives good results in the sedimentary materials encountered (described previously under "Geological Setting"). The effective stagger ratio, V/W (see Figure 7.11*b*) is perfectly balanced (at 1.0), ensuring a high degree of fragmentation of the rock blasted.

While design and operational difficulties in some locations inhibit the use of the modified initiation sequence, efforts to introduce this are being made, particularly in historically unstable areas of the mine.

Designing Overburden Blasts for Improved Stability

Effective integration of the blast features required to optimize pitwall stability, covered in considerable detail in previous publications (McIntyre and Hagan, 1976; Hagan, 1977b; Hagan, 1977c), can best be illustrated by considering the redesign of the hypothetical basic blast shown in Figure 7.12. The unsuitability of

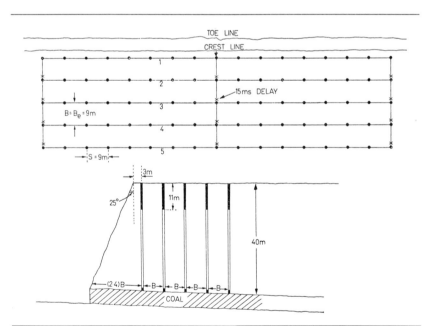

Figure 7.12. Basic square in-line pattern for overburden blasting.

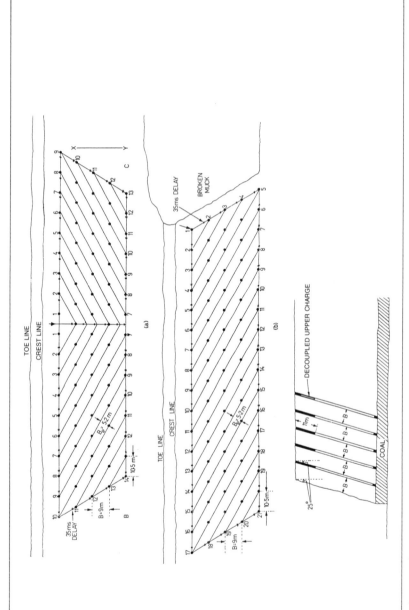

Figure 7.13. Elongated staggered "VI" pattern with inclined blastholes fired (a) to an open face and (b) to a free end.

this basic design is due to the following factors:

1. The excessive toe burden (2.4 times the design value) on the front row (resulting from the use of vertical blastholes).
2. The large number (17) of blastholes detonating simultaneously.
3. The short inter-row delay interval (15 milliseconds).
4. The relatively short overall duration of the blast (60 milliseconds).
5. The relatively small spacing between back-row blastholes, 9 meters (30 feet).
6. The use of vertical blastholes in the back-row.
7. The lack of a decoupled charge within the stemming column of back-row blastholes.
8. The quasi-simultaneous detonation of back-row blastholes.

Because of the lack of blasthole water, the use of ANFO is still maximized in the new blast design (see Figure 7.13). In order to maintain drilling and blasting costs at their present levels, both the powder factor and drilled footage per cubic meter (or yard) of overburden are maintained at their initial values. But the following numerous changes are made:

1. The drilling of inclined rather than vertical blastholes:
   a. Reduces the toe burden on front-row blastholes to its design value.
   b. Helps to delineate the design highwall.
2. The square pattern is replaced by an elongated staggered pattern. This change increases:
   a. The spacing between back-row blastholes by 17 percent and
   b. (as a bonus) the explosive's energy distribution within the burden rock.
3. The use of the "V1" (cf. in-line) initiation sequence:
   a. Reduces the number of blastholes detonating simultaneously by at least 41 percent.
   b. Introduces a delay between the firing of back-row blastholes.
   c. Causes less disruption (as a result of shearing forces during the blast) along the side(s) of the shot by increasing the angle between the principal direction(s) of rock movement and the newly exposed side wall(s).
4. The introduction of a decoupled upper charge in back-row blastholes encourages the formation of an inter-blasthole split at the stemming horizon, thereby reducing the probability of surface overbreak resulting from relative movement of rock at the stemming horizon ("Ecstall Mining's Plastic Pipe Blasting Method," 1973).
5. The use of 35-millisecond rather than 15-millisecond inter-row delays increases the delay interval per foot of effective burden further (in addition to the increase associated with the reduction in B_e), thereby giving each effective row a greater ability to detach its burden from the rock mass before the next effective row detonates. The greater inter-row delay also increases the total blast duration and reduces the intensity of ground vibrations. (It is good practice to increase the inter-row delay to the maximum interval for which cutoffs fail to occur.)

Overbreak created by the failure mechanisms most affecting instability is reduced in the following ways:

1. Because of the reduced effective burden and the longer inter-row delay, explosion gases in back-row blastholes find it easier to displace their burdens in a forward horizontal direction. The period during which gases are confined does not become excessive. For this reason, the gases are not forced to stream into, wedge open, and extend backward-facing cracks and discontinuities. Control of the rate of gas pressure decay is also enhanced by the superior design of back-row stemming columns.
2. The resultant lateral compression of rock beyond the blast boundary is lessened by:
   a. The delayed rather than simultaneous detonation of back-row blastholes.
   b. The fewer blastholes detonating simultaneously.
   c. The longer overall blast duration.
   The amplitude and rate of rebound of the strata beyond the blast boundary, therefore, will be correspondingly less and the extent and magnitude of release-of-load fractures (probably the single most important overbreak mechanism) are appreciably reduced.

It is important to realize, of course, that these numerous advantages can rarely be achieved without a sacrifice in one or more other aspects of blasting. The switch from vertical to inclined blastholes, for example, may reduce the mean drilled footage per shift slightly and, in very weak ground, could possibly increase the incidence of caved blastholes. The collaring of blastholes at their exact intended locations also becomes more necessary if difficulty in seeing effective rows and tying up detonating cord trunk lines correctly is to be avoided.

At most operations, and especially those with severe overbreak/stability problems, however, the advantages of redesigned blasts of this type should far outweigh the limitations. If, as is probable, the muckpile characteristics improve appreciably, it may then be possible to reduce the powder factor by increasing the blasthole spacing (in small increments), thereby reducing overbreak further and lowering overall production costs.

Conclusions

1. Gas extension of discontinuities and release-of-load fracturing are the mechanisms of blast-induced overbreak with the greatest potential of disrupting the newly exposed highwall.
2. Disruption occurs most easily at and/or along natural discontinuities.
3. Field observations indicate that these overbreak mechanisms can be directly and indirectly associated with discontinuities known to encourage instability.
4. With square in-line blast patterns in laminated to medium bedded coal measure sediments, highwall instability is controlled by:

   a. The dip and dip direction of clay-coated bedding surfaces.
   b. The available shear strengths along such surfaces.

    c. The presence of subvertical, release-of-load cracks behind the slope.

    d. The presence of water.

5. Evidence of the association between blast-induced overbreak and 4*b* above is given by:

    a. The rapid increase in frequency of weakened and polished/slickensided bedding surfaces towards the blast perimeter.

    b. The presence of weakened displaced bedding surfaces in the highwall.

    c. The first-time rupture of interfaces between contrasting beds.

    These structures represent recent disturbances in and behind the highwall, their magnitude indicating forces well in excess of those expected from wall relaxation/creep effects alone.

6. Release-of-load fractures represent a direct association between blast-induced overbreak and highwall instability by:

    a. Forming up to 40 percent of the failure surface.

    b. Impounding runoff water to create hydrostatic pressures.

    c. Assisting deep infiltration of runoff water along displaced bedding surfaces.

7. Experience gained in attempting to minimize such overbreak has indicated that overburden shots can be designed to considerably reduce overbreak beyond the highwall without incorporating specialized smoothwall blasting techniques.

8. Overbreak can be reduced and stability promoted by:

    a. Decreasing the peak blasthole pressure in back-row blastholes.

    b. Increasing the blasthole spacing and inter-row delay while maintaining burden constant.

    c. Using a "V1" or "V" rather than in-line initiation sequence.

    d. Drilling inclined rather than vertical blastholes.

    e. Modifying the stemming column of back-row blastholes.

9. The staggered "V1" is the optimum design pattern for producing the least overbreak. However, operational constraints may make this pattern less practicable than the less-than-ideal but highly-successful square "V" pattern.

Acknowledgments

The authors extend their thanks to ICI Australia Ltd. and Utah Development Company for permission to publish this paper. The helpful advice and indirect contributions made by J. K. Mercer and G. Harries of ICI, and W. Komdeur and N. N. H. Godfrey of UDC are also gratefully acknowledged.

References

Anon., 1973, "Ecstall Mining's Plastic Pipe Blasting Method." Canadian Mining & Metallurgical Bulletin, March.

Coates, D. F., 1970, "Rock Mechanics Principles." Mines Branch Monograph 874, Dept. of Energy, Mines and Resources, Canada.

Cook, M. A. et al., 1966, "Behaviour of Rock During Blasting." Transactions of Society of Mining Engineers/American Institute of Mining Engineers, December.

Dally, J. W. et al., 1975, "Influence of Containment of the Borehole Pressures on Explosive Induced Fracture." International Journal of Rock Mechanics, Mining Sciences and Geomechanical Abstracts, Vol. 12, No. 1.

Hagan, T. N., 1977a, "Rock Breakage by Explosives." Paper presented at Sixth International Colloquium on Gas Dynamics of Explosion and Reactive Systems, Stockholm, August.

Hagan, T. N., 1977b, "Smoother Sounder Walls in Surface Operations Through Redesigned Primary Blasts." Queensland Division Technical Papers, The Institution of Engineers Australia, Vol. 18, No. 12.

Hagan, T. N., 1977c, "Overbreak Control Blasting Techniques." Chapter 11 of the Australian Mineral Foundation's 'Drilling and Blasting Technology' Workshop Course, Adelaide, May.

Harries, G., 1973, "A Mathematical Model of Cratering and Blasting." Australian Geomechanics Society's National Symposium on Rock Fragmentation, Adelaide, February.

Hoek, E. and Bray, J. W., 1974, "Rock Slope Engineering." The Institution of Mining and Metallurgy, London.

McIntyre, J. S. and Hagan, T. N., 1976, "The Design of Overburden Blasts to Promote Highwall Stability at a Large Strip Mine." Paper presented at Eleventh Canadian Rock Mechanics Symposium, Vancouver, October.

Underground Mining

The State of the Art in Underground Coal Mine Design

8

H. D. Dahl, Director of Underground Engineering,
Consolidation Coal Co., Pittsburgh, Pennsylvania, USA

Introduction

During the past decade, significant progress has been made by a great number of researchers in applying rock mechanics principles to mine design problems. Some of these new techniques have found their way into the engineering and planning offices of operating mines. In spite of these advances, however, the application of rock mechanics in the day-to-day mine planning and design of U.S. coal mines is relatively rare.

There are a number of reasons for this relatively slow progress, one of which is the question of how to measure the success or cost-effectiveness of a sizable rock mechanics effort in an operating mine. Another concerns the fact that the properties of geologic materials generally vary over a wide range. It is, therefore, difficult to assess whether a change in mining conditions is due to a design change or due to a change in the geologic conditions. A further reason for the slow acceptance of rock mechanics in engineering thinking is the fact that some of the concepts are quite difficult, and that physical evidence of their validity in mine design is still questioned. Also, communication among research, engineering, and applications oriented people appears to function only intermittently.

The Role of Rock Mechanics

In the author's view, however, the most important impediment lies in the "state-of-the-art." Exactly how much can rock mechanics help the mine operator? At the present time, the best that rock mechanics can do as far as design is concerned, is to determine whether one approach is better than another when compared with some design objective (stress, strain, deformation). The absolute best ground control design cannot be determined—only that which is better. If ground control is lost, the statement "it's the best one we looked at" means very little to the harried operator. While the state-of-the-art has advanced a great deal, it has certainly not become an engineering matter in that design handbooks are available—they are not. The success of a rock mechanics program, therefore, can only be determined after a long period of continuous effort.

The only way to achieve an *absolute* assessment of ground stability would be to have an acceptable stability hypothesis. The author does not know of any stability hypotheses that work in underground opening design—only those that appear to work in existing openings such as strain and/or deformation rate analysis.

The purpose of this chapter is to outline some of the design techniques that are being used or tried in the coal mining industry for improved roof and ground control in single, more or less horizontal coal seams.

This chapter covers the basic analysis concept that can be applied to:

1. The design of longwall development headings.
2. The recommendation of pillar extraction schemes.
3. The caving properties of the overburden above longwall panels.
4. The prediction of subsidence.
5. The design of long-term main entries.,
6. The orientation of the mine plan.

Design Procedures (Concepts)

The basic design procedure is quite simple. At first, it is important to clearly define the operational and safety functions that a particular system has to fulfill. Included are size, number of openings, and perhaps of most importance, their required service life.

One important objective in entry design is to minimize maintenance costs. Time dependent effects like weathering of the roof and sloughing of the pillar ribs must be considered. These effects are pronounced for long-term development entries, but may be of secondary importance in the case of short-term panel development headings where coal recovery and high productivity are generally of primary importance.

A good understanding of the nature of the geologic environment is required. This represents, by far, the most difficult step and must include a knowledge of the mechanical properties of the various formations surrounding the opening, the local in situ stress field, and structural features of the rock mass. These quantities are independent variables which cannot be controlled by the design engineer. It is important to recognize them, however, because they are of primary importance in deciding the range in which controllable design parameters may be varied.

Factors affecting mine design

At this point, it is prudent to outline the factors which are within the framework of the designer's control. In the case of coal mining these factors include opening width, pillar width and length, pillar layout, panel width and length, cut sequencing, panel sequencing, and the orientation of all mining activity. The range over which these may be varied is limited; e.g., one cannot mine an opening narrower than that in which existing mining machinery can operate. Panel and development entry orientation cannot easily be changed in an old mine, or in a new mine whose extraction geometry is an overriding consideration.

Design problems in large underground excavations are unique in relation to problems in many other engineering disciplines, because the mechanical and structural properties of the rock mass as well as the precise loading conditions are not very well known. Furthermore, these properties change as mining progresses. Each opening in a strict sense represents a "one of a kind" structure. As a result of all these uncertainties, it is difficult to assess the exact strength of an underground structure and define a meaningful safety factor.

The most common design concept in the past has been to duplicate pillar and opening dimensions, as well as their orientation, from other mining operations under similar conditions. This concept makes use of past operating experience but tends to be very reluctant to introduce changes. This reluctance is partially based on a lack of understanding of the mechanics of failure and stability of underground structures. It often leads, therefore, to loss of reserves and increased maintenance costs because of the tendency to overdesign and the inability to react to changing conditions respectively.

Although this method has led to acceptable operating conditions in the past, the increased demand for coal and more difficult geologic conditions will require a more rational method of mine design.

Modeling Techniques

An alternative approach to mine design consists in the use of various modeling techniques. Laboratory models based on the principles of similitude, together with analog models, have found wide application. The quality of these models for mine design purposes obviously depends on how realistically the model parameters represent those of the prototype, or how accurately the analog relation represents the behavior of the rock mass, respectively. These models are usually quite expensive to construct, and it is not easy to account for changing geological conditions.

Numerical modeling techniques, e.g., finite element models, have proved very effective in the area of rock mechanics. The finite element method is discussed in numerous publications and, therefore, will not be repeated here. This technique has been used by the author to analyze and solve certain mine design problems. Its principal advantages are economy and the capability of modeling a great variety of structural properties and material behavior.

The chief advantage in the finite element technique is that the controllable factors—geometry, sequencing, and orientation—can be varied on paper without resorting to mining. The best variation can then be chosen and recommended to mine management for implementation. It is important to note that this is a *relative* design procedure and does not involve the use of safety or other factors.

Constitutive relations

As mentioned earlier, it is difficult to obtain realistic values to characterize the in situ behavior of the rock mass; it is also recognized that the input parameters that go into a model determine the quality of the model predictions. Field measurements, therefore, are necessary to verify the model predictions *before* any changes in an operating system are made.

At the same time, these measurements provide a way of improving the model through a data matching technique, in which measured displacements are reproduced in the model by suitably adjusting the material properties of the rock. Various continuum theories have been proposed in the literature; since the application of the finite element method is not restricted to any particular theory, a choice can be made according to which theory provides the most realistic results.

Elastic, frictional-plastic model. It has been found that the displacements around longwall and pillar extraction panels in the Pennsylvania and northern West Virginia coalfields are reasonably well described in elastic, frictional-plastic stress strain relations in which the yield condition is a function of confining pressure. This model may not give the best results in other mining districts or other mining systems. It is necessary to determine through experiment which of the constitutive models will give the most satisfactory result in any particular area.

The use of the elastic, frictional-plastic model is based on the observations that follow:

1. Surface subsidence measurements over longwall and pillar extraction systems, as well as measurements underground, have shown that the displacements were essentially time independent; that is, the displacements were controlled by the geometry of the panel and the position of the mining face.
2. It has also been determined by measuring the elongation of a borehole over longwall panels that caving above the seam extends to a distance of 45 to 50 times the mining height. Figure 8.1 shows the results of such measurements. The mining height in this case was approximately 1.7 meters (5.5 feet) with an average overburden of 183 meters (600 feet). The displacements within the caved zone are inelastic because they were caused by partial fragmentation of the rock; this means that the rock behaves more like a granular material which is customarily described by some form of plasticity. Similarly, displacements around the openings in coal measures are known to be largely inelastic in nature.

It is not surprising, therefore, that the elastic, frictional-plastic model provides realistic predictions regarding displacement and stresses around mine openings in sedimentary rock at moderate depth. Where generalized caving does not occur, such as in development headings, the model becomes essentially an elastic one. As indicated earlier, it is necessary to adjust the in situ rock mass material properties before such predictions can be made. These properties may originally have been obtained from laboratory tests; however, the results of these tests generally have to be reduced by a factor of 10 or 20 in the numerical models in order to be able to duplicate previously measured field data.

Once satisfactory agreement between measured field data on an operating system and model calculations have been achieved, the model can be used to analyze the effect of certain design changes. It is not necessary to establish a sep-

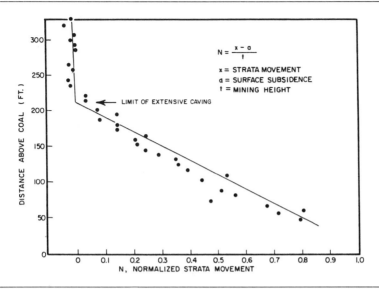

Figure 8.1. Normalized displacement over longwall panel.

arate field program for each model study as long as geologic conditions are sufficiently similar. Every new design based on model predictions should, however, be followed up by a certain amount of fieldwork in order to verify the predictions and to be able to anticipate any possible adverse developments. Such a program will also help to continually improve the quality of the model; it will also serve to establish a data record for a particular mine or mining district which can become the basis for more rational design decisions.

Examples where this concept has been used with considerable success include the design of longwall development headings, room and pillar extraction plans, prediction of surface subsidence, design of long-term main entries, and determination of the optimum roof bolt length.

Longwall design
The most severe problem in longwall design concerns protection of the tailgate entry. Generally speaking, the second or third panel in a series of longwall panels tends to experience the most severe tailgate problems; moreover, it is apparent that these problems are induced by the pressure of the previous panel adjacent to it. It must be decided, therefore, how to design a tailgate development entry so that the future tailgate is protected from the abutment pressures created by the current panel.

There are really only two design philosophies that can be employed. One is to make each panel a "first" panel, leaving sufficient coal between panels to insure that there is no interference between them. This philosophy has the disadvantages that a considerable amount of coal is left unmined and that it requires an inordinate amount of development work per panel.

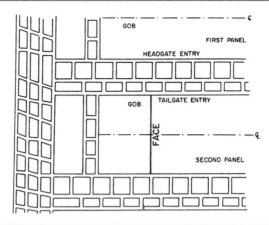

Figure 8.2. Stiff design concept in a longwall system.

The second philosophy involves the design of pillars in tailgate entries so that stresses around the future tailgate are within acceptable limits. There are two ways of achieving this condition; either a stiff or a yielding design can be used. In the stiff method, large pillars are left adjacent to the previous panel, providing a clean break over the overburden. A second small pillar is used adjacent to the tailgate to protect it. The "first panel" concept is a stiff design without use of the yielding pillar for tailgate protection at improved extraction ratios. This method is shown in Figure 8.2. In the yielding method, very small pillars are placed adjacent to both the previous gob and the current tailgate. The idea is that the pillars will carry only as much load as they are capable of supporting and will transfer the abutment load to the large block of coal yet to be mined. Figure 8.3 illustrates such a design.

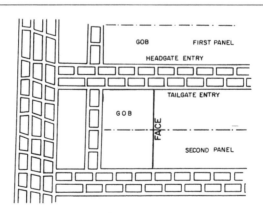

Figure 8.3. Yielding design concept in a longwall system.

The implications in both of these design philosophies in which a first panel concept is not employed is that large differential displacements on either side of an opening to be protected are undesirable. For example, if an opening is subjected to a large displacement on one side with a very stiff pillar on the other, it will likely shear along the stiff edge. In either case, the pillar adjacent to the tailgate cannot act as an independent supporting member—it must deform with the panel block itself.

Room and pillar mining

The same philosophy employed in tailgate protection can be employed in designing room and pillar extraction schemes. An abutment must be provided near the caving zone which is capable of supporting the overburden without causing undue stresses in the active working area. Again, there are two ways of obtaining this result. Either as large a pillar as is practicable is outlined so that the gob again has something very stiff to break on, or very small pillars are outlined on the retreat immediately adjacent to a large virgin block of coal yet to be developed. In either case, the large block of coal acts as a load carrying member, and the smaller pillars outlined are formed as the large pillar is taken out; e.g., wing and pocket or other method of pillar extraction. Figure 8.4 illustrates a system in which large pillars are outlined on the retreat, and Figure 8.5 depicts a system in which yielding pillars are outlined on retreat. Note that the two examples utilize development on retreat. This is due to the importance of time dependent effects in establishing good roof conditions. As few entries as possible should be driven to develop the panel, with the majority of the area being developed as the face retreats. Most openings are then exposed for a period of days rather than months.

For a production panel, the time that an average unit of roof area is exposed should be minimized. A single entry longwall system achieves this result to a very high degree, and development on the retreat approaches this minimum.

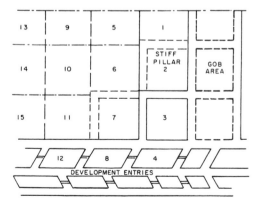

Figure 8.4. Stiff system of pillar mining.

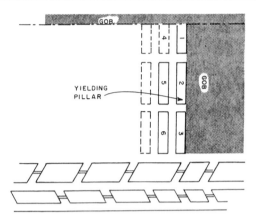

Figure 8.5. Yielding system of pillar mining.

Panel orientation

Most mining areas are characterized by a "good roof" and a "bad roof" direction. If such a condition is present, it is important to obtain optimum panel and main line orientation.

Assume, for example, that best roof conditions are obtained when driving entries in an east-west direction. If possible, main line development entries would ordinarily be placed in an east-west direction. Of course, other complications may preclude this. It is optional in which direction the panels are developed. Since most coal production will be derived from panels, it is important that these be driven in the optimum direction to ensure roof stability; i.e., in this example, the direction should be east-west. First, panel development entries will be in the good roof direction. Second, the break line for either longwall or room and pillar operations will parallel the bad roof direction, resulting in the cleanest possible break and lowest pressure possible in the working area. Third, crosscuts which are needed for development need only be maintained for a short period of time since their utility is significantly reduced after the panel has been developed two or three crosscuts ahead.

It is possible to lay out the mine in such a way that the majority of main line entries and all panel development entries are placed in the optimum direction from a ground control standpoint.

The observation that an optimum direction exists is extremely important. It apparently does not depend on establishing the cause for oriented roof failures in order to take appropriate action for a solution to the problem. Figure 8.6 illustrates proper panel and main line orientations.

Pillar size distribution

Any number of pillar sizes and distributions can be specified. There are many cases in which long-term stability or vertical support capability need to be maximized in order to protect structures from subsidence. The following statement

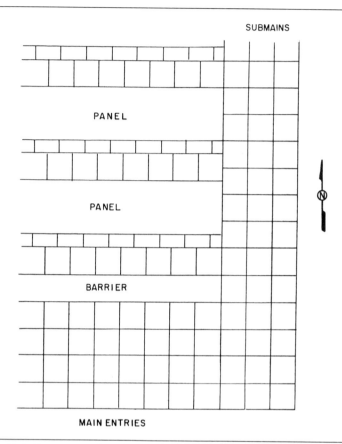

Figure 8.6. Proper panel and main line orientation (north-south is poor roof direction).

can be made: for a given extraction ratio, equal-size pillars minimize the support capability of the pillar layout; or, conversely, the use of equal-size pillars will result in minimizing extraction ratio. This results from the fact that pillar strength increases in proportion to more than the square of its least dimension. This fact is useful in designing both subsidence support areas and, possibly, panel development headings. It may not have an application in designing long-term development headings where time dependent strength and/or material weathering phenomena become important.

Long-term development headings

This is a most difficult problem since deterioration over a long period of time can result in rib sloughing, shale weathering, and a general deterioration of the opening. The author proposes that a long-term development heading design should embrace the following philosophy: all openings should be subjected to stresses

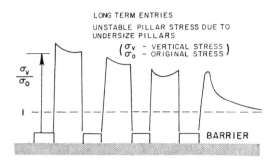

Figure 8.7. Incorrect pillar size for long-term entries.

that are symmetric about its centerline so that differential displacements and strains are minimized over the long term. Moreover, pillars should be sufficiently large so that the absolute magnitude of the stresses reflects a stable design. Barrier pillars should also be sufficiently large so that a virgin stress state exists near the barrier pillar center.

Utilizing the above philosophy, together with observing the optimum orientation in main line development, will result in optimum ground conditions.

Figure 8.7 illustrates how an incorrect pillar design results in increasing stresses toward the panel center. This is undesirable in the long term for two reasons: main line entries are generally located towards the center and nonsymmetrical time dependent deterioration of the opening will occur, possibly leading to shear failure. Figure 8.8 illustrates a design which overcomes these objections.

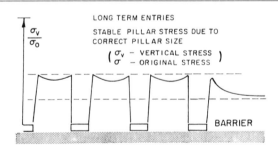

Figure 8.8. Correct pillar size for long-term entries.

The Stability of Underground Coal Mine Workings

9

Neville G. W. Cook, Professor of Mining Engineering,
and M. Hood, Assistant Professor of Mining Engineering,
Department of Materials Science and Mineral Engineering,
University of California, Berkeley, California, USA

Introduction

The advantages afforded by a room and pillar mining system at shallow depths have led to the adoption of this method for extracting coal in most of the underground mines in the United States. However, several problems are inherent in this system; for example, the size of pillars required to ensure safe and economic working conditions increases dramatically as the depth of the mining operations increases. Problems of this nature have led to interest being shown in alternative mining techniques. Longwall mining systems can provide safe working conditions when the stress and geological conditions make room and pillar mining difficult or wasteful, but combine to permit continuous caving of the waste behind the working face. Shortwall mining methods attempt to combine the advantages of flexibility of room and pillar mining, with the advantages of a more complete extraction and safer working conditions offered by longwall methods.

Symbols Used

The symbols used throughout this chapter, and their definitions, are listed below:

o = Average stress across the pillars (megapascals)
H = Depth below surface (kilometers)
e = Ratio of area extracted to that of the original seam
c = Distance between room centers (meters)
w = Width of square pillars (meters)
h = Height of pillar (meters)
$R = \dfrac{w}{h}$
C_p = Strength of the pillar (megapascals)
C_1 = Strength of an ideal cubical pillar of 1-meter side (megapascals)

α and β = Exponents which are determined by experiment
V = Volume of pillar (cubic meters)
S = Safety factor = $C_p\sigma$
L = Minimum horizontal dimension of the excavation (meters)
Z = The height above the center of the excavation at which the transition from tension to compression occurs (meters)
a = Depth to the base of the sill below surface (meters)
t = Thickness of the sill (meters)
ρ = Density of the overburden (kilogram per cubic meter)
E = Young's Modulus (megapascals)
σ_t = Maximum tensile stress in a uniformly loaded sill (megapascals)
δ = Maximum deflection of sill, at center of span (meters)
S_t = Dimensionless parameters to compare stresses
D = Dimensionless parameter to compare deflections
g = acceleration due to gravity (meters per second squared)

Mining Methods

Room and pillar
If the lateral extent of mining is greater than the depth below surface, the average stress on the pillars resulting from the weight of the overburden can be found from:

$$\sigma = 25\,H/(1 - e) \tag{1}$$

and

$$e = (c^2 - w^2)/c^2 = 1 - w^2/c^2 \tag{2}$$

Different investigators (Steart, 1954, Holland and Gaddy, 1957; Holland, 1962; and Salamon and Munro, 1967) have shown that the strengths of square pillars are given by:

$$C_p = C_1\,w^a/h^\beta \tag{3}$$

From the analysis of 125 case histories of South African collieries, Salamon and Munro concluded that $C_1 = 7.2$ MPa, $a = 0.46$ and $\beta = 0.66$ for pillars with ratios of width to height from 0.9 to 8.8.

The difference in the exponents for width and height suggests a slight dependency of strength on volume, given by

$$C_p = 7.2\,R^{0.59}/V^{0.067} \tag{4}$$

From this it follows also that maximum extraction is achieved by mining the maximum safe room width at the full height of the seam, if this can be accomplished. Sometimes, this height may be too great for the equipment in use or it may be desirable to leave top coal to protect the immediate roof. Clearly, if the safety factor S is less than 1, failure of the pillars is likely. If $S = 1$ there should be a 50 percent chance of failure. Depending upon the uncertainties involved, S should be greater than 1 to achieve a low probability of failure. Salamon and

Munro found the mean value of S for that range embracing half the stable situations to be $S = 1.6$.

From equations (1) and (2), and $S = 1.6$, and using $C_p = 1.6\sigma$ it follows that the extraction ratio decreases with increasing mining depth as:

$$e = 1 - 40H/C_p \qquad (5)$$

This relationship is illustrated graphically for typical values of H and C_p in Figure 9.1.

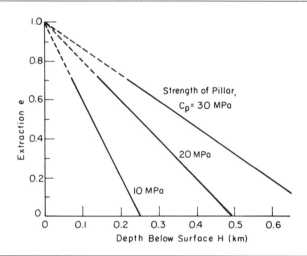

Figure 9.1. Graph showing the percentage of coal extracted from a seam as a function of the depth below surface, with different values for the strength of the pillars.

In underground tests on coal pillars, Wagner, 1974, measured the stress distribution across pillars at various stages of deformation, Figure 9.2. These experiments demonstrated that the perimeter of a pillar is capable of carrying relatively little stress. However, this portion of the pillar provides lateral confinement which enhances the strength at the center of the pillar. The results from tests on pillars of rectangular, rather than square, cross section, indicated that the strength of a long, strip pillar is about 40 percent greater than a square pillar of the same width and height because of the relatively smaller perimeter of such pillars.

Multiple seam extraction. Sometimes more than one seam is mined in the same area. In general, if the vertical separation between seams is more than c, the room center spacing, the effects of the one seam on the other can be neglected. If this separation is less than c, then pillars in adjacent seams should be superimposed upon one another. Where two or more seams are mined, the quality of the strata between them and the width of the rooms also must be taken into account.

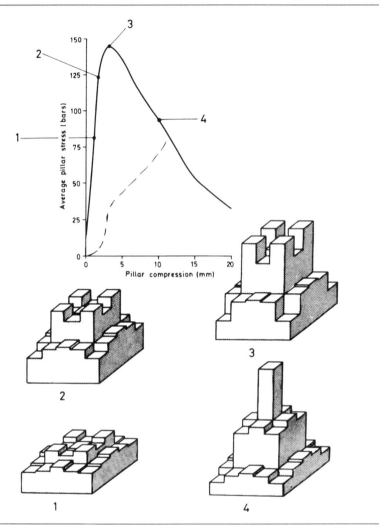

Figure 9.2. A complete average stress compression curve for a coal pillar with a width to height ratio of 2, and isometric histograms (Jaeger and Cook, 1976) showing the vertical stress distribution at various stages of compression.

Only if the strata between seams is particularly competent, can the width of the rooms, $(c - w)$, be made equal to, or greater than, the thickness of the intervening strata. If the thickness of strata separating adjacent seams is less than twice the width of the room, the possibility of failure of this parting must be taken into account. Such a failure may have the effect of increasing the height of the pillars to the combined height of the workings in both seams, with a corresponding reduction in the strength of the pillars. In such situations, it is advisable to ensure that the safety factor of pillars of the combined height is still well in excess of 1, and that the safety factor of the pillars in each seam is greater than ordinarily would be acceptable for working one seam only. Finally, in designing pillars for extracting several adjacent seams, it is good practice to calculate the minimum acceptable safety factor for each seam separately and then to choose as the width of pillar for all seams the greatest of the calculated widths, regardless of safety factor.

The sequence in which multiple seams are mined is not of great consequence. However, since most difficulties are experienced with the roof it is desirable to plan the sequence of extraction so as to protect this as much as is practicable; this may give a marginal advantage to mining the lowest seams first.

Guarding against pillar collapse. Mines should usually be divided into sections by *barrier pillars* to guard against the effects of a pillar collapse or other catastrophes. These are usually long, rectangular pillars, and experimental results (Wagner, 1974) indicate that the width of these pillars should be 10 times their height. Pillars of these proportions are unlikely to fail, except for punching into the roof or the floor. This can occur only if the average stress on the pillar is three or more times greater than the uniaxial compressive strength of either the roof or the floor. Barrier pillars generate pervasive stress concentration in the rock around them and care must be taken in the layout of adjacent seams above or below barrier pillars, and in taking roadways or tunnels through rock in their vicinity.

In an effort to improve the total extraction from room and pillar mining the pillars are sometimes mined on retreat in a second operation. Clearly, this gives rise to an increase in the stresses on the pillars adjacent to the area where the pillars have been mined. The safety factors for pillars where it is intended to recover them in a secondary mining operation should therefore be greater than otherwise, say, about 2. Preferably, the secondary mining operation should be done quickly and methodically using a continuous mining machine. Special supports are installed across each entry and crosscut leading to the pillar to be mined. These prevent the roof from collapsing during the extraction of the pillar. They are intended to be recovered and used repeatedly. The roof over the area where pillars have been mined should cave so as to prevent the weight of the overburden being transmitted fully to the remaining pillars.

Longwall mining

As the depth of mining stratiform deposits such as coal seams increases, so do many of the problems associated with room and pillar mining. These include the sterilization of a large fraction of the reserves in the pillars and difficulties with control of the roof in the rooms. Depth and friable roof conditions are advantageous for longwall mining. In longwall mining, entries or gates are driven parallel to one another in the seam some 100 to 200 meters apart (328 to 656 feet). A connection is established between these entries and the rib of this connection is then mined as the longwall face. Today, the working area adjacent to the longwall face is intensively supported by hydraulic props, and an armored conveyor is used to remove the coal. This conveyor carries the cutting machine. Generally, this is either a planing (plow), or shearing mining machine. Longwall mining possesses many inherent advantages but is a less flexible system than room and pillar mining. In order to optimize the mining operation and for reasons of safety, it is often considered preferable to mine longwalls on retreat.

The key to successful longwall mining lies in the hydraulic prop roof support. This must be capable of maintaining an intact roof between the face and the rear of the support, where the roof should cave consistently. If the roof does not cave steadily the loads on the support may become excessive, and intermittent caving of large overhangs of roof can be dangerous causing the release of methane and large amounts of dust. To cave effectively, the roof should be uniformly friable and the extent of mining should be comparable with the depth below surface.

Stresses induced in longwall mining. Mining induces a vertical tensile stress in the roof of an excavation. The resultant of this induced stress and the original vertical component of the virgin rock stress changes from tension to compression at some distance above the roof. For ratios in the range $0.25 < L/1000\, H < 2$, the ratio of the height of the zone of tensile stress to the minimum horizontal dimension is approximately $Z/L = 0.2$. Likewise the ratio of the height of the zone of horizontal tensile stress to the minimum horizontal dimension is approximately 0.12.

These zones of vertical and horizontal tensile stress are important in causing the roof to cave effectively. If they comprise a large fraction of the thickness of the overburden, caving will occur readily. If $L \ll H$ the vertical extent of this tensile zone is very limited, and adequate caving of the roof may not occur.

Caving can be affected adversely also by the presence of a strong, elastic stratum in the roof such as massive sandstone or intrusive sills. Provided that a sufficiently large area of such a sill is undermined, it is likely to fail and may then continue to break off regularly as mining progresses. It is difficult to determine theoretically how great is the minimum lateral dimension of the excavation necessary to cause this. In addition to the dimensions and proportions of such a sill and the excavation, it also depends upon the strength of the strata and the state of stress in them. However, some theoretical guidance for the interpretation of field observations is available. The maximum tensile stress of a uniformly loaded beam or plate with built-in ends, which approximates crudely a section through such a sill, is given by:

$$\sigma_t = -\frac{\rho g a\, L^2}{2t^2} \tag{6}$$

and the maximum deflection at the center by:

$$\delta = \frac{\rho g a\, L^4}{32E\, t^3} \tag{7}$$

Interestingly, this tensile stress occurs on the top of the sill adjacent to the edges of the excavation. The tensile stress on the bottom of the sill above the center of the excavation has half this value.

Different situations may, therefore, be compared using the values of the following dimensionless factors:

$$S_t = \left(\frac{L_1 t_2}{L_2 t_1}\right)^2 \frac{a_1}{a_2} \quad \text{and} \quad D = \frac{L_1{}^4 t_2{}^3 E_2 a_1}{L_2{}^4 t_1{}^3 E_1 a_2} \tag{8}$$

for stress and deflection, respectively, where the subscripts 1 and 2 refer to each of two different situations. When $S_t = 1$ and $D = 1$ the two situations are comparable in terms of geometry and the stresses induced by mining.

Shortwall mining

Shortwall mining employs parallel entries driven into the seam, separated by distances of about 50 meters (164 feet) or more. Shortwall mining consists of establishing a connection between these entries and then mining a wide web through this connection, using a continuous mining machine. The roof support is provided by hydraulic props similar to those used in longwall operations.

In general, the potential advantages of a shortwall include: a greater flexibility and lower capital costs than a longwall, and better roof control and ventilation than in room and pillar mining.

However, shortwall mining does suffer from several disadvantages, some of which are in the field of strata control. The cut made by a continuous mining machine is much deeper than that usually made by a planing machine or even a longwall shearing machine. This leaves a large area of roof unsupported between the hydraulic props and the face, while the continuous mining machine is making its cut and while it is being trammed back before starting the next cut. The hydraulic props used for roof support must usually be of more robust construction and larger load capacity than often suffices for longwall conditions. Also, the support must be cantilevered a considerable distance ahead of the hydraulic props themselves, to allow room for the mining operation. Shortwalling is likely to work best under a roof which is immediately firm but pervasively friable, so that it caves readily behind the line of face support. Strong massive strata in the roof may pose hazards for shortwall mining. The length of a shortwall is not likely to be sufficiently great to allow such strata to break and cave into the mined-out area. In such situations, therefore, the weight of the overburden must be carried by pillars separating individual shortwalls. Provided that both the roof and floor are strong, such pillars can be very effective in supporting the overburden (see "Room and Pillar" mining). Where the roof caves bodily this problem does not arise.

Subsidence

Underground mining always involves some disturbance of the overburden and surface. This may or may not result in problems, depending on the degree to which the overburden is disturbed and the effects of this disturbance. These effects can be divided into two areas for the purposes of investigation; surface and subsurface. The former, which include changes in surface elevations, may involve damage to buildings or other surface structures. The latter involve changes in the subsurface media which may affect groundwater flows and aquicludes. These latter problems have not been considered to be of major importance in Europe, although this may not be true elsewhere.

Subsidence above underground workings may be principally in the nature of continuous elastic deformation of the overburden, or by clastic caving, involving fracture and settlement of the overburden. The elastic response results in the least disturbance of the overburden and minimum surface subsidence. Clastic caving results in major disturbance of the overburden, but minimum disturbance and maximum stability of the underground excavations below surface.

The limits for elastic deformation of the rock above excavations in a tabular deposit such as a coal seam, depend upon the ratio between the span dimension of the excavation and the depth below surface. If the overburden behaves as a number of beams, theory predicts that the stresses induced in the strata above such an opening will increase as the span is increased (cf. equation 6). This continues until the strength of a stratum is exceeded, at which point it fails and no longer behaves as an elastic material. Also, the depth below surface influences the behavior of the rock mass because, as noted previously, the zone of vertical tensile stress above a mine opening extends relative to the depth as the span increases. Since rocks generally are very weak in tension, the rock in this zone is likely to fail and the strata above an excavation are not likely to behave elastically when the span dimension approaches that of the depth of the workings. Caving, and consequently subsidence at the surface, are very much more probable at shallow depths than is elastic deformation. Consequently, continuous elastic subsidence is limited either to mining at a great depth below the surface or to room and pillar mining at a moderate depth with competent overburden and strong pillars.

Room and pillar mining
One of the advantages of room and pillar mining is that, provided the pillars are not extracted and the roof is strong, subsidence can be limited to elastic deformation so that its effects are negligible. This statement must be qualified, however, because pillars, which have sufficient strength to support the roof when mining operations are in progress, deteriorate with time as a result of weathering; and these pillars sometimes collapse. Strong pillars, such as are needed at any significant depth, can be used only if the roof is capable of withstanding the pillar stress; if not it will fail resulting in uncontrolled subsidence. Alternatively, in conditions where the floor is weak, such as a fireclay which is allowed to become wet, pillars may penetrate into the floor which then results in subsidence in a manner similar to that caused by pillar collapse. It is difficult to predict subsidence with much degree of confidence under these conditions.

Longwall mining
Clastic caving, such as occurs generally when longwall mining is practiced, results in extensive fracture of the overburden, which is potentially damaging to strata acting as aquicludes and, therefore, to aquifers and groundwater flow. However, various techniques have been developed to control the caving process and to limit the strains induced by it. These include the use of temporary and permanent pillars and filling. In some circumstances these techniques may be used to preserve the integrity of an important aquiclude, so as to minimize the effects of mining on groundwater flow.

Theoretical analyses of caving and subsidence have not yielded much useful insight to these problems, on account of the complex nature of these phenomena. However, careful investigations and measurements have been made, particularly in Britain. These data have been reduced to empirical formulations which are of great assistance and which have proved to be remarkably precise in predicting the incidence and the effects of subsidence. Although these studies have been concerned mainly with the monitoring of ground movements at the surface, subsurface movements obviously are related phenomena. Further research is needed to show the extent of clastic failure of the overburden above mine workings. However, the relationships do serve to define the conditions at which inelastic behavior of the overburden begins.

The amount by which the overburden above longwall mine workings will subside depends on:

1. The depth of the workings.
2. The height of the seam.
3. The lateral dimensions of the panel (the width and the length).
4. The inclination of the seam.

Maximum subsidence is achieved when large panels are mined at a depth where the stresses induced cause full clastic caving of the overburden to take place. Under these conditions the maximum displacements at the surface correspond to about 90 percent of the seam height. Maximum subsidence occurs when the minimum horizontal dimension of the excavation is numerically equal to, or greater than, approximately 1.4 times the depth of the seam below surface, that is:

$$L \geqslant 1400\, H$$

When $L = 1400\, H$ this is known as the critical span and when $L < 1400\, H$ this is known as the subcritical span and when $L > 1400\, H$ this is known as the supercritical span.

At shallow depths of less than about 100 meters (328 feet), where the stresses induced in the rock by the weight of the overburden are low, the dimensions of the panel may have to exceed these critical values before maximum subsidence is achieved. Field observations have shown that subsidence effects extend outside the plan area of the seam being mined, to encompass an area described by an angle about $35°$ from the vertical drawn outwards from the edge of the workings, Figure 9.3. This angle is known as the angle of draw (Subsidence Engineers Handbook, 1975). The maximum displacement at the surface takes place at the center of the panel. The profile of the surface displacements, illustrated in Figure 9.3, can be calculated from empirical curves (Subsidence Engineers Handbook, 1975), with an accuracy of about 10 percent.

Formation of a subsidence trough leads to differential horizontal movement of the surface. It is this horizontal strain, together with the curvature of the ground surface, which determines the extent of damage to surface structures. The horizontal strain is a function of the surface curvature and the induced strains. The variations in the strain profiles with the depth and with the critical

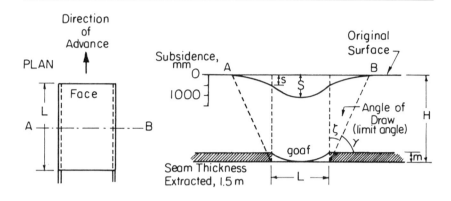

Figure 9.3. Diagram illustrating a profile of surface subsidence above a coal seam and the angle of draw (Thomas, 1973). (Reprinted by permission of Methuen of Australia Pty. Ltd.)

span of an excavation are illustrated in Figure 9.4. This Figure shows that, for narrow spans, high compressive strains are induced in a zone above the panel center. Under supercritical conditions when $L > 1400\ H$, the compressive strain at the center of the caved area returns to zero (see Figure 9.4 and Subsidence Engineers Handbook, 1975).

Two factors are of overriding importance in determining the extent of damage that will be caused to a building as a result of mining subsidence. One of these is the strain induced by the mining operations—and it has been demonstrated that this is a function of the critical span—and the other is the length of the building in the direction in which the seam is being mined, Figure 9.5. The damage caused to surface structures has been classified under five headings by the National Coal Board in Britain in terms of these two parameters, Figure 9.5. The definitions of these five terms are given in the Subsidence Engineers Handbook, 1975.

Techniques Used to Minimize Subsidence

Backfilling of the waste area is a standard technique used in mining to improve the stability of the workings and to reduce the effects of subsidence. Although filling cannot provide active support to the roof, it does limit the maximum amount of subsidence and can be used to strengthen pillars. This can be achieved because dilatation of the pillar compresses the fill in an active manner which enables the perimeter of the pillars to support substantially greater loads than is feasible normally (see Figure 9.2). Since maximum dilatation takes place at the midheight of the pillar (Salomon and Oravecz, 1976), the fill should be placed to a height greater than this. Another of the benefits achieved by filling room and pillar workings is that weathering of the pillars is reduced significantly, and consequently the integrity of the pillars can be predicted with confidence for long time periods.

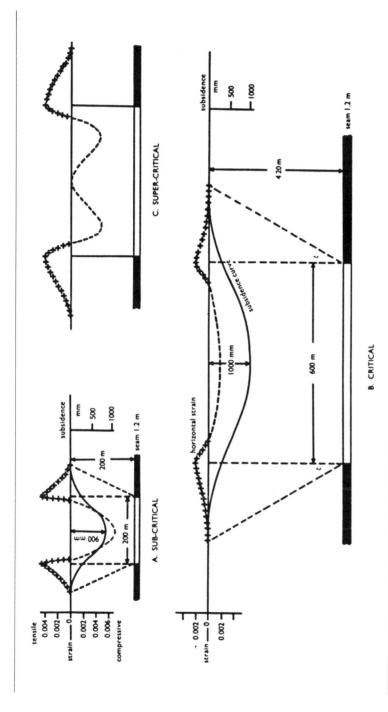

Figure 9.4. Profiles of strain induced in overburden at surface at subcritical, critical, and supercritical spans (Thomas, 1973). (Reprinted by permission of Methuen of Australia Pty. Ltd.)

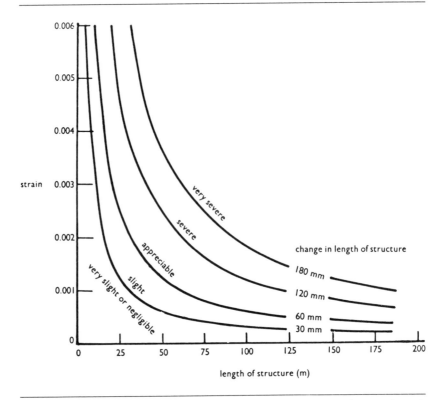

Figure 9.5. Relationship of damage to length of structure and ground strain (Subsidence Engineers Handbook, 1975).

The emplacement of fill under pressure behind longwall and shortwall faces has been shown to reduce the maximum subsidence of the overburden by 90 percent. The use of this technique, therefore, can prevent gross failure of the overburden.

Conclusions

The complex nature of the phenomena which influence both the stability of underground coal mine excavations and subsidence of the overburden above these excavations, has restricted the use of theoretical analysis as a tool to assist in the design of these excavations. However, a large amount of experimental data relating to these subjects has been accumulated. Empirical relationships have been derived from this data and these relationships enable parameters, such as the stability of pillars, to be predicted with a high degree of confidence. These formulations also enable accurate estimates to be made for the likely effects of subsidence in the strata which overlie the mine workings.

References

Holland, C. T., 1962, "Design of Pillars for Overburden Support." Mining Congress Journal, 48.

Holland, C. T., and Gaddy, F. L., 1957, "Some Aspects of Permanent Support of Overburden on Coal Beds." Proceedings West Virginia Coal Mining Institute.

Jaeger, J. C., and Cook, N. G. W., 1976, "Fundamentals of Rock Mechanics." A Halsted Press Book: John Wiley and Sons, Inc., New York.

Salamon, M. D. G., and Munro, A. H., 1967, "A Study of the Strength of Coal Pillars." Journal South African Institute of Mining and Metallurgy, 68.

Salamon, M. D. G., and Oravecz, K. I., 1976, "Rock Mechanics in Coal Mining." Pub. Chamber of Mines of South Africa, PRD Series No. 198.

Steart, F. A., 1955, "Strength and Stability of Pillars in Coal Mines." J. Chem. Metall. Min. Soc. S. Afr., 56.

Subsidence Engineers Handbook, 1975, Pub. National Coal Board, London.

Thomas, L. J., 1973, "An Introduction to Mining." Pub. Hicks Smith and Sons, Sydney.

Wagner, H., 1974, "Determination of the Complete Load Deformation Characteristics of Coal Pillers." Proceedings Third Congress of the International Society for Rock Mechanics, in Advances in Rock Mechanics, Vol. 2, Part B, National Academy of Sciences, Washington, D.C.

Research to Improve Structural Stability and Percentage of Coal Extracted in Australian Coal Mines

10

A. H. Hams, General Manager and Director of Research,
Australian Coal Industry Research Laboratories Ltd.,
North Ryde, New South Wales, Australia

Introduction

In the period from World War II until about 1970, the Australian coal industry, in common with the coal industries of most countries, suffered from a supply of cheap oil in massive quantities to the energy market. In this situation, in order to remain in existence, the coal industry was forced to reduce its production costs to a minimum. To achieve this, coal was mined from only the most favorable areas; if unexpected difficulties arose, that area of the mine was abandoned. Further, there was no incentive to recover a high percentage of coal and an optimum balance developed between ease of mining and amount of coal extracted. Thus, although costs were controlled and output per manshift significantly increased by mechanization, the real casualty in the situation was the high proportion of coal abandoned and lost in the workings.

A recent survey (Smith, 1977) carried out in the state of New South Wales in Australia, in which 48 of the 71 presently operating underground mines were examined, arrived at the following conclusions:

1. Twenty-four mines (half of those examined) are mining less than full seam height, the portion of seam height worked ranging down to 30 percent.
2. Of the seam height actually worked, mining recovery varies between 30 and 80 percent.
3. Considering the whole of all working seams studied in the sample, mining recovery varies between 17 and 80 percent, with the mean at 47 percent.
4. Salable recovery of the seams worked varies between 12 and 75 percent, with the mean at 40 percent.

If only 50 percent of the raw coal is being recovered this means that one ton of coal is lost for every ton of coal mined. Assuming that this sample is representative of the Australian coal industry and the situation continues until the end of the century, then coal losses in the next 23 years could amount to some 1,300 million tons with a present-day value of $A20,000 million. Even if the monetary situation could be ignored (which it cannot), such a loss of a nonrenewable energy resource should not be acceptable.

It is apparent, therefore, that a new balance will have to be obtained between ease of mining and the percentage of coal recovered. The immediate task will be to conceive and develop mining systems which will enable an increased coal recovery, and at the same time minimize or eliminate the cost increases which might otherwise occur.

Such a task will involve predictions of the stability of mine workings. Coal is lost in current mining operations in many ways. These include:

1. Failure of access roadways because they are of the wrong size and/or shape for their environment, resulting in instability manifested by roof falls, floor heave, or excessive maintenance costs.
2. Pillars between roadways being of the wrong size or shape resulting in either crush, spontaneous combustion, or difficulty in final extraction.
3. An excessive number of pillars being formed in multi-roadway developments which are difficult or impossible to extract. The large number of intersections is also a cause of roof failures.
4. Failure to completely extract coal in areas developed for total extraction. This results from: incorrect or careless procedures creating instability and making full extraction impossible; development of spontaneous combustion resulting in the sealing of significant areas of the mine; and deliberately leaving coal in the roof or floor for some purpose.

In Australia, action is currently being taken to improve the situation in each of these areas. Part of this work will be illustrated in this chapter by a description in some detail of two projects designed to improve the stability of access roadways:

1. A project to predict mining conditions when only borehole information is available.
2. A project to improve the stability of access roadway development in unusually bad mining conditions.

The illustrations are presented in case study form and, therefore, no conclusions are drawn by the author.

Prediction of Mining Conditions from Borehole Information

Until recently, it has been difficult to predict the mining conditions likely to result from an underground mining operation before the mining has commenced. It is likely that different procedures might have been adopted in some cases had a prediction of mining conditions been available in advance. So-called bad mining conditions result from a combination of the physical properties of the coal and rock strata, and the virgin stress field in the region, together with the size and shape of the excavation. If bad roof conditions are experienced in a development heading, the situation may be quite different and satisfactory if the heading is driven at different dimensions or shapes.

Accordingly, a project had been initiated by Australian Coal Industry Research Laboratories, Ltd. (ACIRL) in Australia to measure the basic parameters and observe the stability of the opening in a known situation. These parameters are then used to predict by finite element analysis what mining conditions would be expected in a particular case. A comparison can then be made between the predictions and the actual. It is intended that this procedure will be followed in about 20 cases, by which time it is hoped that sufficient data will have been collected to make it possible to predict with more confidence the mining conditions which can be expected in an unknown situation, when only rock core information is available. The following method is adopted.

Specimen collection

A borehole to produce a 50-millimeter (1.97-inch) core is drilled into the roof and cored for about 13 meters (42.7 feet). Similarly the floor is drilled and cored to a depth of about 3 meters (9.84 feet). Block samples of coal are taken from the same site and cored in the laboratory.

Properties

The following determinations are made on the rock and coal specimens:

1. Density.
2. Uniaxial compressive strength.
3. Young's Modulus.
4. Poisson's Ratio.
5. Tensile strength.
6. Triaxial strength.

Typical results are shown as follows:

	Rock (sandstone)	Coal
Density (kg/m^3)	2,456	1,273
Uniaxial compressive strength (MPa)	77.6	33.4
Young's Modulus (MPa)	16,980	3,700
Poisson's Ratio	0.45	0.41
Tensile strength (MPa)	5.0	1.5
Triaxial strength (at 10 MPa)	125	100

Stress measurement

Where suitable strata exist, stress measurements are made using the South African CSIR triaxial cell. This is an overcoring method in which the cell is attached to the rock near the end of a borehole and the installation overcored. Provided the rock is suitable for overcoring, useful results can be obtained. It has been found that fine-grained sandstone is the most satisfactory, but that in laminated shales, tuffs, and mudstones, overcoring is not possible. Various degrees of success can be obtained in intermediate rock types, but if good rock cannot be located within about 6 meters (19.7 feet) of the heading, stress measurement is not attempted. In this case finite element calculations are made using measured rock properties and assumed stress values.

A typical stress pattern obtained by measurement is shown in the table below:

Principal Stress	Magnitude (MPa)	Azimuth	Altitude
Maximum	16	$180° \pm 20°$	$10° \pm 10°$
Intermediate	12	$80° \pm 20°$	$10° \pm 10°$
Minor	8	$330° \pm 20°$	$90° \pm 10°$

Introscope survey

The hole drilled for stress measurement is used for an examination of the roof strata by introscope. This instrument is simply an illuminated periscope capable of being extended to 6 meters (19.7 feet); it enables the strata sequence exposed in the borehole to be examined for bed separation or other fractures. These observations are very useful in forming a judgment about the physical conditions of the strata surrounding the opening.

Mining environment

A close study of the mining environment at the site of the operations is made. Such features as width and height of the opening, nature of roof and floor material, details of roof support and any features such as roof sag, floor heave, or rib spall are noted and recorded.

Mathematical analysis

The data obtained by the work already described is used to conduct a mathematical analysis of the situation using the finite element method. The results are expressed in terms of progressive failure patterns, nodal shifts, and excavation boundary deformations. Processing of the data should produce results which correspond with the conditions already observed and, if it does not, the reasons for the differences must be determined. One area of difficulty is related to the quality of the rock core material. Perfect cores are seldom, if ever, obtained and because the more friable material is at times unsuitable for preparing specimens there is a tendency to overstate the strength of the rocks comprising the roof and floor sections.

Adjustments have to be made, therefore, to allow for the absence of some low-strength material from the stratigraphic sections used for analysis and also for the effects of joints, cleats, and other discontinuities. As the work proceeds it is hoped that the effect of these features can be progressively quantified, so that future predictions made from rock cores in unknown situations can be made more accurately and with greater confidence.

Improved Roadway Stability by Variation of Shape

A colliery operating in the Greta Seam in the Cessnock area of New South Wales in Australia was driving 6 × 2.5-meter (19.7 × 8.2-foot) rectangular development roadways into virgin territory when very bad conditions developed. Although heavy steel roof supports were installed, large falls occurred and one by one the headings were abandoned.

A research project was instituted to seek a solution to this problem, since the long-term existence of the mine depended on the development of access roadways through the area.

The investigation was conducted in accordance with the following procedure:

1. Measurement of material properties.
2. Preliminary assessment of stability of roadways of varying shape.
3. Mathematical analysis by the finite element analysis method of roadways of circular shape and varying size.
4. Driving of a roadway of the selected shape underground.
5. Monitoring the stability of the roadway over a period of 12 months.

Measurement of material properties

A comprehensive program to measure the physical properties of the coal and roof and floor rocks was undertaken. In general the seam is about 6 meters (19.7 feet) in thickness and lies relatively flat under about 300 meters (984 feet) of cover. The roof material is a mixture of coarse sandstone and conglomerate, and the floor is sandstone.

Preliminary assessment of stability
of roadways of varying shape

The preliminary analysis was based on the results of a study made by the U.S. Bureau of Mines (R.I.7297), and was carried out on three different cross sections, shown in Figure 10.1. These sections were circular, horseshoe, and rectangular, and were considered to represent three possible roadways in a thick seam. The maximum closure of the three roadways was compared at a hydrostatic stress of 7 megapascals (MPa). Assuming a normal pressure gradient, this represents the expected in situ stress at a depth of about 300 meters (984 feet). The results of this work are shown in the table below:

Cross section	Maximum closure (mm)	Position
Circular	95	Overall
Horseshoe	200	Horizontal axis
Rectangle	450	Vertical axis

Although this was only a preliminary investigation, it illustrated the disadvantage of the conventional rectangular roadway in thick coal seams. The horseshoe cross section reduced closure by over 50 percent, and the circular cross section was the most stable shape.

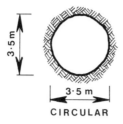

CIRCULAR

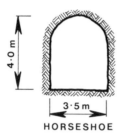

HORSESHOE

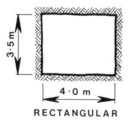

RECTANGULAR

Figure 10.1. Sections of three differently shaped roadways upon which closure tests were carried out.

Mathematical analysis by the finite element method
of roadways of circular shape and varying size

Having reached the conclusion that a circular or near circular shape was likely to provide the best solution, three circular roadways of diameters 3, 4 and 5 meters (9.84, 13.1 and 16.4 feet) respectively were selected for further study by the finite element analysis method, using material properties and stress levels believed to reasonably represent those existing in the mine.

Four bedding planes (partings) were included above the openings, and stress levels of 7 and 10 megapascals were used to compare the results. Material properties were based on previous work and are set out in Table 10.1.

Table 10.1. Properties of Rock and Coal

Coal

Uniaxial compressive strength (MPa)	28
Young's Modulus (MPa)	7,200
Density (kg/m^3)	1,590
Cohesion (MPa)	9.6
Angle of internal friction ($^\circ$)	45.8
Tensile strength (MPa)	1.45

Sandstone

Young's Modulus (MPa)	13,800
Density (kg/m^3)	2,550
Cohesion (MPa)	14
Angle of internal friction ($^\circ$)	31.5
Tensile strength (MPa)	8.62

Bedding planes

Normal stiffness (MPa/m)	1.25×10^6
Shear stiffness (MPa/m)	1.56×10^2
Consolidated normal stiffness (MPa/m)	1.25×10^9
Residual shear stiffness (MPa/m)	4.5×10^{-4}
Cohesion (MPa)	1.7
Angle of internal friction ($^\circ$)	20.1
Allowable joint closure (mm)	1.5

Analyses indicated that the 3-meter (9.84-foot) diameter opening was stable at a stress of both 7 and 10 megapascals. The 4-meter (13.1-foot) diameter opening was generally stable at 7 megapascals. Although the elements in the lower portion of this profile had a factor of safety less than unity, the deformations were believed to be acceptable. The 5-meter (16.4-foot) diameter opening was not stable at either 7 or 10 megapascals because the upper boundary was aligned with a bedding plane in the coal, and this resulted in the failure of the elements in the vicinity. The results of this work for 7 megapascals stress are illustrated by Figures 10.2, 10.3, and 10.4.

Driving a roadway of selected shape underground
In order to test the validity of the predictions, arrangements were made to drive an appropriately shaped roadway underground. Using a Voest-Alpine roadheader a roadway of roughly circular shape and approximately 200 meters (656 feet) in length was driven in the region where the rectangular headings had proved to be unmanageable. Originally it was decided to form a circular, arched roadway of 4.3 meters (14.1 feet) diameter with a flat floor, having a maximum height of 3.3 meters (10.8 feet). The height was subsequently reduced to 3 meters (9.84 feet) because a parting in the coal close to the original roof line caused the top section of the arch to fall away. These sections were supported by steel mesh and roof bolts and the general arrangement is shown in Figure 10.5.

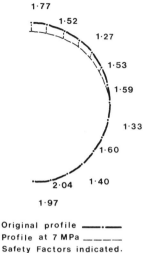

Figure 10.2. Results of analysis on a 3-meter-diameter opening at a hydrostatic stress of 7 megapascals.

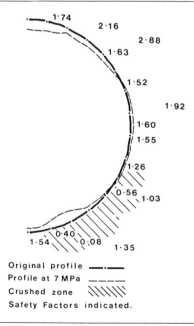

Figure 10.3. Results of analysis on a 4-meter-diameter opening at a hydrostatic stress of 7 megapascals.

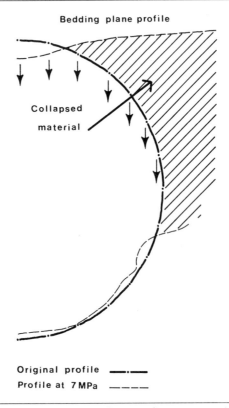

Figure 10.4. Results of analysis on a 5-meter-diameter opening at a hydrostatic stress of 7 megapascals.

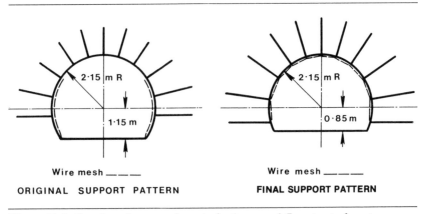

Figure 10.5. Spacing along roadway in both cases 0.7 meter to 1 meter.

Because of the very fractured nature of the coal it was impossible to maintain an accurate profile. Nevertheless, a serious effort was made to drive the selected shape and, in general, this was successful even though the rate of advance was disappointing. The rate of advance was not an important feature at the time, since the primary objective was to drive a heading as near as practicable to the designed shape so that the stability of the opening could be determined. In any case, a high rate of advance could not be expected in an area which had been impossible to develop by other means.

Monitoring the stability of the roadway

In order to test the stability of the new roadway, a monitoring procedure was followed for a period of about 12 months. This was done by leveling from bench marks situated in an area considered to be stable. Several reference bench marks were used so that any movement of the datum bench mark would be indicated by relative movement between them. Levels were taken to the heads of roof bolts along the heading. The results showed that the convergence was independent of the varying length of roof bolts installed from time to time. These varied from 1.1 meters (3.61 feet) to 1.8 meters (5.91 feet) and both chemical and mechanical anchors were used.

The majority of the movement took place within the first two weeks of bolt installation and then continued at a very slow, decreasing rate. At the end of the monitoring period the rates of convergence at all stations were approaching zero and the roadway was considered to be stable.

Reference

Smith, M. J., 1977, "The Inadequacy of Current Technology for the Recovery of Coal Seams by Underground Mining Methods." World Energy Conference, Istanbul, Turkey.

Rock Mechanics Investigation for a Longwall Panel in the Illinois No. 6 Coal

11

Peter J. Conroy, Associate,
Dames & Moore, Park Ridge, Illinois, USA

Introduction

A rock mechanics study was performed as part of the longwall demonstration project which was a cooperative agreement between Old Ben Coal Company and the U.S. Bureau of Mines. The demonstration consists of three longwall panels. This chapter summarizes the studies performed on the first panel. The second panel is in progress.*

Six previous attempts with longwall methods in Old Ben Coal Company's mine No. 21 had been unsuccessful due to roof control problems. A description of these attempts is given by Moroni, 1973. In preparation for the present longwall operation, a study was performed to review existing data on the previous unsuccessful attempts, to establish coal and floor strata bearing capacities in mines Nos. 21 and 24, and to analyze the damage to supports used in previous longwall attempts. The design criteria for the supports to be used in this demonstration were developed from this study.

A summary of the earlier longwall faces is presented in Table 11.1.

The present longwall demonstration consists of three panels in Old Ben Coal Company's mine No. 24. The mine is located 3.2 kilometers (2 miles) northwest of Benton, Franklin County, Illinois, and is producing coal from the Herrin (No. 6) Seam. Panel No. 1 measured 140 meters (460 feet) wide by 529 meters (1,735 feet) long with a seam height of 2.4 to 2.6 meters (8 to 8.7 feet). The equipment used in this operation included 95 Rheinstahl 4.5-meganewton (MN) (500-ton) shield-type supports, an Eickhoff double drum, bidirectional shearer, and an Eickhoff single-strand armored face conveyor with head and tail drive units.

---

* At the time of writing this presentation for the First International Symposium on Stability in Coal Mining, the author intended to compare the results of monitoring panel 2 with those of panel 1 which had been reported earlier (Wade and Conroy, 1977). Panel 2 was delayed by the 1977/1978 U.S. coal miners strike and data were not available.

Studies Performed Prior to Mining

A study was performed prior to mining to establish design criteria for the proposed support system. The study included: a review of the data from the previous longwall faces, including a damage survey of supports which were available from face No. 5, and a series of in situ bearing capacity tests to obtain the strength of the roof and floor.

The review of previous data indicated that the supports used for faces Nos. 4, 5 and 6 were not capacitated. Roof control problems were generally associated with stability of the chocks rather than load-carrying capacity. Based on the review of available data it was estimated that an average support density of 863 kilonewtons per square meter (kN/m²) (125 psi) would provide adequate support for the roof in mine 24. This capacity would provide for some contingency in the event that the roof in mine 24 did not behave similarly to the roof in mine 21.

The damage survey performed on the supports used on face No. 5 indicated that structural damage to those supports could occur at the measured loads if they were set at inclinations in excess of 7°. It was reported that supports were loaded at inclinations up to 15° from the vertical. It was recommended that the supports proposed for mine No. 24 be capable of withstanding a lateral thrust equivalent to the lateral component of the roof load inclined 15° from the vertical.

The bearing strengths of the roof and floor in the vicinity of the proposed panels were established by in situ bearing tests. These tests established that the bearing capacity of the floor increased with thickness of floor coal and ranged from 2.1 MN/m² (315 psi) on wet fireclay to in excess of 43.6 MN/m² (6,320 psi) on 36 centimeters (14 inches) of floor coal. Roof strength exceeded the test equipment capacity of 43.6 MN/m². Comparison of the results of the bearing tests performed in Old Ben No. 24 with the tests performed by the U.S. Bureau of Mines in mine No. 21 indicated that the strengths of the roof and floor with floor coal were comparable between the two mines. The low bearing strength measured on the fireclay in mine 24 was due to softening of the clay caused by moisture. Based on the evaluation of bearing test results it was recommended that a minimum of 152 millimeters (6 inches) of floor coal be left on the fireclay. Based on the test results, 152 millimeters of floor coal would provide a bearing capacity of approximately 6.9 MN/m² (1,000 psi).

Studies Performed During Mining

Surface and underground monitoring was performed during mining of panel 1. The monitoring program was designed to:

1. Provide the capability of observing unusual situations during mining so that remedial measures could be instituted quickly, if required.
2. Provide design guidelines for future longwall systems.
3. Evaluate surface subsidence to develop criteria to assist in predicting subsidence for other mines having similar roof conditions.

Both subsidence and development of roof cave were monitored.

Table 11.1. Summary of Longwall Mining Operations in Old Ben Mine No. 21

Face	Operation dates	Face Length (ft)	Support type	Spacing (ft)	Production (tons)	Retreat (ft)	Yield load (set) (tons)	Area roof support (ft^2)		Support problems
								B.C.	A.C.	
1	6/12/62 to 8/15/62	460	4 prop set R-W 45 ton/prop articulated cap	4.1	1,119	24	180	44.98	90.08	Lack of capacity
2	8/21/62 to 10/25/62	460	Do. plus Lamella friction props	4.1	17,240	150	238	44.98	90.08	Lack of capacity
3	7/15/63 to 11/13/63	460	6 prop set R-W 45 ton prop articulated caps, connected by semi-flexible connector	4.1	23,823	220	270	58.69	66.87	Roof support insufficient; decision made to redesign chocks; decide to use 100-ton leg and solid roof bar.
4	1/04/65 to 8/14/65	460	4 prop set R-W 100 ton/prop (Main) in between each was a 2 leg "pilot"	(main) 5	50,866	585	470	50.90	61.40	Some semblance of roof control could be and had been established. The forward newly exposed roof needed to be supported as promptly as possible. Stability of paramount importance.

	Dates		Equipment								Remarks
5	7/15/66 to 8/01/67	600	4 prop set R-W 120 ton/prop, solid roof bar	4.1	115,000	1,070	480	(41.74)	(50.35)		Chock stability had become a paramount problem, in fact face was discontinued for this reason.
								53.30	61.91		
6	6/09/69 187 days to 88 days 4/10/70	300 210	Joy-Gullick 7 leg chock sets, 6 @ 85 ton, 1 @ 50	3.5	70,000	924	510				Constant massive roof falls at head and tailgates, much pressure exerted in advance of face most of time.

Source: Moroni, E.T., "Longwall Experiences in the Herrin No. 6 Coal Seam," proceedings of the Illinois Mining Institute, October 11, 1973.
Note: R-W = Rheinstahl-Wanheim
B.C. = Before cut
A.C. = After cut

Surface monitoring

Monitoring of surficial subsidence. The magnitude and extent of surficial subsidence were monitored by a series of monuments placed in three rows over both panels. The monuments were installed as benchmarks with horizontal control established to third order, Class I accuracy. Vertical control was established to second order, Class I accuracy. The monuments were installed and monitored by Old Ben Coal Company personnel with direction and assistance from the Bureau of Mines. Subsidence contours two weeks after the completion of the panel are shown in Figure 11.1.

Maximum surficial subsidence observed over panel 1 was 1.34 meters (4.39 feet). This is approximately 63 percent of the extracted seam height of 2.1 meters (7 feet). The angle of draw was calculated to 3.1 millimeters (0.12 inch) of subsidence. This angle was computed to be 14° at the south end of the panel, 30° at the north end of the panel, and 23° at the west end of the panel. The draw angle was not calculated over the east side since this area was subject to subsidence from conventional mining operations in an adjacent section.

Monitoring of roof cave. The monitoring of roof cave was accomplished by means of a 75.7-millimeter- (2.98-inch-) diameter borehole drilled from the surface over panel 1. Geophysical logging of the borehole was performed by Bureau of Mines personnel from the Denver Mining Research Center. Finally, a 22.2-millimeter- (0.88-inch-) diameter coaxial cable was grouted into the borehole. Subsequent collapse of the strata caused damage to the cable which could be detected by electronic analysis of the cable using Time Domain Reflectometry (TDR). The results of the TDR monitoring are shown in Figure 11.2.

The TDR study was particularly valuable in analyzing the behavior of the rock strata above the mine after removal of the coal. The analysis permitted observation of the strata failure with respect to face advance and stratigraphy. By examination of the TDR records, it was possible to observe initial strata separation and development of the separation until failure occurred.

Subsurface monitoring

The subsurface monitoring was designed to measure:

1. Pressure in the support hydraulic system.
2. Changes in stress in coal and floor.
3. Differential strata movement.
4. Roof-floor convergence.

Particular attention was given to stress changes and strata movements with respect to face position.

Support monitoring. Instruments were placed on 10 shields spaced at regular intervals along the face. Each instrument installation consisted of two hydraulic pressure gauges and a continuous seven-day, two-pen recorder. One pressure gauge and one pen of the recorder measured pressure in each of the two shield legs making the system capable of measuring differential loads on the shield. During mining operations, the gauges were read and canopy heights measured daily. The recorders were inspected daily and serviced weekly.

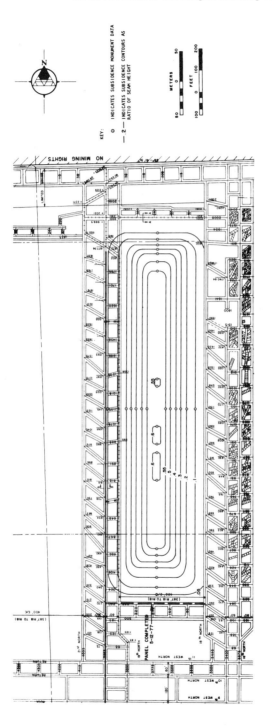

Figure 11.1. Surface subsidence contours two weeks after completion of panel.

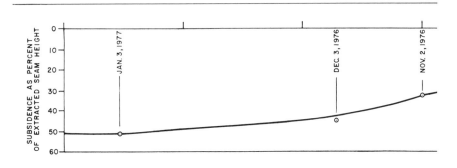

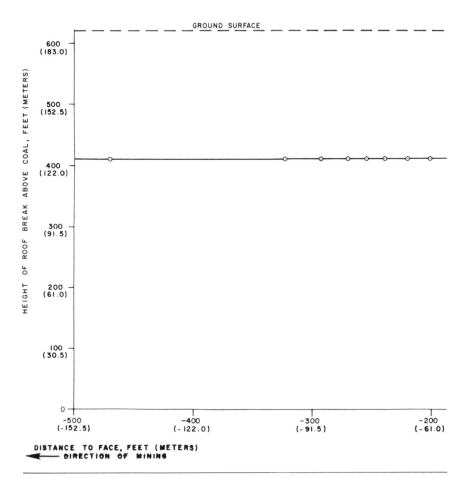

Figure 11.2. Roof break versus face advance.

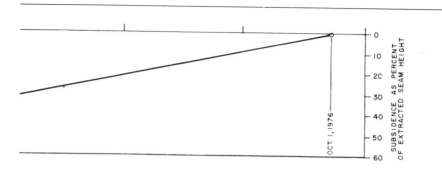

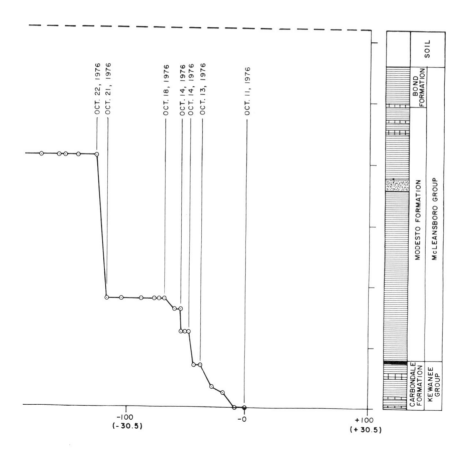

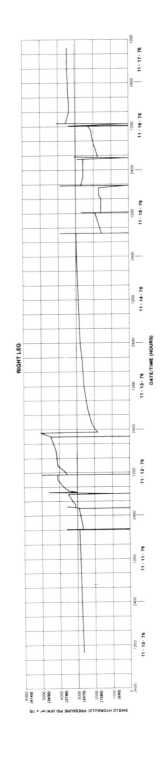

Figure 11.3. Seven-day hydraulic pressure record (typical).

Figure 11.3 illustrates a typical seven-day record of hydraulic pressure for one shield. The graph shows that the hydraulic pressure increases at a higher rate during the mining cycle than during idle periods. The record also shows a characteristic pressure increase in each leg immediately prior to unloading. It is believed that this characteristic is the result of a temporary load increase caused by the unloading of the adjacent leg.

The capacity of the shields used in panel 1 was approximately 4.5 MN (500 tons) at a seam height of 2.1 meters (7 feet). This capacity is less than that of the chocks used unsuccessfully on longwall face 6 in mine 21. In general, shields were loaded below their rated capacity. The hydraulic pressures used to set the shields were erratic and commonly less than the nominal setting pressure of 30 MN/m^2 (4,350 psi). The shields did not approach yield pressures until they reached the recovery room at the end of the panel. Shield hydraulic pressures remained generally below nominal setting pressure until approximately 3.1 to 4.6 meters (10 to 15 feet) of coal remained in the panel. During the last few passes of the shearer, pressures increased rapidly and continued to rise after the last coal had been removed. In the recovery room several shields showed extremely high hydraulic pressures often above nominal yield pressure.

Throughout the mining operation, there was no observable cyclic pattern in shield loading. Likewise, there was no observable reduction in canopy loads after the first or second major roof falls. Differential loading was observed on the shields. In several cases the difference in hydraulic pressure between the left and right shield legs exceeded 6.9 MN/m^2 (1,000 psi). No consistent pattern was observed in the differential loading.

A review of the typical seven-day shield pressure record indicates the highest rates of loading occur during mining cycles. During idle periods, such as weekends, canopy loading occurs gradually or may decrease.

Stress changes measured. Stress changes in the coal seam and in the siltstone floor were measured at various locations in and around panel 1. Stresses were measured using vibrating wire stressmeters manufactured by Irad Gage, Incorporated of Lebanon, New Hampshire. The stressmeter consists of a cylindrical gauge containing a highly tensioned steel wire mounted diametrically across the gauge. A portable readout device vibrates the steel wire at its resonant frequency to monitor changes in the length of the steel wire which result from stress changes. The readout device displays the period of the wire at its resonant frequency. The readout value (period of the wire) must then be converted to the change in stress; this is accomplished by using a calibration curve developed for the type of rock in which the stressmeter is installed.

The stressmeters were used to monitor pressure changes in the coal seam, in the longwall panel, in the pillars surrounding the panel, and in the siltstone floor under the longwall panel. The stressmeters were placed in 38.1-millimeter- (1.5-inch-) diameter boreholes drilled both horizontally and at an incline into, around, and under panel 1. The setting distances ranged from 7.1 to 84.7 meters (24 to 278 feet). The locations of the stressmeter installations are shown in Figure 11.4. Figure 11.5 shows typical results for stressmeters that were installed in the panel and Figure 11.6 shows typical results for stressmeters that were installed under the panel.

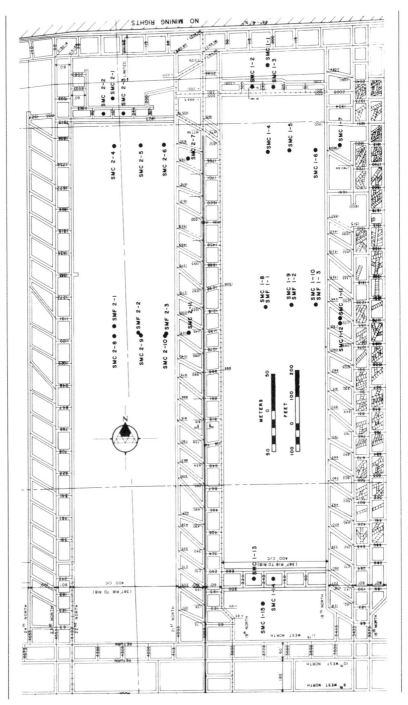

Figure 11.4. Stressmeter locations in panel 1.

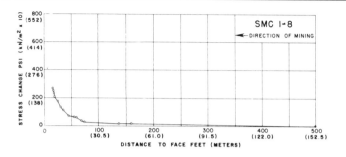

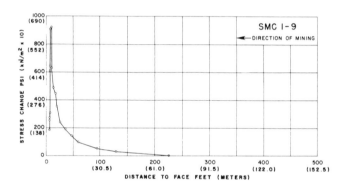

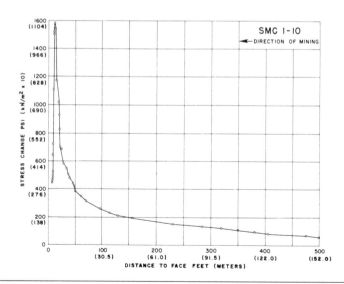

Figure 11.5. Stress change versus distance to face: stressmeters SMC 1-8, 1-9 and 1-10. Typical results for stressmeters in the panel are shown.

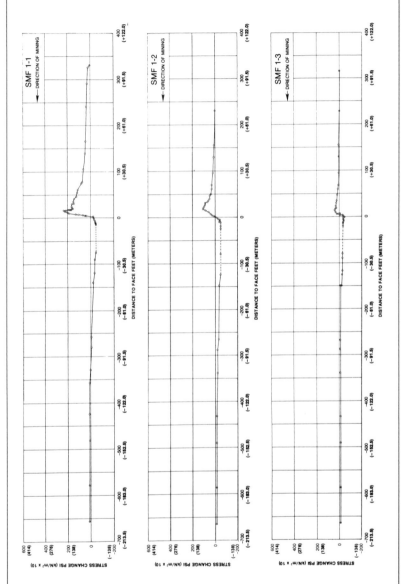

Figure 11.6. Stress change versus distance to face: stressmeters SMF 1-1, 1-2 and 1-3.

A review of stressmeter data indicated that stresses in the panel began to increase when the face was approximately 61 meters (200 feet) away. The rate of increase accelerated when the face was approximately 30.5 meters (100 feet) away and accelerated to a maximum rate 15.2 meters (50 feet) from the face. The maximum stresses occurred in the panel approximately 2.7 to 3.1 meters (9 to 10 feet) in front of the face.

Stressmeters installed in pillars at the bleeder entries showed a marked increase in stress of 2.8 to 4.1 MN/m^2 (400 to 600 psi) when the face was advanced to approximately 45.7 meters (150 feet). At that point, a sharp reduction in stress was noted. Stress then increased to exceed 4.1 MN/m^2 (600 psi) above initial stress. The sharp stress reduction was accompanied by large roof falls behind the shield and by failure of the underclay beneath the bleeder entry pillars. The failure of the underclay was observed as heave in the floor of the bleeder entries. A similar failure of the underclay was observed at pillars that were installed in the recovery area when the pillar stresses reached 4.1 to 5.5 MN/m^2 (600 to 800 psi) above overburden stress.

Conclusions

The completion of panel 1 was the first successful application of longwall mining in the Illinois Basin. Roof control was achieved primarily due to the inherent rigidity of the shield supports. In general, the supports were loaded below their rated capacity. Yielding of supports was not observed until they approached and entered the recovery room at the end of the panel. A cyclic pattern of shield loading corresponding to a pressure front was not observed, and there was no observable reduction in canopy loads after the first or second major roof falls.

The necessity of leaving 152 millimeters (6 inches) of floor coal was demonstrated during mining. Whenever the floor coal was removed the shields would push into the underclay and would often require timber blocking to resume advance.

The results of rock mechanics monitoring are summarized in Figures 11.7 and 11.8. Figure 11.7 summarizes the data through a centerline section of the panel. The stress change due to mining is shown, and is related to the measured development of caving and subsidence. This figure indicates that at distances of approximately 76.2 meters (250 feet) from the face, loading on the goaf resumed overburden pressures and that at distances greater than approximately 91.4 meters (300 feet) from the face most of the subsidence was due to consolidation of the goaf. Figure 11.8 summarizes the data obtained from stressmeter data showing the generalized stress contours in the panel and adjacent pillars. The contours indicate an increase in stress due to mining and are expressed as a ratio of overburden stress.

References

Moroni, E. T., 1973, "Longwall Experience in the Herrin No. 6 Coal Seam." Proceedings of the Illinois Mining Institute.

Wade, L. V., and Conroy, P. J., 1977, "Rock Mechanics Study of a Longwall Panel." AIME Fall Meeting, St. Louis, Missouri.

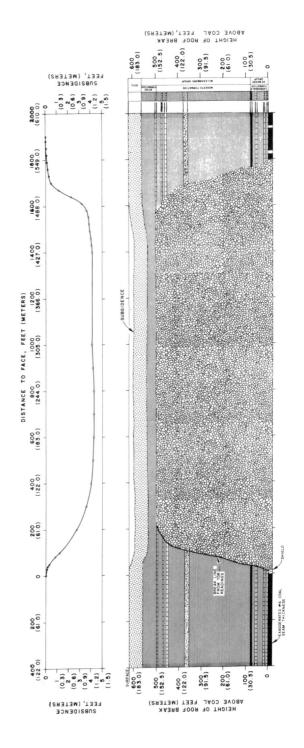

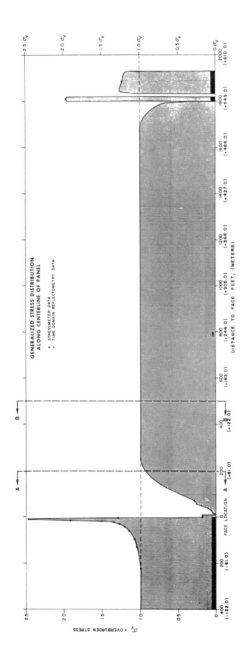

Figure 11.7. Summary of data along panel centerline.

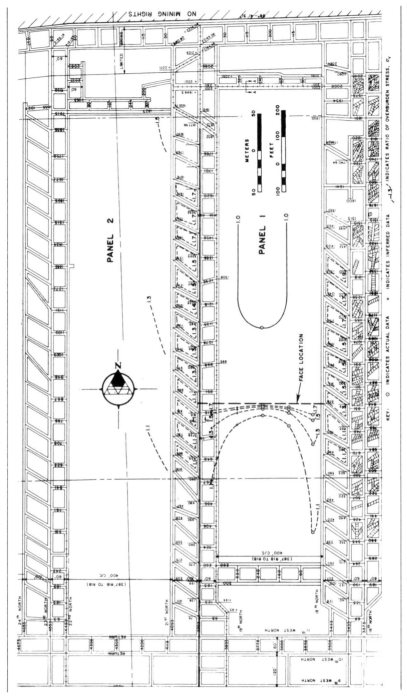

Figure 11.8. Generalized stress contours.

Working of Coal Seams Which Are Not Suitable for Mechanization

12

Serafettin Ustunkol, Mining Engineer,
Eregli Coal Company, Zonguldak, Turkey

Introduction

It is a fact that, at the present time, the world is faced with an energy problem. Nations are searching for new energy sources and increasing their production from existing resources. Coal, which is essential for the steel industry, is one of the most important energy resources mankind possesses. Because coal is a non-renewable resource, seams which were previously considered to be economically unworkable are now being explored. Such seams are especially important in developing countries which have limited coal resources.

Among coal seams which previously were not thought economical to mine, there are seams which are not suitable for full mechanization. It is now possible to work these seams economically by applying special methods. Although these methods rely greatly on manpower, partial mechanization is also utilized. Such methods can be used most economically in countries like Turkey, where cheap manpower is available and coal sources are poor.

Characteristics of Coal Seams Unsuitable for Mechanization

Seams requiring special methods have been squeezed, crushed, and broken by faults, and their dips have been irregularly distorted by extreme tectonic stresses. In places, these seams thin out or even disappear completely. In other places they have become very thick. Figure 12.1 is a coal seam showing these characteristics.

The coal seams of the Zonguldak Basin of Turkey are examples of distorted seams. Coal mining in this basin, which is the earliest known coal basin in Turkey, is carried on in the Namurian and Westphalian Carboniferous Coal Measures which are overlain by Jurassic and Cretaceous formations. Mining is accomplished in very broken and folded structures caused by the effects of the Hercynian and Alpine orogenies.

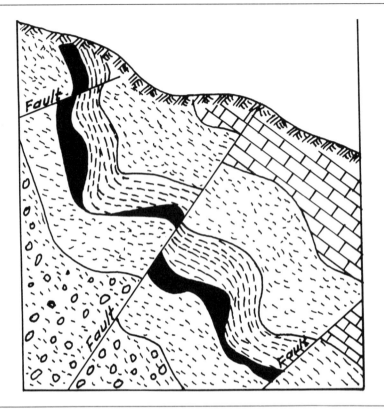

Figure 12.1. A typical, very disturbed coal seam found in the Zonguldak Basin.

Geological and stratigraphical features of the basin are very complicated because of the intense tectonic stresses to which it was subjected. The strata have been folded to form many synclines and anticlines, and have been displaced by numerous faults. From about 25 workable coal seams in the basin, 4,989,600 tons of salable coal are produced annually using semi-mechanized mining methods. Attempts to apply full mechanization by various engineering firms have failed.

Mining Methods

Coal seams which are unsuitable for mechanization can be divided into two groups according to their angle of dip: gently sloping or inclined seams up to 45°, and steeply inclined seams over 45°.

Gently sloping or inclined seams up to 45°
In Zonguldak, gently sloping and inclined seams are being mined by the back-caved or backfilled retreating longwall method. Mining is carried out on levels. The vertical distance between levels is from 50 to 100 meters (164 to 328 feet).

Thick seams are being worked in slices, and tailings from washeries are used as filling material.

Geological obstructions (squeezed and broken zones) at the face can be circumvented by driving curved raises, and small faults can be passed through by small shafts supported with timber. Figure 12.2 shows a plan view and section of such a longwall.

While negotiating geologically disturbed ground, coal production sometimes has to be stopped. It is therefore necessary to have spare longwall faces ready for production.

The coal is cut by pneumatic picks. On longwalls, timber is used to support the face. For this purpose pine is preferred because it is easy to cut, is light to handle, and makes a warning noise when it starts to yield under pressure. If backfilling is not carried out, then the rear of the face is supported by installing chocks made of robust-structured timber, like oak. Support systems in the raises and small shafts on the longwalls are made of three- and four-piece timber sets respectively. The distance between sets may be about 0.91 meter (3 feet), and lagging completes the supporting. The raises and shafts are divided into two compartments: manway and coalway.

Full mechanization can be applied in all kinds of tunnel drivages, and supporting may be accomplished by using rigid, two-piece steel arches.

Steeply inclined seams over 45°
In steeply inclined seams, methods other than current worldwide longwall systems are being used in Zonguldak.

If the seams are thin, the following methods may be applied: diagonal twin longwall mining with backfilling, and sawtooth longwall mining.

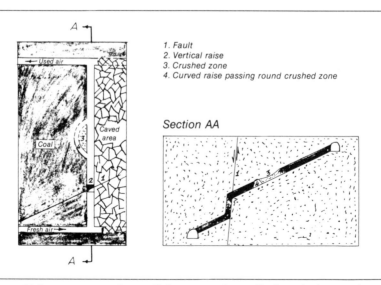

1. Fault
2. Vertical raise
3. Crushed zone
4. Curved raise passing round crushed zone

Figure 12.2. A retreating longwall face in geologically disturbed ground.

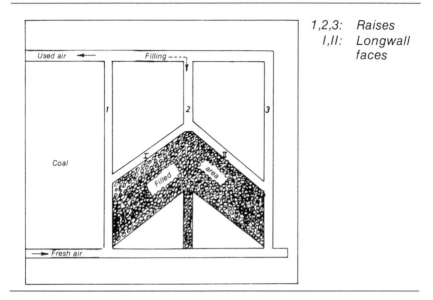

Figure 12.3. Diagonal twin longwall mining.

Diagonal twin longwall mining. The vertical distance between the lower and upper gate roads is about 60 meters (197 feet). Three raises are driven from the lower to the upper gate road, as can be seen in Figure 12.3. From the bottoms of the first and third raises, two diagonal raises at about 42° are driven towards the middle raise, and so, two longwall faces are formed.

From these two longwalls, production is carried out in a direction of advance at right angles to the faceline. Filling materials are sent down by gravity through the middle raise, and the backs of the longwalls filled. Rectangular timber sets, reinforced with lagging, are used for supporting the faces.

Sawtooth longwall mining. The sawtooth method of longwall mining is illustrated in Figure 12.4. The vertical distance between the lower and upper gate roads is about 60 meters (197 feet). A secondary road is driven 5 meters (16.4 feet) above the lower gate road; a pillar of coal is left, affording protection for the lower gate road. These two roads are connected by small shafts driven through the coal pillar at intervals of 5 meters (16.4 feet), supported by square set timber. From the secondary road, a diagonal raise is driven at about 43° towards the upper gate road and teeth are formed in the coal at the top of the raise. There will be as many faces as there are number of teeth. The height of each tooth is about 2.6 meters (8.53 feet). A team, consisting of a picker and his helper, works at every tooth face.

A rectangular timber set is installed as support every 1.3 meters (4.27 feet), shown in Figure 12.5. For reinforcement, a timber bar is inserted horizontally between two posts, and set in the middle. Lagging is applied to the sides and roof between sets.

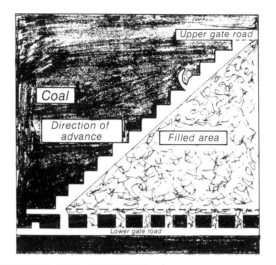

Figure 12.4. Sawtooth method of mining.

1. *Prop*
2. *Cap*
3. *Sill*
4. *Lagging*
5. *Middle bar*

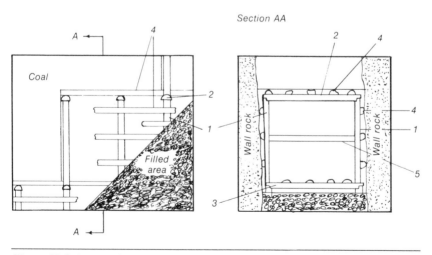

Figure 12.5. Sawtooth mining method. Sections show system of supporting teeth.

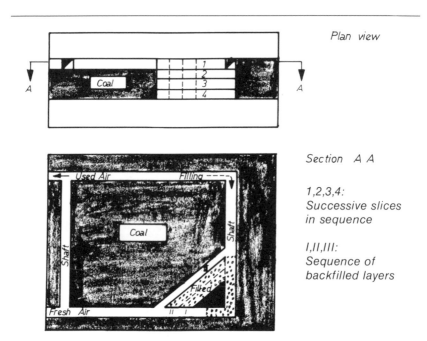

Figure 12.6. Shaft method of mining.

The filling shift follows the production shift, and filling materials, consisting of fine and wet washery tailings, are sent down by gravity from the upper gate road.

If the seams are thick, then the shaft method of mining can be applied.

Shaft method of mining. Mining is carried out on levels. The vertical distance between levels is about 50 meters (164 feet). Shafts, supported by square set timber, are driven through the coal at intervals of about 35 meters (115 feet) along the true dip of the seam, from the lower gate road to the upper gate road. As can be seen in Figure 12.6, a diagonal raise is driven at about 42° from the lower gate road toward the shaft; for example, following the roof rock. A short longwall is then formed, which is worked toward the floor rock in slices. After each slice has been mined, the back area is filled with material sent through the shaft. When a layer is completely worked out, a second raise just above the first is driven in the same direction, and so another longwall is formed. This procedure continues until the whole panel between two shafts has been mined out.

Face supporting is accomplished by three-piece timber sets reinforced with lagging.

Gas Problems

When mining very disturbed coal seams, especially in deep mines, gas problems will increase considerably, since methane, which is dispersed homogeneously in undisturbed seams, will have accumulated in certain zones within the disturbed seams. For example, the probability of gas accumulation is higher in the zones of faults and breaks, and in the thicker parts of a seam where the two ends have been pinched by the action of compressive tectonic stresses.

Accumulations of compressed methane may outburst with great force and may cause fatal accidents. Many underground workers have died because of methane outbursts in Zonguldak.

During the driving of roads or raises, more care must be taken when approaching gassy zones. The most successful way to prevent methane outbursts in Zonguldak, especially when driving raises through coal, is to drill boreholes to locate gas. As a result, zones with gas accumulations are discovered, and it is possible to draw out the gas, or at least relieve the pressure of the compressed gas.

Pillar Extraction on the Advance at Oakdale Colliery

13

B. Nicholls, Colliery Manager,
Clutha Development Pty. Limited,
Narellan, New South Wales, Australia

Introduction

Oakdale colliery is one of a group of coal mines operated by Clutha Development Pty. Limited.

The mines are situated adjacent to a deep gorge called the Burragorang Valley to the southwest of Sydney.

For many years, other mines in the area have worked destressed areas adjacent to the outcrop giving mostly good roof conditions.

The mine has been in operation for 25 years. None of the boundaries of the lease outcrop in the Valley gorge. Several attempts have been made in the past to develop east into the deeper reserves within the colliery lease, but all these development drivages were very difficult and have ultimately proved impossible. All attempts to mine east, particularly at depths in excess of 430 meters (1,411 feet), have failed resulting in high production costs while mining and the sterilization of substantial quantities of coal. For many years, coal has been won using continuous miner and shuttle car systems. Massive roadway failure has occurred mainly in the north-south direction, irrespective of the type or amount of support used.

Because of the failure to develop the major reserves to the east, and the rapidly depleting reserves to the west of the shafts, it has become critical that a solution to the problem be determined quickly. Failure to develop toward the east in the present development sections would mean that access to the substantial eastern reserves would be lost from the present shaft entries. The inability to maintain permanent roadways in the eastern reserves could involve costly alternative shaft or drift entries farther to the east of the existing mine site.

Obviously, the success of such a drastic measure would still depend on a solution being found to the mining of the deeper and more difficult coal to the east. All indications are that the present strata behavior can be anticipated in these eastern reserves.

Geology

Coal seam

The Bulli Seam being worked is the top seam of a sequence comprising the coal measures of the Southern Coalfield within the Sydney Basin. The measures are of Permian age.

The Bulli Seam in the Burragorang Valley area is a medium coking coal. Within the Oakdale lease, the seam has low ash, medium swell characteristics resulting in a high yield after preparation. The average approximate analysis of the Bulli Seam coal in the Oakdale colliery, excluding bands, is as follows:

Moisture	2.1
Volatiles	25.8
Fixed Carbon	62.1
Ash	10.0
Swelling Index	$3\frac{1}{2}$
Rank (R_0[max])	1.03

The seam varies in thickness from 1.8 meters (5.9 feet) to 2.7 meters (8.9 feet).

Roof and floor

The immediate roof consists generally of laminated shale/mudstone. There are areas of sandstone-infilled stream channels running through the lease area which do create considerable mining problems when encountered. The upper roof measures consist of massive thicknesses of sandstone known as the Hawkesbury Sandstones.

The seam floor consists of a reasonably hard, dark grey shale. Figure 13.1 shows a section through the Bulli Seam.

Zone of monoclinal flexure

A major structural geological feature of the Sydney Basin is present in the Oakdale area as a monoclinal flexure zone passing through the lease to the east of the shafts on a north-northwest strike, as shown in Figure 13.2. This is seen underground as a local increase in seam dip from the average 1 in 20, to in excess of 1 in 11. Difficult mining conditions prevail when extracting coal from within this zone in either solid development or pillar extraction. The roadway stability problems are aggravated within and to the east of this zone.

It is considered that entry into the deeper coal on the eastern side of the monoclinal flexure zone has proved difficult, because headings are entering strata subjected to higher stresses than those in the western areas. The flexure zone apparently effectively locks in the stress field to the east. The western areas of the mine, however, may be relatively destressed by the presence of the Burragorang Valley.

Igneous intrusions

Several minor dike and sill structures are encountered within the lease. None are difficult to handle, and they create only minor mining problems.

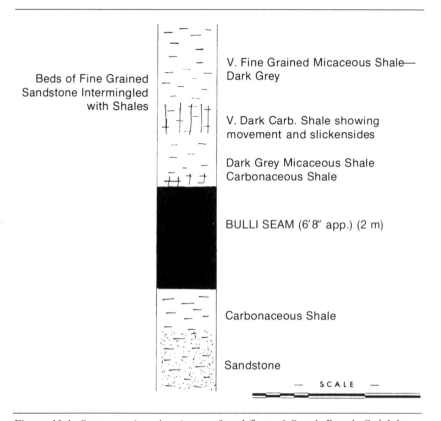

V. Fine Grained Micaceous Shale—
Dark Grey

Beds of Fine Grained
Sandstone Intermingled
with Shales

V. Dark Carb. Shale showing
movement and slickensides

Dark Grey Micaceous Shale
Carbonaceous Shale

BULLI SEAM (6'8" app.) (2 m)

Carbonaceous Shale

Sandstone

—   SCALE   —

Figure 13.1. Seam section showing roof and floor, 6 South Panel, Oakdale.

Roadway Stability Problems

Historical

All attempts made in the past to develop down the dip to the east have failed. The problem of roadway stability, particularly in the north-south direction, has increased in and to the east of the monoclinal flexure zone as the workings have become deeper.

Massive roadway failure in the north-south direction has occurred irrespective of the density and type of support. Roof support methods previously used, in an attempt to prevent roof falls, included combinations of wood props and half-round crossbars, pointed heavy props, crossbars roof bolted to the roof, and heavy steel girders on props.

Directional mining was attempted after geological analysis and recommendations as to the preferred mining direction, but this also proved unsuccessful. Pillar sizes have also been varied, and panel designs using staggered and conventional intersections have been attempted. The addition of timber chocks and, in some instances, brick walls, have still not prevented failure.

Present support system

Recently, roof bolted W straps with rib props have been adopted as the means of support throughout the mine. The normal roof bolt being used is the 1,500 × 24-millimeter (59 × 0.94-inch) slot and wedge bolt, the numbers of bolts per W strap being determined by mining conditions. Bolt anchorage is usually adequate with pullout tests showing anchor competence in excess of 15 tons.

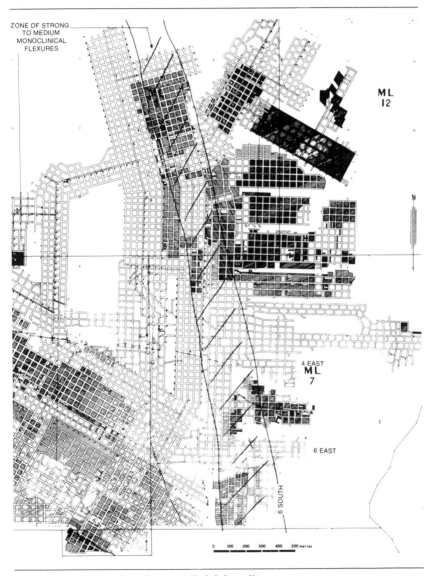

Figure 13.2. Plan of workings at Oakdale colliery.

In the past, the top band of high ash coal was left in the roof. The total seam section is now being extracted in an attempt to prevent crushing-out of top coal from over the supports. Again, despite this intensive support system, massive failure in the north-south direction has occurred.

Implications on Future Mine Development

Previous attempts to mine east with the subsequent failure has determined that eastern development can now only be achieved from the 4 East and 6 East areas, shown in Figure 13.2. These areas are at depths of 440 meters (1,444 feet) and 430 meters (1,411 feet) respectively.

With only three years of coal left for one unit in the western areas, failure to resolve the problems of development to the east would have a drastic effect on the future of the mine. Some 50 million tons of reserves could be cut off from the existing shaft entries.

It then became imperative that a solution to the problem of the eastern development be determined quickly and implemented immediately.

Investigations into the Stability Problem

Australian Coal Industry Research Laboratories

During 1973, investigations were carried out by the Australian Coal Industry Research Laboratories, Ltd. (ACIRL), into the possibility of using sacrifice pillar extraction to protect development headings. (R. T. Hall, 1974.) These investigations using U.S. Bureau of Mines stress measuring cells, were conducted in the Main East development at the colliery.

It was believed at the time that some relief was afforded by the total collapse of an adjacent extraction area. However, no major collapse occurred adjacent to the area being monitored, and the results of the investigation were inconclusive. The investigation did, however, indicate that the major stress was horizontal and in an east-west direction.

As the results of the investigations were considered inconclusive, no further work or trials were carried out.

University of N.S.W. (School of Mines)

More recent investigations into the problems of stability in coal mine roadways have been carried out at other mines in the Southern Coalfield. (R. A. Yeates, 1977.) The results of these latter investigations again indicated a very high horizontal stress field.

By adopting the standard procedure for calculating overburden stress (stress = depth of cover × rock density), a value of approximately 10 megapascals (MPa) was obtained for the working depth at Oakdale colliery.

By using the finite element method to determine the stress patterns around the roadways and theoretical horizontal to vertical stress relationships, failure patterns were predicted. The failure pattern using a 4 to 1 horizontal to vertical stress relationship, closely approximated the pattern of failure occurring in the

Oakdale workings. The horizontal stress in the eastern area of the mine is then estimated to be between 40 and 45 megapascals.

Practical interpretation of the problem

From these investigations and a practical in situ assessment of the magnitude of the problem, it became obvious that ground stability could only be achieved by accommodating these stresses in collapsed areas. Accommodation of the horizontal stresses had to be achieved immediately after, or, if possible, before the roadways were developed.

It appeared that no practical support system would be adequate to contain and control strata movement due to these high horizontal stresses.

Failure Pattern

Roof guttering

Generally, mining conditions at the working face were not difficult, except when mining under wet sandstone or localized geological anomalies.

Roof deterioration usually began with roof guttering on the high side of the roadways. Some floor heave was also experienced. Rib spalling was directly associated with the roof guttering and, in fact, aggravated the problem by widening the guttered area. The roof guttering progressed into a major problem of roof fall from the affected area and heavy rib spalling.

The floor heave generally increased and, in some instances, became quite considerable before actual roadway failure occurred. At this stage, roof bolts were under high tensile stress with considerable deflection of roof bolt washers and W straps between the supporting roof bolts.

The ultimate result was that the guttering worked up into the roof on the pillar rib line. This continued, usually almost vertically, until it reached above the level of the support anchors. At this stage, massive collapse of the roof strata in the roadway occurred, shearing off vertically on both ribs and traveling considerable distances along the roadway.

Before intensive bolting was used, the sheer deadweight of stone and continuing pressure destroyed the supports. Several hundreds of meters of roadway have been destroyed in this manner. Figure 13.3 shows the sequence of roadway failure.

Relief headings

The driving of a relief heading adjacent to a collapsed roadway, leaving only 6 meters (19.7 feet) of coal between the two roadways, showed the ground around the collapsed roadway to be destressed and stabilized. In none of these relief headings were there any indications of roof guttering, rib spall, or floor heave. Indeed, it was indicated that these relief headings could be driven with absolute minimum support.

This has been proved where, of necessity, development roadways have been driven through old collapsed workings by splitting the pillars, and minimum support was required in the new drivages (see Figure 13.4).

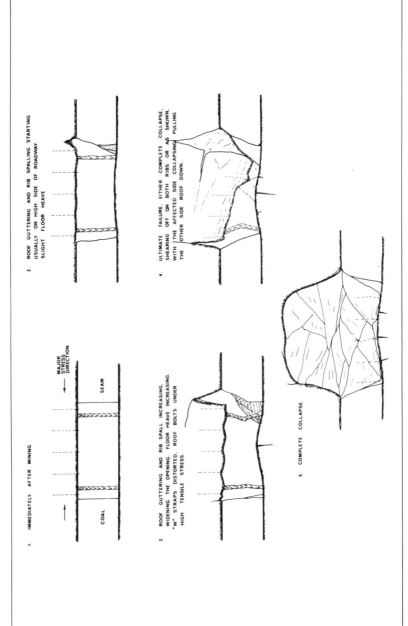

1. IMMEDIATELY AFTER MINING

MAJOR STRESS DIRECTION

SEAM

COAL

2. ROOF GUTTERING AND RIB SPALLING STARTING USUALLY ON HIGH SIDE OF ROADWAY. SLIGHT FLOOR HEAVE

3. ROOF GUTTERING AND RIB SPALL INCREASING, WIDENING THE OPENING. FLOOR HEAVE INCREASING. "W" STRAPS DISTORTED. ROOF BOLTS UNDER HIGH TENSILE STRESS.

4. ULTIMATE FAILURE, EITHER COMPLETE COLLAPSE, SHEARING OFF ON BOTH RIBS OR AS SHOWN, WITH THE AFFECTED SIDE COLLAPSING PULLING THE OTHER SIDE ROOF DOWN.

5. COMPLETE COLLAPSE.

Figure 13.3. Sequence of roadway failure at Oakdale colliery.

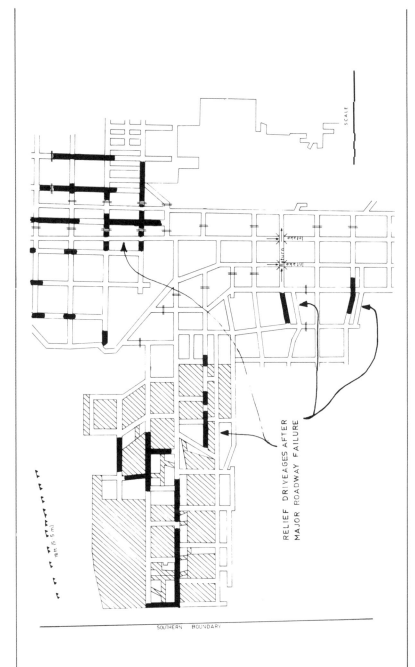

RELIEF DRIVEAGES AFTER
MAJOR ROADWAY FAILURE

SOUTHERN BOUNDARY

Figure 13.4. Plan of 5 South Panel, Oakdale colliery showing partial extraction of coal.

Several attempts have been made at other mines in the Southern Coalfield to design and implement *sacrifice heading* techniques to protect development headings. Only minimal success was achieved as the single collapsed heading did not offer sufficient stress relief to protect more than the immediately adjacent heading. At times, the sacrifice heading failed to collapse, thereby completely negating the system.

Extraction of pillars
At other locations, some pillars have been extracted when advancing a panel, but usually because of economic necessity. However, the total concept of "Pillar Extraction on the Advance" had never been utilized as a method of destressing strata to enable panel development to be successfully achieved.

Pillar Extraction on the Advance

Stress relief requirements
It was obvious that following the collapse of a roadway, the adjacent strata became stabilized after the stresses were relieved.

The question was: By extracting pillars on the advance, could sufficient stress relief and, thereby, ground stabilization be effected to enable the successful development of production panels under increasing cover?

A panel design for pillar extraction on the advance was developed at Oakdale colliery, and it was decided to implement the designed system immediately. This was to be an attempt at stabilizing the ground within which the panel was being developed. It was to be effected by allowing the high inherent horizontal stress to be relieved into the large goaf area being formed as the panel was developed.

Operating unit
The panel in which the system was introduced was 6 South. This panel was actually within the monoclinal flexure zone. The mining unit consisted of a Jeffrey 120H Heliminer and two Noyes Hydracars. All production machines were operated at 1,000 volts.

The roof was supported by 1,500 × 24-millimeter (59 × 0.94-inch) slot and wedge roof bolts and W straps. Roof bolting equipment consisted of Atlas Copco Falcon stopers. A row of wooden props was set on each side of the roadway.

Initial drivage into the panel proved difficult and two major roadway failures occurred before any pillars could be extracted. Extensive rib guttering had again occurred in the panel headings, and a high density of support was required to hold the roadways open.

Output was low owing to the necessity for extra support, and additional labor was required to effect roadway support to try to contain the roof guttering outbye of the working faces.

Sequence for drivage
Obviously, roadways have to be driven to form pillars before pillars can be extracted. The design for pillar extraction on the advance was such that minimum solid drivage was done before pillar coal was extracted (see Figure 13.5).

The requirements of the system were:

1. The system must give the required stress relief and strata stability.
2. The system must be safe to operate and easy to ventilate.
3. The system must be acceptable to the men involved and to the New South Wales Department of Mines Inspectorate.
4. Existing mining units must be usable with the system.
5. Fender widths should not exceed 9 meters (29.5 feet).
6. Wheeling distances to the conveyor boot end must be kept to a minimum in order to maximize mining time and output of coal.
7. Roadways farthest from the goaf must not be driven until the goaf is formed as close as practicable to those roadways.

When pillar extraction was completed as part of the sequence of events, solid drivage again commenced to form the next row of pillars. At this stage, a stepped

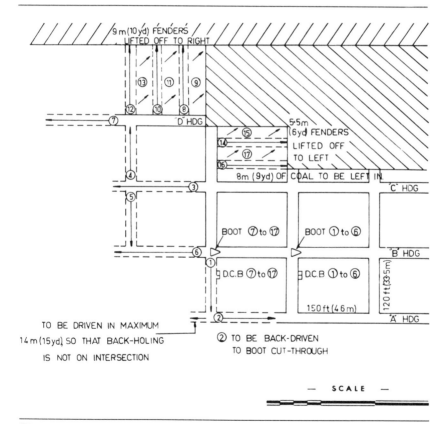

Figure 13.5. Sequence of drivage for pillar extraction on the advance, 6 South Panel, Oakdale colliery.

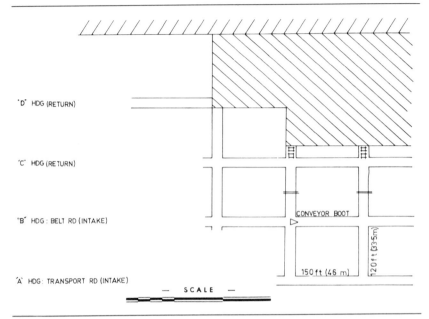

Figure 13.6. Formation of stepped goaf line, 6 South Panel, Oakdale colliery.

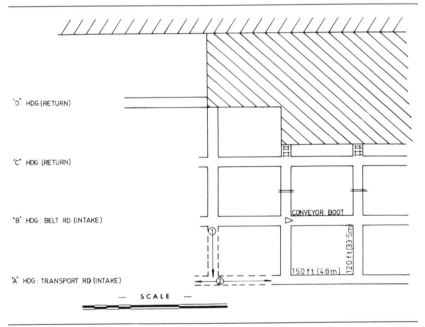

Figure 13.7. Formation of conveyor boot cutthrough, 6 South Panel, Oakdale.

goaf line was established to the farthest inbye completed cutthrough and the most inbye cutthrough of the panel (see Figure 13.6). After a short machine flit, the next inbye cutthrough was completed and the transport heading *A* was back-driven to the conveyor boot (return) end cutthrough. This roadway was the one farthest from the goaf, as shown in Figure 13.7.

The heading was back-driven to eliminate moving machines around the conveyor boot end and simplified the machine move. The machines were then moved to drive *C* heading and the next inbye cutthrough right and left of *C* heading. *D* heading was holed first to establish the return airway. The machines were then moved to drive *B* heading, the conveyor belt heading, which holed into the already formed cutthrough.

At this stage, the conveyor was extended one pillar distance, the boot end having been sited on the intersection. The last return *D* heading was then driven inbye one pillar distance before pillar extraction was commenced. This drivage was required as the intersection of the cutthrough and *D* heading was destroyed as the block of coal was extracted (see Figure 13.8).

Extraction commenced at the outbye end of the pillar and the goaf edge was taken inbye for 45 meters (148 feet) with a split and fender system, and splits at

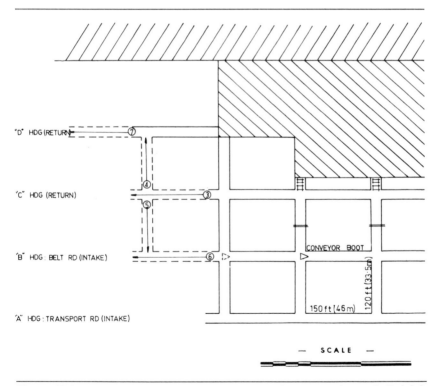

Figure 13.8. Formation of the last return "D" heading, 6 South Panel, Oakdale colliery, before pillar extraction was commenced.

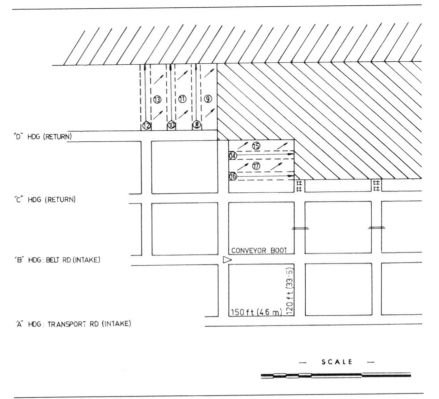

Figure 13.9. Partial extraction of pillars, 6 South Panel, Oakdale colliery.

90° to the headings. Support in the splits was by roof bolt and W strap and wood props were used in the lifts. The unit was then moved outbye one pillar along *C* heading and the pillar between *C* and *D* headings was partly extracted. Only 9 meters (29.5 feet) of coal were left in to protect the single remaining return, as shown in Figure 13.9.

Shuttle car wheeling for the first block of coal extracted was around the B to C pillar. Wheeling for the second block was along the boot cutthrough to the conveyor boot end.

Additional support, wooden chocks, were erected in the cutthrough adjacent to the goaf edge to prevent the goaf running into the return heading. Extra wooden props were set in the return to ensure that the single return could not fall in due to general roof weighting.

Mining conditions were good at the face area, and production rates when on pillar extraction were high. Panel development including the extraction on the advance of some 84,000 tons of coal in 16 pillars has now been completed.

Total development time for the panel was 14 weeks and the unit has been put on full pillar extraction. No major roadway failure has occurred in the panel since the system was adopted.

Conclusions

Pillar extraction on the advance has been accepted and adopted by the men at the colliery operating the 6 South unit as a safe and effective system.

A high degree of ground stabilization has been achieved by forming the large goaf area as the panel was developed. This was evident in that, as the goaf area was increased in size, the normal symptoms of highly stressed strata in development headings were gradually eliminated. Toward the end of the development of the panel, there was no evidence of stress conditions in any heading, despite the panel being within the monoclinal flexure zone with its known high stress problems.

The system has ensured the successful development of a production panel using conventional continuous miner and shuttle car equipment accompanied by roof bolting techniques.

This has been achieved for the first time in the Oakdale colliery when mining at a depth in excess of 430 meters (1,411 feet). There has been no major roadway failure due to unstable ground created by high stress conditions. This has also been achieved with the panel within the highly stressed monoclinal flexure zone.

Productivity from the panel has been reasonably good, but cyclic, in that pillar extraction outputs per shift were higher than solid development outputs per shift.

With the pillar size adopted development of a completed row of pillars including the extraction of the associated pillars for stress relief takes approximately eight days.

The system minimized roof support requirements, yet owing to the strata being stabilized offered a safe method of operation with minimum support costs.

The system is to be adopted in other development panels in the deeper eastern area of Oakdale colliery.

Owing to the deterioration of the return airway in 6 South, because of its close proximity to the goaf, the sequence to be adopted for the next panel will be changed slightly. A larger pillar will be left to protect the return airway, and all the block of coal forming the advancing goaf is to be taken by longer splits to the existing goaf. One unit move will be eliminated in this new sequence (Figure 13.10).

The company is confident that the system has proved a major breakthrough in panel design and development techniques. The mining of deeper, highly stressed coal measures in the Oakdale/Burragorang Valley areas of the Southern Coalfield, with continuous miner and shuttle car mining units, now seems a more practical proposition than was previously anticipated.

Acknowledgments

To Clutha Development Pty. Limited, in particular J. W. Brown, general superintendent of collieries, and J. Slater, assistant superintendent of collieries, for their support and assistance in the implementation of the system and preparation of the paper. To the men and mine officials operating the 6 South unit for their enthusiasm and cooperation in adopting and operating the system of "Pillar Extraction on the Advance." To Dr. Ross Blackwood and the University of New South Wales, School of Mining, for assisting with the practical analysis of the problem at Oakdale colliery.

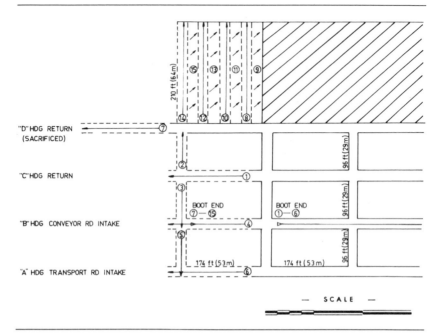

Figure 13.10. Plan of proposed modified pillar extraction on the advance, 7 South Panel, Oakdale colliery.

References

Hall, R. T., 1974, "Underground Monitoring of Strata Control Procedures." Australian Coal Industry Research Laboratories, Ltd. Report No. 413.

Yeates, R. A., 1977, "Types of Underground Roadway Failures Occurring in the Southern Coalfields of New South Wales." School of Mining Engineering, The University of New South Wales, Rocks Mechanics Investigations Report, No. 1/77.

Underground Strata Control with Resin Grouted Roof Bolts in McIntyre Mines' Coal Operations

14

F. Grant, Research Scientist, Canada Centre for
Mineral and Energy Technology, Calgary, Alberta, Canada,
R. M. Wigelsworth, Assistant Chief Engineer, Underground Mines,
and K. Charlton, Senior Planner, Underground Mines,
McIntyre Mines Ltd., Grande Cache, Alberta, Canada

Introduction

McIntyre Mines commenced coal operations in 1968 with two mines in the 6.1-meter- (20-foot-) thick No. 4 Seam, one on each side of the Smoky River Valley. The mines were in the front ranges of the Rocky Mountains, in disturbed Cretaceous strata where the depth of cover increased rapidly. Roof support problems were encountered almost immediately. The mining company with the cooperation of the Mining Research Centre of the Department of Energy, Mines and Resources, initiated a cooperative support research program to improve strata control.

The mines were originally designed for longwall panels developed by three-entry systems. One longwall face was started in each mine, then the system was abandoned due to the thick seams, soft floor coal, heavy pressures, and mechanical problems with equipment. Large openings had been created by the longwalls, and the heavy pressures caused major floor heaving. Within a short time the No. 5 mine was closed due to strata problems. This was followed by the start of mining in No. 2-11 Seam, then recently in No. 2-10 Seam, with both operations stratigraphically above the original No. 2 mine in No. 4 Seam. The stratigraphic correlation of the coal seams mined by McIntyre is shown in Figure 14.1.

The roof control research established that the use of expansion shell roof bolts with various auxiliary aids was the most compatible roof support system to match the rapid mining cycle. Effective expansion shell anchors, roof bolt types and lengths, and auxiliary aids such as wire mesh, installation techniques and equipment were used and established. In some mines or in some areas, the roof bolt support was not sufficient and additional support methods had to be used; these included posts, timber sets, steel sets, arches, cogs, or changes in entry or pillar dimensions. The extra support for weak strata when required is expensive, interrupts the mining cycle, and may lead to roof failure if not installed quickly.

Research on polyester grouted roof bolts was started with the objective of improving strata control in areas of weak roof. The new mine in No. 2-10 Seam will

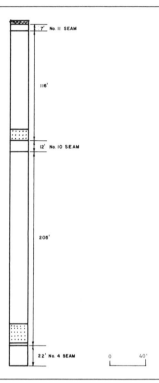

Figure 14.1. Stratigraphic correlation of coal seams mined by McIntyre Mines.

have soft roof conditions with which to contend. The Mining Research Laboratories had conducted research in other mines using polyester resins with some success. The method has been tried in other mines in the USA and in Europe with good results.

General

Mine No. 2A-4 Seam started production during June 1975, with adverse strata conditions being encountered almost immediately. From start-up until the polyester resin grouted roof bolt program was initiated there were over 30 roof falls. These ranged from small nuisance falls to large ones causing significant lost production and/or complete road realignment or loss.

Most large falls had one outstanding anomaly in common: they were all breaking at the expansion shell anchor horizon or a short distance above. This problem resulted in Canada Centre for Mineral and Energy Technology (CANMET) being asked to conduct pull tests in late 1975, and the results indicated that the bolts would take an anchor load of 9 to 11 tons before movement started. However, as the roads advanced, the overburden increased and the roof continued to deteriorate to the point where expansion shell bolts were difficult to tension. The mine at this time reverted to the system of installing timber booms on wood legs at 1.5-meter (5-foot) centers or less. Changes in weight pressing onto the wooden booms were visually noticeable; and the loss in production, due to the slower support system, was significant.

Once the decision to test polyester resin bolts was made, an extensive literature search was started to familiarize the McIntyre personnel with the types of resin cartridges available, bolt styles, and the problems encountered by other mines using polyester resin.

Polyester resin cartridges

The manufacturers of the polyester resin grouts have made intensive studies concerning the development of a product capable of easy installation and possessing great strength. The components of the resins have to be mixed in the precise proportions to develop sufficient strength after a short period of mixing, gelling, and curing. The polyester resin and catalyst have to be packaged in a casing so that they are protected in transit, stay separated until mixed, and hold together while being mixed to form a strong grout binding the bolt and strata in place when set. The installation must be made quickly with a minimum of trouble using standard mine equipment. This operation and that of placing the flexible cartridges in roof holes 3.05 to 4 meters (10 to 13 feet) above the floor, has been achieved fairly successfully to date.

It was decided to test three different resin brands. They were as follows:

1. Resin A: 25.4-millimeter-diameter × 559-millimeter-long (1 × 22-inch cartridge with a gelling time of 45 seconds at +15 degrees Centigrade and with an average curing time of 5 minutes (code A 25/550).
2. Resin B: 32-millimeter-diameter × 305-millimeter-long (1.25 × 12-inch) cartridge with a gelling time of 1 minute (code 3212 0001).
3. Resin C: 32-millimeter-diameter × 305-millimeter-equivalent-length cartridge (1.25 × 12-inch equivalent) with a gelling time of 1 minute (code T 1.25 12 0001).

In conjunction with the above, three different rebar bolts were used as follows:

1. No. 6 with a diameter of 19.1 millimeters (0.75 inch).
2. No. 7 with a diameter of 22.4 millimeters (0.88 inch).
3. No. 8 with a diameter of 25.4 millimeters (1 inch).

The bolts were all 1.8 meters (6 feet) long with a 28.7-millimeter (1.13-inch) forged square head and washer attached. The forged head bolts were selected because a mechanized-type roof bolter was being used in the No. 2A-4 Seam and also because, if the trials were successful, this type of bolt offered the fastest installation time in 3-meter- (10-foot-) high cuts made by standard continuous mining machines. Also, the use of forged head bolts does not require a man to follow the bolting operation to attach a plate and nut to each bolt or to put up steel mesh screen.

Trials

No. 2A-4 seam

From the literature search and check calculations performed it was noted that the best economic conditions using polyester resin were with a No. 6 rebar bolt in a 25.4-millimeter (1-inch) hole. However, every hole drilled for roof bolting by McIntyre Mines Limited is 35 millimeters (1.38 inches) in diameter so that the expansion shell will slide into the hole. Therefore, a drill steel and bit improvement program was incorporated into the polyester resin tests.

Initially, the 25.4-millimeter- (1-inch-) diameter bits with matching drill steel were checked. Problems developed during the first day of the tests with this new steel and the 25.4-millimeter bits were not used again until the last test hole. The three main problems encountered were:

1. The hole "puffed" continuously while drilling.
2. Steel wore quickly at the throat of the rod where the bit was coupled to it.
3. Excessive wobble and bending occurred while drilling.

Due to the results experienced here, and at other mines in the United States, the drill steel company altered its bits and rods so that the above problems were minimized. In fact, due to the modification of the bit, McIntyre Mines is now achieving longer bit life and a drilling time reduction of up to 50 percent.

Because of the initial problems with the 25.4-millimeter bits most of the test holes were 35 millimeters (1.38 inches) in diameter, therefore, giving an over-sized-diameter hole for all the No. 6 and No. 7 rebar bolts installed.

The installation trials with polyester resin indicated that the roof bolting technique in No. 2A-4 Seam would require modification and improvements for this type of strata support.

The problems encountered varied from high room heights ranging from 2.74 to 4.57 meters (9 to 15 feet); varying seam dips of 10° to 20° which approached the limit of the roof bolters' tramming power; heavy, blocky roof with weak joint and fracture planes; and the No. 2A-4 Seam roof lay between geological folds on the high side and No. 2-4 Seam goaf area on the low side.

Because the drill hole was slightly larger than required, more resin than was desirable had to be used to fill the larger annulus around the bolts. Difficulties arose while placing the flexible cartridges up into the holes in the high roof and holding them in place until the rebar bolt could be inserted. Control of the roof bolt penetration and mixing speeds was also a problem. This was compounded by having to hold the bolt and wire screen in place until the resin set. The penetration time had to be closely controlled since Resin C "gelled" before the bolt was fully inserted if the penetration was slow. The overall time tolerance for insertion, mixing, and gelling was limited.

These problems were solved in a reasonable fashion with close control of installation technique and perseverance of the bolting crews. New bolting crews encountered the same problems and had the same difficulties adapting to them.

Table 14.1 gives the installation times recorded at Location *A* as seen on the plan in Figure 14.2. The average setting times for each polyester resin are shown at the bottom of the chart.

Table 14.1. Data for Bolts and Resin Used in Tests at Location A

Bolt no.	Test no.	Bolt size	Hole size (in.)	Resin type	No. cart.	Time (min)			
						Drill	Insert	Spin	Hold
1	—	#7×6'	1-3/8	B	5	2:45	0:30	0:35	2:25
2	—	#7×6'	1-3/8	C	5	2:38	0:45	0:17	—
3	—	5/8×6'	1	A	2	2:45	0:48	0:50	3:20
4	—	#7×6'	1-3/8	B	4	3:02	0:45	0:20	0:57
5	1	#7×6'	1-3/8	C	5	3:45*	0:55	0:10	0:10
6	2	#7×6'	1-3/8	B	5	4:10	0:55	0:25	1:15
7	3	#7×6'	1-3/8	C	5	3:14	0:50	0:10	0:12
8	4	#7×6'	1-3/8	B	5	N.A.	0:45	0:30	—
9	5	#7×6'	1-1/4	A	3	5:21	0:30	0:35	0:40
10	6	#7×6'	1-1/4	A	3	10:04	0:40	0:35	0:43
11	7	#7×6'	1-1/4	A	3	6:35	0:54	0:45	—**
12	8	#7×6'	1-3/8	C	5	3:38	1:03	0:09	—
13	9	#7×6'	1-3/8	B	5	3:09	0:53	0:36	0:59

Average	Insert	Spin	Hold
Resin A 25/550	0:43.0	0:41.25	1:15.75
Resin B MV3212	0:45.6	0:29.2	1:07.2
Resin C 1-1/4"×12"	0:53.25	0:11.5	0:05.5

*Adjusted roof bolter thrust during drilling.

**No holding after spinning, approximately 152 millimeters (6 inches) of relaxation by bolt.

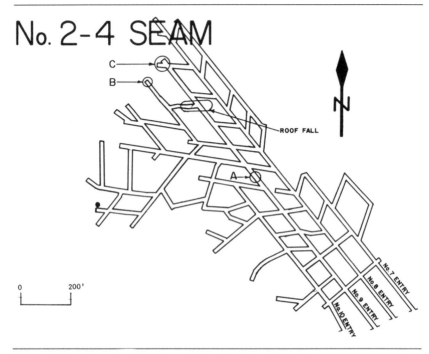

No. 2-4 SEAM

C
B

ROOF FALL

A

0   200'

Figure 14.2. Plan of the No. 2-4 Seam, McIntyre Mines.

Further testing was then carried out on Resin C to determine if varying the amounts of resin in the test holes would still give the desired results. The amount of resin varied from one cartridge per hole to five cartridges per hole. The results of the tension tests done on all bolts are shown in Figures 14.3 through 14.6.

Two wire cables were installed diagonally across the roof junction to assist in support at Location *C* in Figure 14.2. Tension at installation was relatively taut. On inspection two weeks later, however, these cables were found to be exceptionally tight indicating that a load was acting on the intersection. The rest of the intersection at Location *C* had been resin-bolted before the two wire cables were installed. The No. 9 Entry road was bolted to the face and the crosscut to No. 10 Entry was bolted to within 4.6 meters (15 feet) of the face. At the time of the cable inspection, it was also noted that 4.6 meters of "free" roof (i.e. not supported) had fallen. This fall, however, stopped exactly at the first row of resin bolts. It must be pointed out that the resin bolts were installed into roof that had stood unsupported for approximately three weeks due to a roof fall in the supply roof (No. 9 Entry) at the third crosscut inbye, stopping all production until it was cleaned up.

While using resin bolts, a roof fall occurred in No. 2A-4 Seam, shortly before the suspension of operations. It was found that most of the bolts were improperly installed. The faults included: wrong bolt diameter to hole diameter; insufficient resin to compensate for wrong diameter ratio; and insufficient spinning to create proper mixing. Pictures of the improperly installed bolts are shown in Figure 14.7.

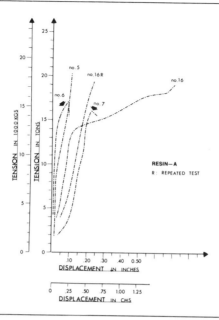

Figure 14.3. Tension test results using Resin A in No. 2A-4 Seam.

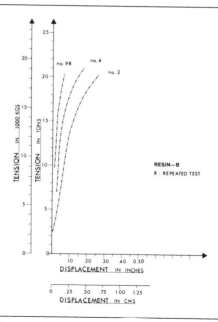

Figure 14.4. Tension test results using Resin B in No. 2A-4 Seam.

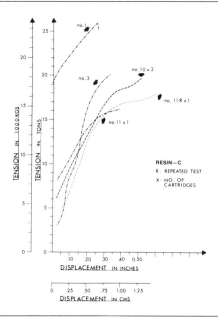

Figure 14.5. Tension test results using Resin C in No. 2A-4 Seam.

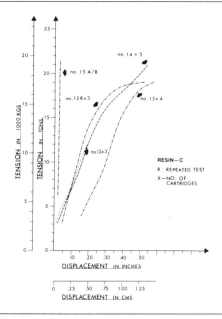

Figure 14.6. Tension test results using Resin C in No. 2A-4 Seam.

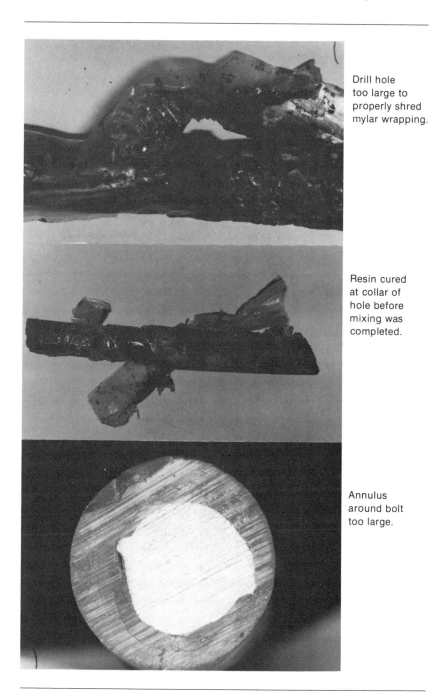

Drill hole too large to properly shred mylar wrapping.

Resin cured at collar of hole before mixing was completed.

Annulus around bolt too large.

Figure 14.7.: Photographs showing results of improperly installing bolts.

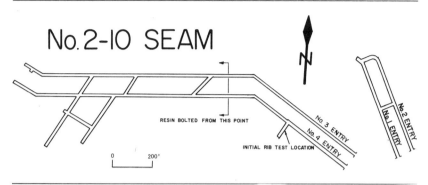

Figure 14.8. Plan of No. 2-10 Seam development, McIntyre Mines.

No. 2-10 seam

In the last two days of the week, an attempt to resin-bolt the coal ribs in No. 4 Entry of No. 2-10 Seam was made. However, due to equipment problems, the test bolts could not be satisfactorily installed. No. 2-10 Seam development is shown on the plan in Figure 14.8.

After the trials in No. 2A-4 Seam and the small attempt to bolt the ribs in No. 2-10 Seam, a decision to reinforce the arch support with polyester resin bolting in No. 2-10 Seam naturally followed.

A contracting firm was employed initially to advance No. 3 and No. 4 Entries in No. 2-10 Seam, and a new varying set of problems was encountered with the preceding decision to install resin roof bolts. Two previous attempts to form crosscuts between these entries had failed, and in both cases large falls had occurred despite the use of closely lagged W8 × 24 steel sets.

Installation of the resin bolts by the contractor was performed with a stoper-type drill. The stoper bits first used were the standard 35-millimeter (1.38-inch) bit, but this created an annulus of up to 16 millimeters (0.63 inch) and once again it was found that correct mixing was not being achieved. This, in conjunction with a slow setting resin, caused excessive slow bolting.

To remedy this, the stoper bits were changed to 29 millimeters (1.13 inches) in diameter, the smallest in stock, and the roof bolts were also changed from No. 6 to No. 7. This resulted in the reduction of the annulus to approximately 6.4 millimeters (0.25 inch). Resin B was also changed from MV0510 to MV0001, and Resin C was introduced. Resin C was not used initially because it was thought that the resin would set too quickly for stoper use. However, the slower rotation of the stoper drill compared to that of the mechanized roof bolter seemed to contribute directly to the gelling time of the resin.

When the McIntyre crew replaced the contractor in No. 2-10 Seam, a mechanized rotary bolter was introduced to the mine. Three objectives were achieved by this move as follows:

1. When using the stoper, water was being inserted into roof cracks and making the already fragile roof even more unreliable. This problem was solved.

2. The quantity of roof bolts set per shift was increased thus enabling more production to be attained.
3. The use of a vacuum drilling system became available and testing of the 25.4-millimeter- (1-inch-) diameter bits could begin.

As indicated, the 25.4-millimeter system tested in No. 2A-4 Seam had not performed to the expected results. In No. 2-10 Seam both system 1 and system 2 bits were subjected to testing.

Different grades of carbide content, modified bits, and incremental bit diameters were experimented with over a few months.

At the time the 25.4-millimeter drill bit system was introduced, McIntyre had changed to using Resin C exclusively. The polyester resin used with the narrow-diameter holes was 25.4-millimeter × 0.62-meter equivalent (1 inch × 2 feet), 25.4-millimeter × 0.91-meter equivalent (1 inch × 3 feet), and 25.4-millimeter × 1.22-meter equivalent (1 inch × 4 feet). These longer cartridges resulted in a shorter installation time per drill hole.

Due to the roof conditions of No. 2-10 Seam, longer polyester resin cartridges with a two-minute gelling time and rebar bolts 2.4 meters (8 feet) long with 29-millimeter (1.13-inch) diameter holes were used. Table 14.2 gives a cost rating comparison of the different support methods available for use in No. 2-10 Seam.

Discussion of Results

General

The standard roof bolt tension and torque tests have been used on resin grouted bolts in this mine. The measurements in anchorage strengths and torque with the grouted bolts to some extent differ in manner from test results with the mechanical roof bolt anchors. The results may in some respects be misleading.

For example, each graph in Figures 14.9 through 14.11 shows tests results from three mines, including McIntyre, that are average for each mine. These are compared with one sample test of an expansion shell roof bolt tested in the same

---

Table 14.2. Comparison Costs in No. 2-10 Seam

	Cost rating/ft. ($)
Steel arch	20.0
Steel on wood legs	7.9
Wooden set	1.0
Expansion shell bolts	2.4
Resin bolts	5.5
Combined 5 & 2	13.5
Combined 5 & 3	7.4

Note: Support based on entry cross section
3.05 x 5.5m (10 x 18 ft.) at 1.5m (5 ft.) spacing

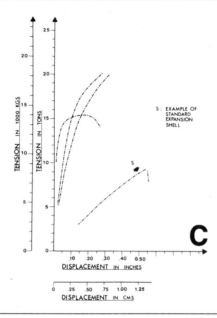

Figure 14.9. Anchorage strength test results at mine C.

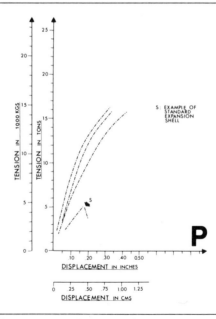

Figure 14.10. Anchorage strength test results at mine P.

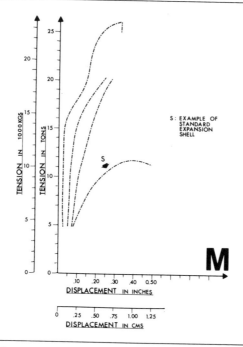

Figure 14.11. Anchorage strength test results at mine M.

type of strata. It will be noticed that the standard bolts indicate a fair amount of deformation for reasonably low loads. These results indicate that the tension tests are reacting well within the limit of the strain or elongation qualities of the steel or, to put it another way, the anchorage strength of the expansion shell is less than the strength of the steel rod or bolt. In nearly all cases, with the polyester resin grouted bolts, despite different resins and amounts, the tension tests indicate deformation in the same order as the yield values of the steel. It appears that the tension applied is just on the steel and if carried to failure it follows the stress strain curve of the metal until failure, which is usually at a stress concentration point at the neck of the forged head and pulling stud. There has been the odd case where yield has been noticeable due to slipping of the metal-to-grout and grout-to-wall rock bond weakness. However, these tension tests definitely indicate a bad or poorly installed bolt. Usually, poor results indicate an improper mix, holes too large for the material in it, or poor installation.

Torque tests are similar. If sufficient torque is exerted with the wrench, the head may tend to twist, then return to normal on removal of force due to the elasticity of the metal. If the bolt does happen to rotate, it is an immediate indication of a poor installation.

When a standard expansion shell roof bolt is tested by either pull tests or torque, there is an immediately noticeable transmittance of the applied torque into the wall rock through the short expansion shell and the roof bolt plate. The polyester grouted roof bolts do not give this reaction as the applied force or tension appears to be taken up in the grout filling the annulus between the hole perimeter and the rod. The reaction may be spread through any amount of grout from a few inches at the bottom, to the height of the column in the hole. Although the applied tension has at times been very high, generally the stress concentration point near the collar of the head of the rod fails. There have been a few cases where with large holes, soft rock, and high tension, the anchorage has been made to yield. It is debatable which is stronger or weaker in the grout bond to the metal or rock. There is also some doubt whether the grout forms a chemical bond, like an adhesive, or a mechanical bond.

No. 2A-4 seam

As mentioned earlier, the roof conditions in No. 2A-4 Seam were considered to be worse than those encountered in the other producing mines, and the locations chosen to install the resin bolts were representative of the roof as a whole. The faces of No. 9 and No. 10 Entry were bolted in the hope that these two roads could be extended to allow more coal to be released for future pillaring. However, a decision to depillar the mine was made within a short time after the resin test was concluded.

The location shown as *A* in Figure 14.2 was especially chosen as a convenient place to learn installation techniques. The loads obtained on all test bolts were substantial, and more than had been obtained on earlier expansion shell bolt tests in No. 2A-4 Seam. These previous tests showed that an 11-ton pull was required before the expansion shell began to fail or slip in the hole. Even the holes that had one and two resin cartridges, respectively, loaded to a greater extent than the expansion shell bolt. The interesting fact about the resin bolt slipping was that it held a constant load as the bolt was pulled out.

As the graphs in Figures 14.3 through 14.6 indicate, all the test bolts loaded with little deformation except No. 16. However, on the repeat test it loaded as the other bolts had, the reason possibly being that Resin A had not fully cured before a load was applied. In the representative pull test results, shown in Figure 14.12, the bolts were approximately 152 to 254 millimeters (6 to 10 inches) from the washer to the roof, and not one showed evidence of "necking" from the applied load.

It is, therefore, stated from the evidence of the tests, even though only 16 bolts were pulled, that a polyester resin roof bolt will take substantially more load than an expansion shell bolt in No. 2A-4 Seam roof, if it is installed correctly. However, from the problems observed in installation, it will be necessary to concentrate on improving roof bolting techniques, thus ensuring a consistent installation of resin bolts.

The first 13 bolts installed as recorded in Table 14.1, show the average times required to insert, spin, hold, and drill. The insert time for each type of resin can be meaningfully compared to show that the time can be reduced for longer cartridges. That is to say, the time varies directly with cartridge length and number. The spin time is a function of the resin's chemical composition and the Resin C was significantly faster since it set at 17 seconds, like Bolt No. 2, which stopped the roof bolter rotation. The other two resins remained sufficiently viscous to deter this stoppage. As in the spin times, hold time is again a function of chemical composition.

It must be stressed that all mechanical machinery should be in good operating condition. The type of roof bolter utilized by McIntyre was found to be ideal because the thrust could be adjusted to compensate for the lighter 25.4-millimeter- (1-inch-) diameter steel and, also, the vacuum is adjustable. It is fast and powerful and, although the bolter used did not have an articulating mast for angled bolt installation or corner work, a "nut runner" on the steel worked well for the tight locations.

MINE: McIntyre #2A Mine   LOCATION: 9 Entry   10 Entry   DATE: Nov. 8, 1976

TEST NO.	#9		#10		#11		#12	
SHELL TYPE	B		C		c		C	
SHELL DIA. "	32mm x 12" x 5		1¼" x 12" x 2		1¼" x 12" x 1		1¼" x 12" x 3	
ROD DIA."xLGTH"	7/8" x 6' (#7)		7/8" x 6' (#7)		7/8" x 6' (#7)		7/8" x 6' (#7)	
HOLE DIA. "	1 3/8"		1 3/8"		1 3/8"		1 3/8"	
SETTING TORQUE								
TEST TORQUE			300+		300+		300+	
TENSION APPLIED TONS	DISPLACEMENT INCHES		DISPLACEMENT INCHES		DISPLACEMENT INCHES		DISPLACEMENT INCHES	
	INITIAL	REPEAT	INITIAL	REPEAT	INITIAL	REPEAT	INITIAL	REPEAT
0	0.0	0.00	0.000		0.000		0.00	
2	0.0	0.00	0.010		−0.003	0.00	0.00	0.00
3	0.023	0.00	−0.002		−0.005	−0.001	0.013	0.054
4	0.037	0.000	−0.009		−0.006		0.029	0.053
5	0.040	0.002	0.001		−0.006	0.055	0.048	0.064
6	0.040	0.002	0.021		−0.006		0.079	0.085
7	0.005	0.003	0.045		0.002	0.105	0.096	0.099
8	0.004	0.005	0.067		0.035		0.114	0.109
9	0.004	0.007	0.089		0.052	0.138	0.133	0.121
10	0.004	0.011	0.111		0.090	0.156	0.146	0.133
11	0.004	0.015	0.135		0.106	0.176	0.169	0.147
12	0.004	0.019	0.171		0.132	0.193	0.187	0.160
13	0.004	0.023	0.192		0.170	0.214		0.178
14	0.000	0.027	0.216		0.198	0.233		0.195
15		0.028	0.248		0.239	0.264		0.219
16		0.029	0.278		0.360	0.309		0.253
17		0.028	0.311			0.497		0.288
18		0.048	0.351			0.599		0.336
19			0.468					0.527
19+			0.523				21 tons	0.547
O ELASTIC REBOUND		None	0.390		0.622	0.330	Ø	0.438
REMARKS	Moving wedges & out of alignment		Washer bent. Rod pulling out		Washer bent rod pulling out on repeat rod pulled ½ in. Load holding steady		Washer bent ¼ in. rod pulled out approx. ¼ in.	

Figure 14.12. Results of representative roof tests at No. 2A mine.

The three resins were basically advertised correctly. Resin A is a slow resin, and would be ideal for horizontal or down holes. It was found to lengthen the cycle of bolt installation considerably, in addition to being more expensive to purchase initially. Resin C was ideal, since it minimized the bolting cycle mix time to 20 seconds with a short hold or cure time to prevent bolt relaxation. However, this quick gelling time allows no room for error or inexperience. The cost of Resin C was the same per cartridge as Resin B, but Resin C and hardener are separated by Mylar and not by a chemical interface as in Resin B. This should give Resin C a much better shelf life. Resin B was the mid-range resin tested. It was classified as MV-Medium Viscosity, and this still was considered to be too viscous. Consequently, 200 cartridges of HV-Heavy Viscosity cartridges were ordered, to see if a high viscous resin would help to improve the hold time of Resin B.

The size of hole drilled will directly affect the economics of the resin bolt, because hole size varies directly with bolt and resin size. The 25.4-millimeter (1-inch) hole will give the best cost for rebar bolts and resin, provided the previously mentioned drilling problems are completely overcome. The 25.4-millimeter system is currently being thoroughly tested in No. 2-10 Seam.

No. 2-10 seam

Earlier tests on drill bit life showed that System 1, 25.4-millimeter bits were capable of 17 to 30 holes per bit. With minor changes to the bits, the number of holes recorded on some of the 25.4-millimeter bits was increased to as high as 50 holes per bit.

The System 2 bits have proved they are capable of drilling up to 75 holes per bit. However, based on an average performance, 45 holes per bit is a more realistic average.

The significant disadvantage between the two types of bits is that they are not interchangeable and thus require stocking separate sets of vacuum rods.

The overall advantage of the improved bits tested, in addition to the good bit life, is that the drill time per hole is reduced by up to 50 percent.

During the course of the drill bit tests and polyester resin tests, it is estimated that a total of approximately 2,800 2.44-meter (8-foot) rebar roof bolts were installed. There were no roof falls in No. 2-10 Seam after polyester resin bolts were used as auxiliary support on a regular basis. The initial bolting pattern used to match the timber supports was a 0.91 × 0.91 meter (3 × 3 foot) pattern. However, when the timber supports were enlarged, with respect to centerline distances, the bolting pattern was increased to 0.91 × 1.22 meters (3 × 4 feet).

Conclusions

It is evident from past observations and results that No. 2A-4 Seam was in an area of high tectonic stresses and required an effective and efficient roof control method. The cost of arches, booms, or steel sets and other supports is substantial, but these supports are economic when compared with coal abandoned through roof failure. The main problem with using one of the three roof support methods mentioned, is the "backbye" work required to keep them in effective support.

Because all the mechanized equipment used by McIntyre Mines is large and powerful, it does not take much to damage a wooden chock or knock out a post. This problem is complicated by the possibility of broken posts and misaligned chocks caused by floor heave as well as by rib movement. The support method chosen must be safe, inexpensive, out of the way of equipment, and able to resist, within reason, the tectonic forces found in the coal seams operated by McIntyre. The study of resin roof bolts indicates that this type of roof support may possibly control the strata found in the No. 2A-4 and No. 2-10 seams.

It is also noted that, compared to an expansion shell bolt, a resin bolt is expensive. However, when the expansion shell will not anchor effectively, cost becomes basically irrelevant between the two bolts, if the strata is to be supported with roof bolts.

Roof Fall Prediction at Island Creek Coal Company

15

B. Rao Pothini, Mining Research Engineer,
and H. von Schonfeldt, Manager, Mining Research,
Occidental Research Corporation, Lexington, Kentucky, USA

Introduction

Roof falls seriously affect safety, cost, and productivity in coal mines. Rockfalls constitute the single largest cause of fatal injuries in the nation's underground coal mines. About 30 percent of the fatal injuries in the first 11 months of 1977 were attributed to roof falls (Mining Enforcement and Safety Administration, MESA, 1977). They are, thus, a safety concern. The operating division in which the following work was carried out has been fortunate not to have had any roof fall fatalities in recent years. Nevertheless, the potential danger of serious injury is always present.

Cleanup of rock falls is a labor intensive operation, and the cost for labor, materials and equipment is high. Frequently, production is lost for part of the cleanup period. The cost due to lost production is even more significant for highly mechanized mining methods like longwall systems, for example. A large proportion of Island Creek Coal Company's high-grade metallurgical coal comes from longwalls. Unit face production of these systems is 5 to 10 times higher than that of continuous miner sections. On the other hand, capital requirements for longwall systems are significantly higher. It is, therefore, necessary to keep their production and productivity high in order to pay for the enormous capital investment required.

Figure 15.1 illustrates the negative impact of poor roof conditions on cost and productivity in a number of mines.

The high roof support costs in mine Nos. 4, 5, and 6 are due to a higher than average roof fall frequency. Because of the large disparities of cost and productivity in the various mines, even a small improvement in the performance of the worst mines will lead to significant savings.

An important factor, which cannot be expressed in numbers directly, is the psychological effect of difficult and potentially unstable roof conditions on the work force.

For these reasons, the company's management authorized a program to develop an effective roof fall prediction method. In the following, some of the results of this effort are reported. The program is still in a development stage, and some of the conclusions will have to be viewed as preliminary. The basic concept of the technique, however, is cost effective and has met with the approval of the operating division. It is, therefore, considered to be successful.

Development of Base Data

Cost savings potential

The coal measures are of Pennsylvanian age. The mines are working the Pocahontas No. 3 Coal seam which is part of the Pocahontas Formation (Miller, 1974). The overburden thickness varies between approximately 457 and 762 meters (1,500 and 2,500 feet). The roof strata consist of horizontally bedded, hard sandy shale, and sandstone layers.

Roof falls were generally considered a problem in the operating division. Very little quantitative information was available, however, regarding the extent and cost of the problem.

The first task in this program, therefore, was to develop base data which would supply answers to these questions.

To determine the extent of the problem and the cost savings potential, a detailed study of all roof falls that occurred over the history of one of the mines in the division was made. This particular mine had continuous miner units only

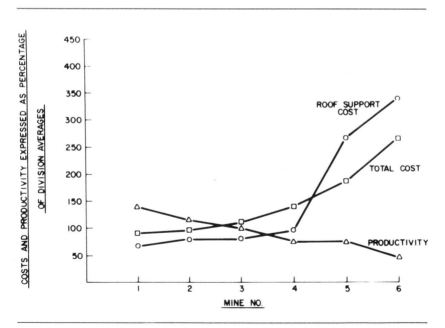

Figure 15.1. Impact of roof control on productivity and costs.

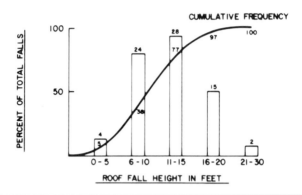

Figure 15.2. Fall frequency versus height for four-way intersections.

during the period covered. Over 100 falls were counted. The fall frequency was calculated to be approximately one fall for every 26,000 tons of clean coal.

The volume of a typical rockfall was determined by measuring height, width, and length of each fall and then taking the average of all falls. Thus, typically 397 cubic meters (14,000 cubic feet) or 1,080 tons of rock fell in each fall, resulting in a fall rock to clean coal ratio of 1 ton in 22 tons. Figure 15.2 shows the distribution of falls by fall height.

Under the (conservative) assumption that on a ton by ton basis it costs as much to clean up a fall as it does to mine coal, a savings potential of 5 percent of the mining costs in development exists if all falls could be prevented. An even higher savings potential exists in longwall units when the cost of lost production is taken into consideration.

Possible causes of falls

To determine possible causes of the roof falls, the influence of tectonic forces, geologic conditions, mining geometry, and roof support techniques were considered in planning the program.

No correlation between roof falls and possible high lateral stress in the mine could be established. The existence of the Keen Mountain fault in the area (McCulloch et al, 1975) has no appreciable effect on roof conditions except in the immediate vicinity of the fault zone.

Geologic disturbances like slickensides and channel sands are more commonly associated with the roof falls, as are transition zones between slate and sand rock. Little can be done in the short term to anticipate and bypass these zones since they occur at random.

The correlation between topographical features like creek beds or mountain tops with roof falls is inconclusive. It appears that in one particular mine a higher incidence of falls occurred under creek beds. This result could not be confirmed in any of the other mines, however.

The influence of the mining geometry on roof fall frequency is significant as illustrated in Figure 15.3. All mines are developed with multiple entry systems in

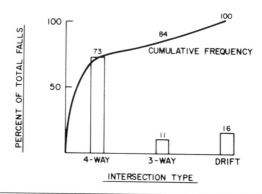

Figure 15.3. Roof fall history of a mine.

room and pillar fashion. During one year, about 500 four-way, 300 three-way, as well as 1,200 entries and crosscuts are developed in each mine. The fact that 73 percent of all falls occur in four-way intersections—that is in only 20 percent of all mine openings—will later on allow selective monitoring techniques to be developed.

It is interesting to note that deformation measurements described below determined that four-way intersections which have an average roof span of 12.5 meters (41 feet) converge more than twice as much as the entries and crosscuts which have a nominal span of 5.5 meters (18 feet).

To ensure that the conclusions developed from one mine over its entire history are valid for the other mines in the division, a second survey was made covering all mines for a period of one year. Figure 15.4 suggests that the conclusions regarding fall frequency in one mine may be generalized for the entire division.

In this latter survey, it was also established that 79 percent of all falls occur within 305 meters (1,000 feet) of the active mining front (Figure 15.5). Recalling that most falls occur in four-way intersections, and combining this with the fact that only a few hundred four-way intersections are excavated within 305 meters (1,000 feet) of the face, enables 60 percent of all roof falls to be detected and predicted by restricting the surveillance to only a fraction of the exposed total mine roof.

In the foregoing, possible causes of roof falls and base data for a roof fall prediction technique were discussed. While no clear correlation between various geological factors and roof falls could be determined, it was established that most falls could be detected with a suitable technique by monitoring only a small portion of the mine.

The method described herein is based on precise observations of floor and roof deformations. Its objective is to recognize unstable roof areas early, before visual signs in the form of breaks and cracks occur. This enables the mining engineer to take preventive measures before the roof instability becomes uncontrollable and results in significant reduction of vertical clearance of the mine entry.

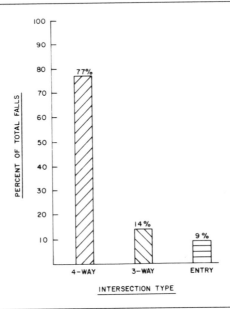

Figure 15.4. Distribution of roof falls by intersection type for one year (all mines).

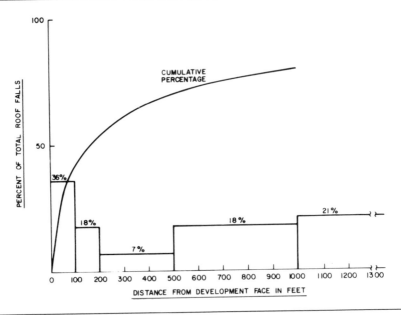

Figure 15.5. Distribution of roof falls versus distance from mining face.

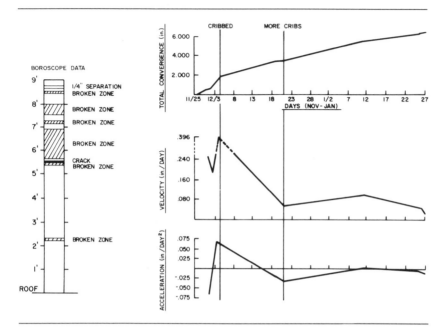

Figure 15.6. Convergence and borescope data for an unstable area in a development unit.

While the principle of using deformation measurements to monitor mine opening stability is well established (Schuermann, 1975; Jacobi, 1976; Pothini et al, 1976), the present method is a more cost-effective, specially adapted technique for large single seam coal mines using room and pillar or multiple entry longwall systems.

The essential features of the method are explained with the help of Figure 15.6.

To evaluate the roof stability of a point or area in the mine, the following criteria are applied. A roof is considered stable if a maximum permissible value of any of the following is not exceeded:

1. Total roof to floor convergence.
2. The average daily convergence rate (velocity).
3. The change of the convergence rate (acceleration).

Instability is suspected otherwise. In this case, the area in question will be more closely monitored; if any two criteria, e.g., convergence and convergence rate, suggest a "weak spot," a borescope analysis (Figure 15.6) of the roof strata is carried out to confirm and define the extent of the unstable area. Although more time-consuming and difficult to install, roof sag meters may also be used to determine the size of the unstable zone.

The primary purpose of independent roof sag measurements in the context of this program was to determine the amount of floor heave. It was calculated as 61 percent of the total entry convergence; the assumption was made, however, that the strata 3.05 to 4.6 meters (10 to 15 feet) above the entry into which the deepest sag meter was installed did not move appreciably. Finally, once the volume of unstable rock is known, the supplementary support requirements necessary to stabilize the area can be estimated.

It was found that these criteria can be successfully applied to advancing development units, retreating longwalls as well as long-term entries. It is important to note, however, that each of these units will generally have a different characteristic convergence, velocity, and acceleration which signal instability as demonstrated below. These numbers will have to be developed for each mine or division with the help of measurements.

Criteria for Roof Fall Prediction

Convergence and roof sag data were taken over a period of approximately 12 months. The measurements were made accurate to the nearest 0.025 millimeter (0.001 inch) with a specially designed "convergence rod." Expansion bolts, 51 millimeters (2 inches) long, installed in 10-millimeter- (0.38-inch-) diameter holes in the roof and floor were used as reference points for the convergence stations. The sag stations consisted of three to four expansion shell roof bolts installed in close proximity at various depths into the roof, usually 0.61 meters (2 feet), 1.22 meters (4 feet), and 3.05 meters (10 feet).

The guidelines and results that follow were drawn from these measurements:

1. It was found that a distinction had to be made between mining-related convergence as a result of the redistribution of forces and "time dependent" deformation. The former is limited to the first 30.5 meters (100 feet) outbye the face in advancing entry development systems.
2. On advance, the average convergence for the first 6.1 meters (20 feet) behind the face is 0.42 millimeter per meter (0.005 inch per foot) and 0.08 millimeter per meter (0.001 inch per foot) for the next 6.1 meters. It should be noted that intersections are mined on 30.5-meter (100-foot) centers.
3. As pointed out earlier, the majority of falls occur in four-way intersections, and deformation monitoring should concentrate on those areas to secure optimum cost efficiency. The following results pertain primarily to intersection behavior.

   The average total stable convergence measured over a period of 61 days was 30.5 millimeters (1.2 inches); three-way intersections converged only half as much. Figures 15.7 and 15.8, respectively, give a measure for the variation of convergence in three-way and four-way intersections. The average maximum stable convergence rate, therefore, is 0.5 millimeter (0.02 inch) per day.
4. Convergence rates in excess of 0.5 millimeter (0.02 inch) per day more than 30.5 meters (100 feet) outbye the face indicate potentially unstable points which require closer surveillance. Although this number was derived specifically for four-way intersections, it would indicate "weak spots" in three-way intersections and entries as well. Generally, velocities in excess of 2.54 millimeters (0.1 inch) per day in any opening type 30.5 meters (100 feet) outbye

the face or more than one month old are considered a warning signal.
5. Data indicate that four-way intersections can withstand a total convergence of up to 25.4 millimeters (1 inch) without detrimental effects to the integrity of the roof. It is important to note, however, when measurements are made in older entries much of the "permissible" deformation may have already taken place. In this case, it is considered more useful to closely observe the convergence rates.
6. Any point in a freshly mined entry experiencing more than twice the conver-

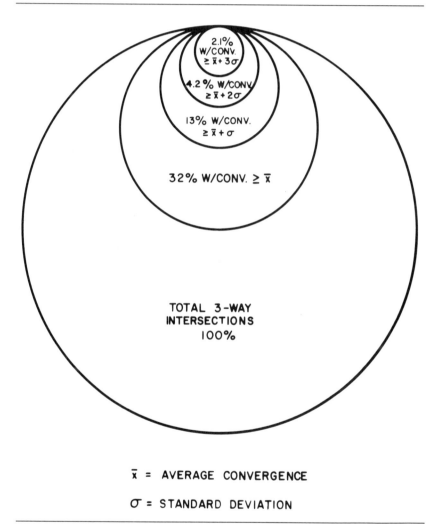

$\bar{x}$ = AVERAGE CONVERGENCE

σ = STANDARD DEVIATION

Figure 15.7. Diagram showing the statistical distribution of convergence data for three-way intersections.

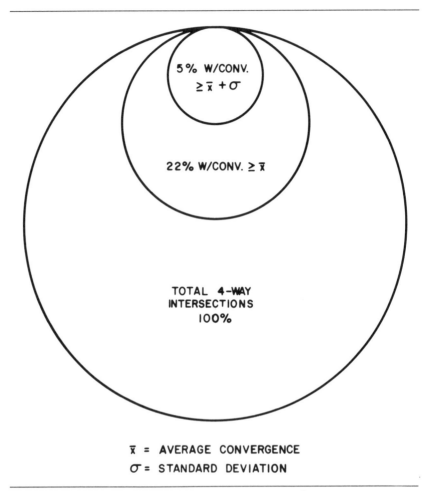

5% W/CONV.
$\geq \bar{x} + \sigma$

22% W/CONV. $\geq \bar{x}$

TOTAL 4-WAY
INTERSECTIONS
100%

$\bar{x}$ = AVERAGE CONVERGENCE
σ = STANDARD DEVIATION

Figure 15.8. Statistical distribution of convergence for four-way intersections.

gence than neighboring stations in the same time period must be considered potentially unstable.

7. Acceleration (greater than zero), i.e., an increasing rate of convergence (Figure 15.6) is a sign of instability.

8. In old entries such as the longwall headgate entry in Figure 15.12, for example, total convergence as well as convergence rates relative to other points in the same entry have proved most valuable in detecting an unstable top.

9. Longwall gate roads show a substantially different behavior in the vicinity of the mining face as illustrated in Figure 15.9. Much of this movement, which often exceeds 127 millimeters (5 inches) at the tailgate, is due to floor heave. A detailed analysis of the convergence behavior in this area is beyond the scope of this chapter.

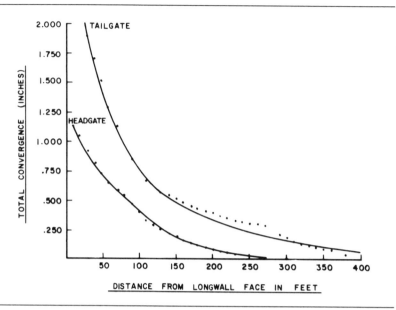

Figure 15.9. Normal convergence curves for longwall gate entries.

In the foregoing, various quantitative criteria that warn of possible roof falls were discussed. The authors wish to point out that the numbers derived herein are not expected to apply in general to all coal mines. They are characteristic of a specific mine or division and might vary with different geologic environments and mining geometry. The method, however, is expected to be generally valid.

Case Histories

The following case histories show how unstable areas have been detected and cite some of the corrective measures that have been taken.

1. Figure 15.6 illustrates a case history from a development entry where total convergence over 25.4 millimeters (1 inch) and velocity far in excess of 2.54 millimeters (0.1 inch) per day aided in detecting the unstable area. Subsequent borescoping confirmed the presence of extremely broken and slickensided surfaces in the roof above the 1.52-meter (5-foot) bolt anchors. The total convergence of more than 152.4 millimeters (6 inches) also verifies the degree of fracturing in the roof. Supplementary supports in the form of cribs were installed to bring velocity and acceleration to normal levels. At such high convergence levels, wide open extension fractures in the roof were also evident.
2. The next example is taken from a future longwall supply entry within 30.5 meters (100 feet) of the mining face in an intersection. Unstable roof was suspected because of the factors on the following text page (Figure 15.10):

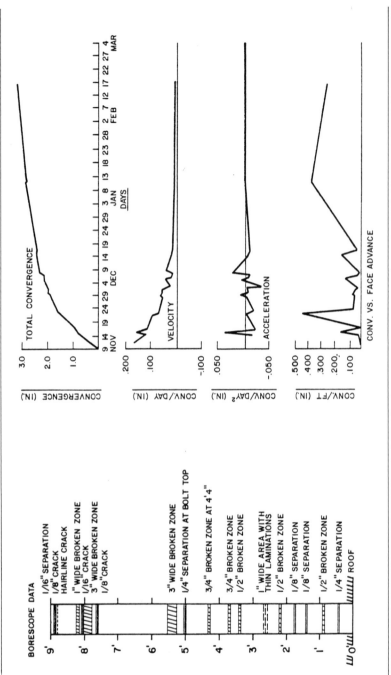

Figure 15.10. Convergence and borescope data for an unstable area in a development unit.

a. Convergence total of over 25.4 millimeters (1 inch).
b. Velocity of more than 2.54 millimeters (0.1 inch) per day.
c. Erratic accelerations instead of monotonic deceleration.
d. Extremely high convergence of over 0.8 millimeter per meter (0.1 inch per foot) of face advance instead of the normal rate of 0.42 millimeter per meter (0.005 inch per foot).

    Borescope analysis verified broken rock to a height exceeding the 1.52-meter (5-foot) bolt anchor horizon, and steel I-beams on 1.83-meter (6-foot) centers were installed and propped up by 152.4-millimeter (6-inch) circular steel columns to arrest the movement.

3. Figure 15.11 illustrates a classic example in which the cumulative convergence and convergence rate at one point (B6) are so much higher than at any of the other stations that unstable roof is suspected.

4. The final example illustrates how the roof fall prediction technique may be applied in an entry that is several years old and where total measured convergence is actually less than 25.4 millimeters (1 inch). Figure 15.12 shows convergence records superimposed on the plan of a longwall panel. The high total convergence near the face is considered normal (Figure 15.9). There are several other points, H24 and H12, for example, with a considerably higher than normal velocity. Borescope inspections confirmed broken rock zones above the bolt anchors. Several alternatives to stabilize these areas are currently being considered.

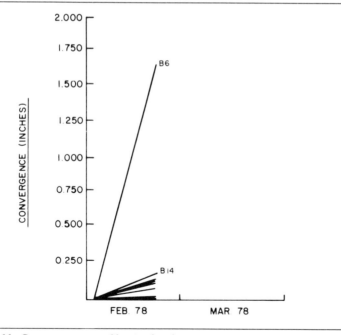

Figure 15.11. Convergence profiles in development units.

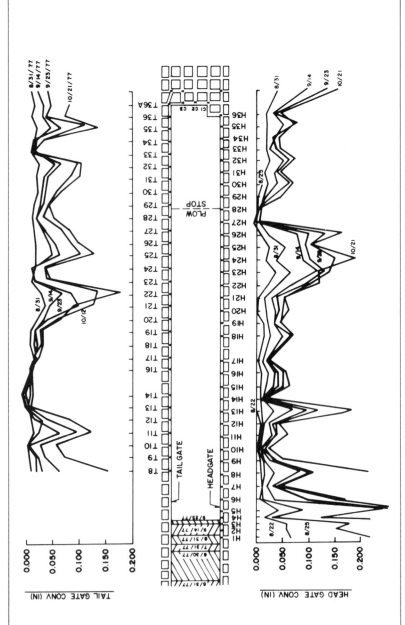

Figure 15.12. Total convergence profiles along the gate entries of an active longwall face.

Summary and Conclusion

A method to predict roof falls in coal mines with the help of precise deformation measurements has been described.

The criteria which determine whether or not a roof is stable are total convergence, convergence rate (velocity), and change of convergence rate (acceleration).

The criteria were derived with the help of a carefully planned field program, which included regular measurement of roof to floor convergence and roof sag in mine development units and on retreating longwalls of six mines. Borescope inspection of the roof strata was used to confirm unstable roof and determine the extent of the affected zone.

Extensive mapping of past roof falls was used to determine the causes, location, and frequency of the falls.

It is concluded that roof falls in the deep mines of Island Creek Coal Company in Virginia occur primarily in four-way intersections within 305 meters (1,000 feet) of an active mining face.

Local geologic disturbances such as slickensides and facies changes are often associated with the falls.

Convergence over 25.4 millimeters (1 inch) and/or convergence rates greater than 2.54 millimeters (0.1 inch) per day indicate that the roof is unstable and will fall without additional support.

Numerous falls have been prevented with the help of this technique at a modest cost. Since the inception of the program, no roof fall occurred in the areas under surveillance, which contain over 600 monitoring stations in 22 development units.

Acknowledgments

The authors wish to thank the management of Island Creek Coal Company for permission to publish; and Mark Christianson, Jerry Lee Haynes, Steven Carter, and Robert H. Moffitt for their assistance in field data collection.

References

Jacobi, O., 1976, "Praxis Der Gebirgsbeherrschung." Verlag Gluckauf, Gmbh, and Essen.

McCulloch, C. M., D. W. Jeran, and C. D. Sullivan, 1975, "Geologic Investigations of Underground Coal Mining Problems." U.S. Department of the Interior, RI 8022.

M.E.S.A., 1977, "Safety Reviews." U.S. Department of the Interior, Table No. 8.

Miller, M. S., 1974, "Stratigraphy and Coal Beds of Upper Mississippian and Lower Pennsylvanian Rock in Southwestern Virginia." Virginia Division of Mineral Resources, Bulletin 84.

Pothini, B. R., J. H. Thaler, and W. L. Finlay, 1976, "Rock Mechanics Considerations in Mine Design for a Deep, Bedded-Deposit Under the Influence of High Tectonic Forces." Proceedings, 17th U.S. Symposium on Rock Mechanics.

Schuermann, F., 1975, "Das Messen Von Gebirgsbewegungen Mit Langmessankrern." Glückauf, Volume III, No. 13.

Working Thick and Steep Coal Seams in India: A Case Study

16

A. K. Gulati, Project Officer/Deputy Chief Mining Engineer,
and A. K. Singh, Assistant Colliery Manager,
Bharat Coking Coal Limited, Dhanbad, India

Introduction

Coal production (excluding lignite) in India was 100 million tons in 1977-78, and is planned to be increased to 137 million tons by 1981-82. This means an annual increase of about 9 million tons, which is a big challenge for all connected with the coal mining industry in India. It calls for a new look at the trends of mining methods and the exploitation of difficult coal deposits, particularly of coking coal, which were earlier abandoned. Bold, scientific steps in this direction have been possible with the nationalization of coking coal mines in 1971, and all other coal mines in 1973. All Indian coal mines are now placed under the centralized control of Coal India Limited (a public undertaking), making it easier to take coordinated steps for the development of new mining methods. It is proposed here to deal with experiments carried out during the last few years in the mining of steeply dipping seams (over 27°) more than 3 meters (9.8 feet) thick, with hydraulic sand stowing at Sudamdih. Mining with caving has also been successfully tried elsewhere in India.

In India, seams more than 3 meters thick account for nearly 70 percent of the total coal reserves, and 80 percent of the present output comes from thick seams. At the present time, the maximum depth of workings is about 600 meters (1,969 feet). In some new projects it is intended to mine to a depth of 1,000 meters (3,281 feet).

Generally, seams dip from 5° to 15°, but seams as steep as 80° from the horizontal are also being worked. The overburden consists mostly of sandstone and shale, with sandstone accounting for about 75 to 80 percent of the total overburden. In the Singarauli coalfields of Madhya Pradesh, a seam 131 meters (430 feet) thick has been found, and is being mined by open cast methods.

Methods of Mining

Both room and pillar and the longwall method of mining are being practiced for extraction of India's coal deposits. The room and pillar method is extensively used, while longwall mining has only recently been introduced to the Indian coal mining industry. However, longwall mining is becoming more and more popular, with new faces being opened in various coalfields. The first powered support face in India has been commissioned at the Moonidih mine project of Bharat Coking Coal Limited. Equipment is supplied by Dowty Mining Equipment, Ltd., U.K.

Room and pillar mining

All the extraction techniques have been associated with stowing. Seams as thick as 8.53 meters (28 feet) are being extracted in one slice. A seam 14.63 meters (48 feet) thick was extracted by room and pillar* working by developing it in three sections, with 3-meter partings left in between the sections. This resulted in heavy losses of coal, and invariably led to the occurrence of fires due to the spontaneous heating of coal.

Longwall mining with caving

Longwall mining with caving is the universally practiced method in all advanced coal mining countries, and is likely to be the most popular method in India during the next ten years, because of its inherent advantages. Longwall mining is being applied more often because of the shortage of stowing materials. In India, the method has not been practiced on a large scale, primarily due to the nonavailability of equipment such as friction/hydraulic props, armored conveyors, and so on. With the manufacture of this equipment by Mining and Allied Machinery Corporation (MAMC) Ltd., Durgapur, longwall mining is being applied more often. A field survey has yet to be undertaken in India for determining the support resistance required at the face, but generally 35 to 40 tons per square meter (3.26 to 3.72 tons per square foot) at the face, and 80 to 100 tons per running meter (24.4 to 30.6 tons per foot) at the goaf edge have been found suitable for caving roof strata. In a recent experiment for the extraction of a thick and steeply inclined seam, the following longwall caving methods have been successfully tried.

Successive slices using wire netting. Imported from France, the technology of using wire netting in successive slices can be successfully used in seams between 5 and 6 meters (16.4 to 19.7 feet) thick, where development has already been carried out by room and pillar mining.

Longwall sublevel mining. In seams thicker than 7 to 8 meters (23 to 26.3 feet), longwall sublevel mining has found better application than longwall caving, with successive slices using wire netting. A 12-meter- (39.4-foot-) thick seam has been successfully worked in four slices below wire netting. A 2.5-meter- (8.2-foot-) thick coal parting between the longwall faces on the third and fourth slices was taken by undermining. This method is finding application in other mines in various parts of the country.

Longwall mining with stowing

Up to the present time, only hydraulic sand stowing has been practiced in India. Limited experimental work was carried out with pneumatic stowing, but it was

* Room and pillar mining is commonly known as bord and pillar in India.

found far more expensive and the experiments were discontinued. In flat seams (below 7°), multilift extraction has been carried out successfully after mechanizing the face with a shearer. Thick seams of varying thickness, which have been developed by room and pillar mining, have been extracted with stowing, but the percentage recovery has been very poor. It is the steep (over 27°) and thick seams of from 3 to 20 meters (9.8 to 65.6 feet), which are posing a challenge. As huge deposits of such seams containing coking coal are available, a bold venture was undertaken to establish the various technologies available for mining thick and steep seams.

In the Sudamdih mine in Bihar, seams from 3 to 22 meters (9.8 to 72.2 feet) in thickness with gradients between 27° and 50° are present. Various methods for mining these seams have been tried, and have become fairly well established. They are now being extensively adopted by other mines. In this chapter, the details of these trials at the Sudamdih mine are described.

Production

The mine is served by two shafts, respectively 7.2 and 6.5 meters (23.6 and 21.5 feet) in diameter, and has a planned output of 6,000 tons per day. Coal is wound in two 10-ton-capacity skips. Mining is being carried out in two phases. During the first phase, coal seams are being mined on three horizons at depths of 200, 300 and 400 meters (656.2, 984.3 and 1,312.4 feet), and during the second phase, work will be carried out to a depth of 850 meters (2,878.9 feet). Total extractable reserves of coal in the mine are 90 million tons; 40 million tons of these reserves are being worked in the first phase.

Geology

The Sudamdih holding is located in the southeastern corner of the Jharia Coalfield in eastern India, partly on the northern bank of the river Damodar, and partly on the southern bank. The strata dip at an angle varying from 27° to 60°. It has been established that there are 24 coal seams within the Barakar sequence of the Jharia Coalfield, all of which occur in the Sudamdih area. The property lies south of, and close to, the great Patherdih horst, and also north of, and very near to, the great southern boundary fault believed to be of post-Gondwana age. Due to the proximity of the Patherdih horst and the great boundary fault, the area is highly disturbed geologically. The coal seams are very gassy and steeply inclined. Other geological disturbances in the form of faults and igneous intrusions, such as dikes and sills, are common in this area. Faults hade at low angles, and some are almost horizontal in nature, resulting in large areas of barren ground. These unusual faults cause many problems in the development and extraction of the coal seams.

Method of Development

The deposits have been worked on the outcrop by open cast mining and from shallow inclines to a depth of 150 meters (482.2 feet). These old workings are waterlogged, and fires are known to exist. Due to the presence of these old work-ings, fires, and the river Damodar, the Sudamdih Shaft mine was planned to be worked from a depth of 200 meters on horizons leaving a barrier of 50 meters. The upper horizon at a depth of 200 meters constitutes the main ventilation level, and the lowest horizon at 400 meters is the main transport level, in which there is a tippler and skip loading point. Main drivages have a cross-sectional area of 14 square meters (150.7 square feet), and the main drivages in the seams are driven from crosscuts to form coaling blocks of 100 meters along the strike. All the main drivages are supported by steel arches.

Methods of Extraction

The various extraction methods being practiced at Sudamdih are as follows:

Jankowice system

The Jankowice method has been practiced at Sudamdih over the last eight years, and is very suitable for seams 3 to 12 meters (9.8 to 39.4 feet) thick, and on gra-dients from 27° to 45°. A seam exhibiting the following characteristics has been worked extensively by this method:

Thickness of seam	7 to 7.5 m (23 to 24.6 ft)
Inclination	30° to 40°
Roof	Sandstone
Floor	Shaly-sandstone
Crushing strength of coal	350 kg/cm^2 (4,977 lb/in.^{2})

The details of the method are shown in Figure 16.1.

Method of development. The length of the face is 110 meters (361 feet) and one 6-meter (19.7-foot) slice is mined at a time. The seam is extracted in two lifts, each 3.5 meters (10.9 feet) thick. The face moves in a direction along the strike, while the longwall advances from dip to rise. A coal rib 20 meters (65.6 feet) thick is left above the main lateral road for its protection, and another gallery is driven connecting two rises, supported with timber, to start the face.

The face is divided into small subfaces by driving two stables, giving six short-moving fronts, each 6 meters long. Sixteen holes are drilled in each front, 1.2 meters (3.94 feet) in length, and charged with 500 grams (1.75 ounces) of explosive per hole. After blasting, 60 percent of the coal is loaded onto a conveyor by gravity, and the balance of 40 percent by shoveling.

On the faces, a double-chain armored face conveyor is used, while along the direction of dip a retarder conveyor is used for transporting the coal away from face. The coal is loaded into 2.5-ton-capacity mine cars in the main lateral roads, and transported to the pit bottom by battery locomotives.

Material supply. Materials are supplied by monorails at the tailgate end of the faces. Timber is moved on the faces over the conveyors.

Stowing. The maximum span of exposed roof allowed from the sand pack is 8 meters (26.3 feet); 6 meters (19.7 feet) of void is stowed at a time, leaving 1 meter (3.3 feet) to start the next face. Before stowing begins, the conveyor is dismantled

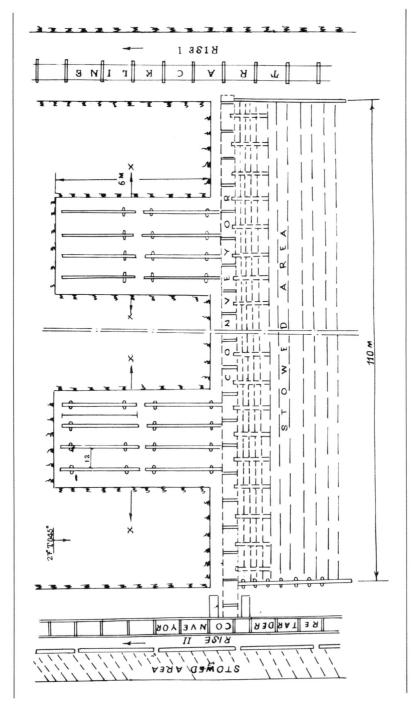

Figure 16.1. Plan showing Jankowice longwall method of mining–Sudamdih mine.

and repositioned. Side barricades are constructed along the rise and another barricade is built along the length of the face. Galvanized iron pipes 150 millimeters (5.9 inches) in diameter, with valves placed at intervals of 10 meters (32.8 feet), are laid at roof level supported by the front props. Sand stowing is completed in about 9 hours, and rates of 250 to 300 cubic meters per hour (147.2 to 176.6 cubic feet per minute) have been achieved. The side barricades are made of wooden planks and bamboo matting, while the face barricade is constructed of hessian cloth. The entire stowing cycle takes about 15 working shifts or 140 manshifts.

Support. Faces are supported by timber with crossbars set over props. The distance between each row of props is 1.2 meters (3.9 feet), and the distance between each prop in a row is 2.5 meters (8.2 feet). The maximum distance allowed between the last row of props and the face is 2.5 meters.

Organization. Twenty miners work on each face and the whole mining cycle takes 10 days. An output of 3,000 tons is produced with a face output per manshift (OMS) of 5 tons.

Second lift extraction. After the first lift has advanced 30 meters (98.4 feet), a second lift is developed. The second lift is approached by the same rise of first lift in the manner shown in Figure 16.2.

Comments. The Jankowice method is not suitable for seams with gradients greater than 45°, because it becomes difficult for miners to stand and accomplish various operations. This method is also unsuitable in seams dipping below 27° because shoveling of coal becomes excessive. This becomes a controlling factor for all operations when face productivity falls sharply.

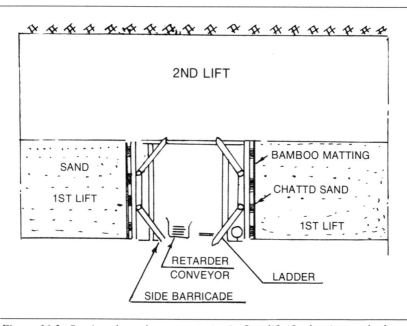

Figure 16.2. Section through transport rise in first lift (Jankowice method).

Kazimierz method

The Kazimierz method—horizontal slicing in ascending order—is suitable for mining thick seams having a gradient of more than 27°. At Sudamdih it has been used in a seam with the following characteristics:

Thickness	22 m (72.2 ft)
Gradient	27° to 35°
Floor	Shaly sandstone
Roof	Sandstone

Method of development. Development is almost the same as in the Jankowice system, except that a crosscut is driven from the floor to the roof of the seam, and this forms the longwall face. In the seam having a thickness of 22 meters, the face length is about 47 meters (154.2 feet). For each lift, an independent gate road is driven. The details of development are shown in Figures 16.3, 16.4 and 16.5. Each lift is 3 meters high, and subsequent lifts are extracted over the sand fill of the previous lift. It is proposed to extract coal between two horizons at Sudamdih in 33 lifts.

Method of extraction. In the face and gate roads, double-strand chain armored conveyors are used. The face conveyor is advanced mechanically by a pusher using the electric coal drill machine. The maximum distance allowed between the pack and the face is 6.5 meters (21.3 feet). Blast holes are drilled along the length of the face, 1.4 meters (4.6 feet) deep. Solid blasting results in a pull of 1.2 meters (3.9 feet); the same advance is obtained in each cycle. Again, the conveyor is moved, supports are set, and the same cycle is repeated.

Support system. The distance between the last row of props and the face is 1.8 meters (5.9 feet). The roof is temporarily supported by timber cantilever bars. The distance between each row is 1.2 meters, and the distance between props is 1.5 meters (4.9 feet), and all the props are lagged. After the extraction of four slices, the face is stowed. The front barricade is constructed of hessian cloth stretched over timber props laid 0.75 meter (2.5 feet) apart. Barricade props are strengthened by two ropes stretched along the face. The gate roads are supported by timber, and the main rises are supported by steel arches or goalpost supports.

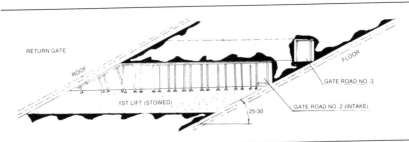

Figure 16.3. Vertical section through a face showing the Kazimierz (horizontal slicing) method of mining.

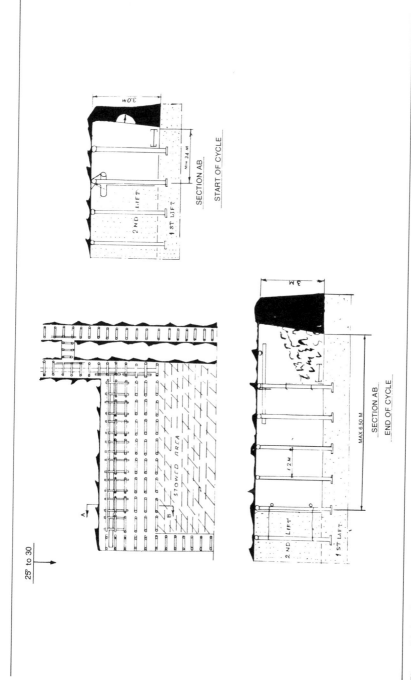

Figure 16.4. Plan and sections showing stages of extraction in a cycle (Kazimierz method).

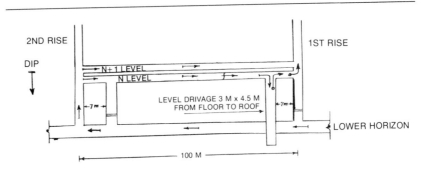

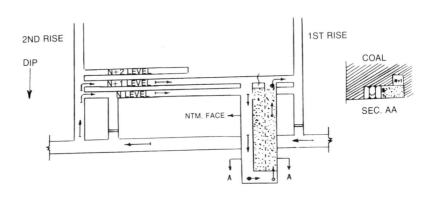

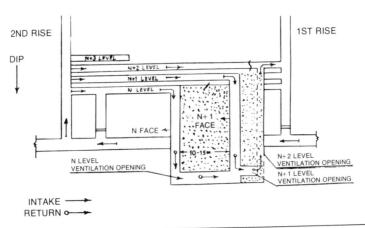

Figure 16.5. Plan showing stages of development of successive lifts (Kazimierz horizontal slicing method).

Organization. Work is carried out on three shifts, two of which mine coal. A cycle of 1.2 meters advance is completed in one day by 40 miners. Each cycle results in an approximate output of 160 tons, with a face OMS of 4 tons.

From a central rise, three lifts are worked on either side, giving a daily production of 900 tons at one loading point.

Multilift inclined slicing system

An alternative Kazimierz method is the multilift inclined slicing system. During extraction, when using the horizontal slicing system, the following difficulties have been experienced:

1. Spontaneous heating.
2. Frequent roof falls.
3. Ventilation problems.
4. Extensive development work.
5. High consumption of timber.
6. Inefficient stowing resulting in poor packing.

An equivalent material model study was carried out by the Central Mining Research Station (CMRS) Dhanbad, and the following observations were made:

1. Horizontal slicing is likely to be associated with large-scale deformation and dislocation of the immediate coal roof. The magnitude of the movement depends on the efficiency of the support and sand pack.
2. In the case of subsequent working, timber props normally set on the stowed pack are likely to penetrate it, resulting in a decrease in its efficiency and more movement.
3. Where the sand packing is very poor, or where it has not been done, even single horizontal slicing could cause heavy movement and deformation in upper layers of coal.
4. Effective sand packing is likely to reduce coal break and delay subsequent fall.

Some of these disadvantages were eliminated by using cross inclined slicing in the 1X/X Seam. The following points were made during investigation of the model:

1. Roof stability is likely to improve with an increase in the slope of the face.
2. Roof conditions improve with a steeper gradient of cross inclined slicing, at the cost of face length and operational difficulties.
3. The load on supports is likely to increase on flatter gradients.
4. Stowing is very effective.
5. Better pack density is likely to avert surface subsidence.

Taking into account all the disadvantages of the horizontal slicing method, multilift inclined slicing was tried in the 22-meter-thick seam. It was proposed to extract the total thickness of coal in six lifts, each 3.5 meters (11.5 feet) thick, leaving a parting 0.5 to 1 meter (1.6 to 3.3 feet) thick between subsequent lifts.

Essential features of the method. The pattern of development in multilift inclined slicing is much the same as in the Kazimierz method, except that a central entry rise is driven at a gradient of 60° from the middle rise, to serve as a supply road for the upper slices, as can be seen in Figure 16.6. Two slices were proposed to be worked simultaneously in one lift. For each lift, individual development work has to be carried out. The first lift is composed of numbers one and two slices, and the second lift of numbers three and four slices. The third lift will be composed of number five slice. Details of this method are shown in Figure 16.7.

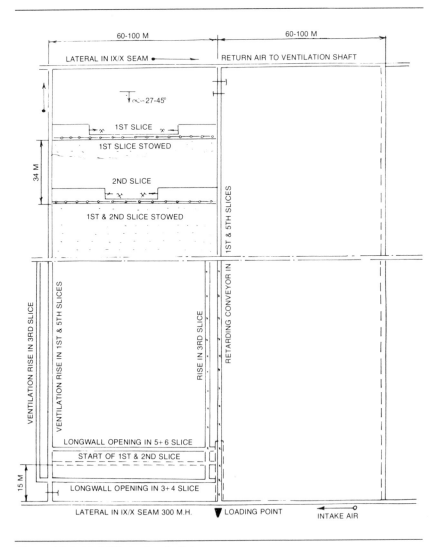

Figure 16.6. Multilift inclined slicing.

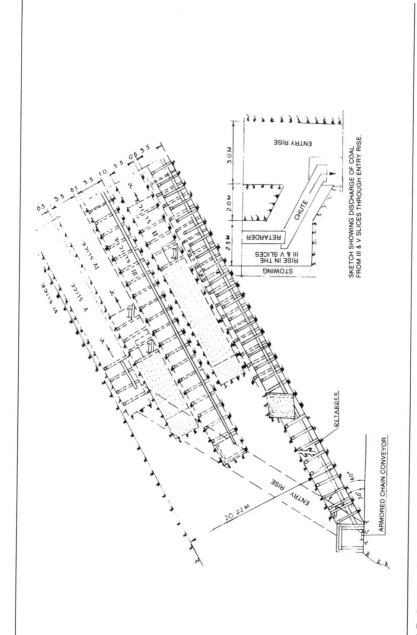

Figure 16.7. Section showing sequence of the multilift mining operation.

Between consecutive lifts, a coal parting with a thickness of about 1 meter is left, and between slices, a parting 0.5 meter thick is kept. The middle rise is utilized for intake ventilation purposes, while the peripheral rise is used as a return airway.

After extraction of the first lift and stowing of the mined-out ground, new development work is carried out for the next lift. Four faces will be worked along the central rise, resulting in an output of 600 tons of coal per day through a single loading point. The entrance to the peripheral rise is supported with steel, while the faces are supported with timber, as has been explained in the Jankowice method.

At the present time, only two lifts are being mined, and development work is proceeding for subsequent lifts. It has still not been established whether the fifth and sixth slices can be mined in ascending order, since spontaneous heatings in the worked-out lifts cannot be ruled out. It is being considered whether to take the top two slices first in descending order and the bottom three later in ascending order.

For the successful outcome of this method, extraction progress has to be fast, the support of the faces has to be good, and stowing must be carried out in a systematic manner. The extraction order of the coal blocks must be carefully followed. Total extraction must be aimed for, without leaving pillars or coal ribs in the slices, and a continuous watch must be kept for any sign of spontaneous heating, especially during the opening up and extraction of the top slices.

Room extraction method or Komara System

Application. This method is suitable for thick seams with steep gradients of more than 40°. The method has been adopted in the XV Seam, which has the following characteristics:

Thickness	4 to 12 m (13.1 to 39.4 ft)
Gradient	40° to 60°
Roof	Sandstone
Floor	Shaly-sandstone

The greatest advantage in working steep seams is that gravity eases the flow of coal.

Development. Formation of coal blocks in this seam is similar to that in other sections of the mine. Each block is divided into two by a middle rise, which is partitioned to form a coalway and a manway. It also serves as an intake airway. The rises on the flanks of the blocks act as transport roads for materials and as return airways.

Galleries are first developed from the rises and have dimensions 3 × 3.5 meters (9.8 × 11.5 feet). A coal barrier is left, 15 meters (49.2 feet) from the main lateral. The drivage of further galleries (with a maximum parting of 40 meters or 131.2 feet), depends on the geological conditions for forming sublevels in the block. From the lowest gallery to the next one higher up, rises, 3 × 2 meters (9.8 × 6.6 feet) in dimension, are driven 10 meters (32.8 feet) apart. The method is shown in Figures 16.8 and 16.9.

Extraction. After driving two rises (rooms) adjacent to the peripheral rises, each room is widened on either side for a distance of 2 meters (6.6 feet), leaving a chute at the entrance. The total width of each room after extraction of the coal is 7 meters (23 feet), and the height to the main roof is about 7 to 8 meters. A solid barrier of coal, 3 meters thick, is left between adjacent rooms.

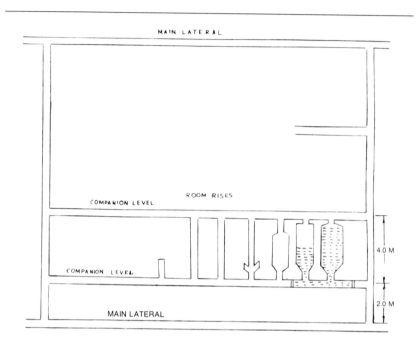

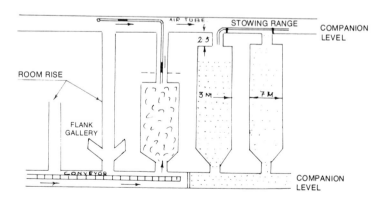

Figure 16.8. General layout of Komara system.

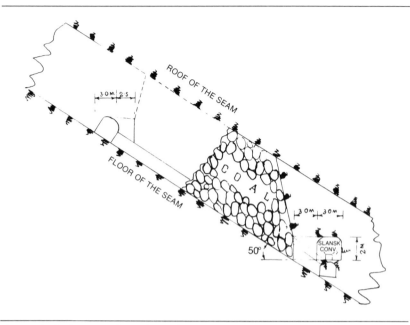

Figure 16.9. Section along dip in Komara system.

Solid blasting is carried out for extracting the coal, during which some coal falls through the chute leaving enough coal to enable drilling to be continued in the roof (the miners stand on the pile of coal left to reach the higher places). After blasting has been completed, the remaining coal is passed through the chute and onto a conveyor in the gallery. The conveyor moves the coal to the middle rise, where it gravitates down the coal compartment. Ventilation of the rooms being mined is by auxiliary fans.

Stowing. As soon as the extraction of coal has been completed, and the room is free of blasted coal, a strong barricade is constructed at the narrow entrance to the room. A second barrier is constructed in the gallery between two rooms, and the conveyor is shortened back. For clearing stowing water, holes are made from the adjacent rise, at intervals of 1 meter, to the extracted room. Stowing one room takes about 8 hours, at an average stowing rate of 250 cubic meters per hour (147.2 cubic feet per minute).

Organization. In each coal block, four gangs of miners work simultaneously, each gang consisting of eight men. A room is normally extracted in about 10 days, and approximately 2,000 tons of coal are produced.

Modified Komara method

Two alternative methods are used at Sudamdih: The first involves driving two rises at invervals of 10 meters and enlarging them, instead of enlarging a single rise, as previously described. This results in a higher percentage of extraction, better ventilation and safety.

The second alternative overcomes the problems experienced in blasting coal more than 5 meters (16.4 feet) thick in one round, by driving one central rise along the roof. This also prevents the accumulation of gas in the void created after blasting.

Conclusion

In this chapter, the recent trials carried out at Sudamdih mine have been emphasized. The methods used have been practiced extensively in Poland and have been suitably modified to suit different conditions at Sudamdih. The methods are being applied increasingly in other coalfields having similar seams.

Acknowledgment

The authors wish to thank the management of Bharat Coking Coal Limited, a subsidiary of Coal India Limited, for giving permission to make this contribution.

Preventing Karst Water Inrushes in Hungarian Coal Mines

17

Dr. Paul Gerber, Chief Geologist,
Tatabánya Coal Mines, Tatabánya, Hungary

Introduction

The Transdanubian mountains mainly consist of Triassic limestone and dolomite but, in some places, coal seams of large thickness occur. These seams were formed in the karst marshlands which developed at the beginning of sedimentation in the Eocene period. Before the formation of coal, however, on the uneven karstified Triassic landscape, thinner sediments of continental and freshwater origin had been deposited.

In the majority of cases, the basement of the Eocene coal basins is formed of Dachsteinian limestone and dolomite which have been karstified. The thickness of the Upper and Middle Triassic sediments is about 3,000 meters (9,834 feet), and the whole series form a homogeneous karst aquifer along the full length of the mountain chain, about 180 kilometers (112 miles) long.

The karst water level in various parts of the mountains varies from 100 to 230 meters (328 to 755 feet) above sea level, depending on the base level of erosion and the surface of infiltration. Originally, the karst aquifer was drained by several springs with a large discharge along the boundary of the mountains, as is shown in Figure 17.1.

Since the early 1950s, considerable mining development has taken place in this part of Hungary, and coal seams liable to water inrush have been explored over extensive areas. In 1961, the total quantity of water pumped in mining operations was 3,364 liters per second (53,710 U.S. gallons per minute), 80 percent of which was pumped from coal mines.

By 1975, the total quantity of water pumped had risen to 10,429 liters per second (165,313 U.S. gallons per minute) and, as a consequence, the natural springs in the mountains have dried up. A large quantity of water is pumped from some mines in the Tatabánya area where, in individual cases, specific pumping raises 37,000 liters per ton (9,775 U.S. gallons per ton). A graph showing the development of pumping in the Tatabánya Basin is shown in Figure 17.2.

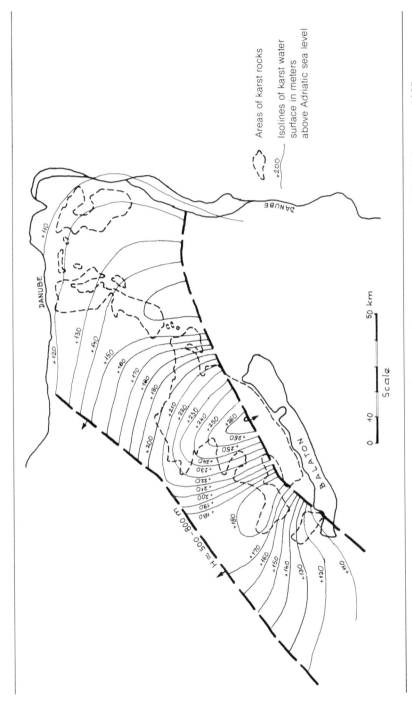

Figure 17.1. Map of the main karst water levels in the Transdanubian central mountain chain of Hungary (1957).

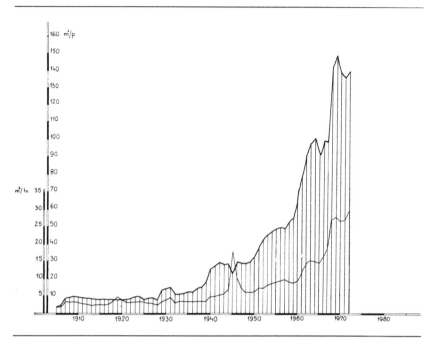

Figure 17.2. The increase of pumping in the Tatabanya coal basin.

Methods of Protection

In Hungary, the methods of protection from water inrush in coal mines have greatly developed as the danger has increased. Methods differ in certain mines, according to geological and hydrogeological conditions.

At first, the passive-preventive method of protection was the most widespread. This method involves the preparation for pumping water which may inrush, despite previous protective measures. Reducing water discharge by post-stopping of larger inrushes results in more favorable pumping costs and subsurface protection.

The possibility of recognizing the relationships between the protective layer effect, structure, and the inrush of water, plays a major role in determining a suitable method for protection. In some areas, the expected water discharge can be determined on this basis by increased exploration. An optimum water protective system can be developed with an order of mining, as well as by making preparations for expected pumping in good time.

Based on many years of experience, Hungarian experts have obtained good results in:

1. Determining protective layers.
2. Establishing more effective and economical methods of forming barriers (pillars) along faults, and between adjacent mines liable to water inrush.

3. Reliably forecasting expected water discharge.
4. Diverting the most important underground streams having a regional effect, thereby reducing the danger of inrush in certain areas. The reduction in the supply of water ensures a small or nonexistent water pressure over a much larger area.
5. Post-stopping of water inrushes achieved by packing with sand-cement materials. Other materials are also used efficiently, for example, fly ash, bentonite, fiber glass, cement, and other plastic materials.

Passive-preventive method

The passive-preventive method relies on safe exploration and mining procedures, the optimum utilization of pumping capacity, the improvement in pumping efficiency, and the automation of pumping stations.

At the present time, work is being carried out in developing the technology for constructing stoppings with cheaper materials, and with better efficiency. Adequate mine drainage systems are also being formed to protect coalfaces from water. This helps mining continuity and makes possible a more effective synchronization between pumping and water economics.

It must be understood, however, that this method can only be applied in those areas where protection layers of varying thickness are found. Earlier, the method was applied successfully in a number of coal mines in the Transdanubian mountains, such as the Oroszlány coal mines, and in the Tatabánya and Dorog coal basins. Since then, the method has become less important in the Dorog and Tatabánya areas more liable to water inrush because of the forced cut back in mining.

Because the method was not applicable in some mines with a small or nonexistent protection layer, a few shafts were closed. In the Tatabánya district, other methods of protection were chosen, in the early 1960s, in places where there was a great deal of danger from water inrush. Passive water protection is applied when extracting small blocks of coal, and where there is a large protective layer providing considerably greater safety.

Active water protection method

Safe and economical mining is achieved by planning and regulating the drainage of water from the aquifers, relieving stress, and depressing the water level. In the Tatabánya area, this is applied with success in the southeast of the basin where the basement is formed of dolomite. Here, 1,480 to 1,497 liters per second (23,461 to 23,730 U.S. gallons per minute) of water are pumped from an area of 0.52 square kilometers (0.2 square miles), which has resulted in a drop in the water level of 80 meters (263 feet).

This area, which was very susceptible to water inrush because of the high pressure of karst water, has been safely mined by caving methods. Figures 17.3 and 17.4, a plan and cross section respectively, illustrate the method of draining water from the dolomite.

In the western part of the basin there exists a horst bordered by a fault with a throw of 150 to 200 meters (492 to 656 feet). The coal seam lies on a basement of Dachsteinian limestone and is protected from water inrush by a thin layer of impermeable rock. A shaft was sunk on the downthrow side of the fault and, after

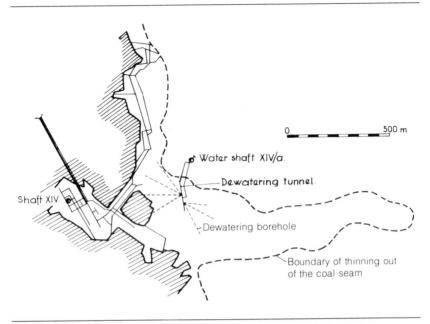

0 _____ 500 m

Water shaft XIV/a.

Dewatering tunnel

Shaft XIV

Dewatering borehole

Boundary of thinning out
of the coal seam

Figure 17.3. Plan showing area dewatered by the water shaft XIV/a.

constructing a pumping station, a tunnel was driven into the limestone behind bulkheads. After driving the tunnel for a short distance, boreholes were drilled in different directions from it which intersected an underground stream resulting in a water flow of 673 liters per second (10,688 U.S. gallons per minute).

The line along which the water flowed was known, and so a decision was made to change the direction of flow to form a more favorable depression. It was not possible, however, to accomplish this because of the urgent demands of mining operations. In order to protect a spring used as a recreational facility, regular pumping was started only after ensuring that water was available in wells, and after swimming pools were constructed.

Coal mining can be safely carried out by driving a water tunnel—a third entry—at the deepest point along the dip. When boreholes are drilled from the third entry under the face, they drain any underground water or streams intersected by the holes, thereby reducing the possibility of water inrush. This has also been successfully proved in other mines. In Hungary, the active protection method is being successfully applied in bauxite mining operations where safe and economical working is dependent on the observance of efficient dewatering practices.

Dewatering the deeper brown coal and bauxite areas would result in the formation of a depression causing significant environmental harm in the central mountain area. It is not intended to apply this method here, except to a limited degree, even though it is the safest method from a mining point of view, and it also provides drinking water.

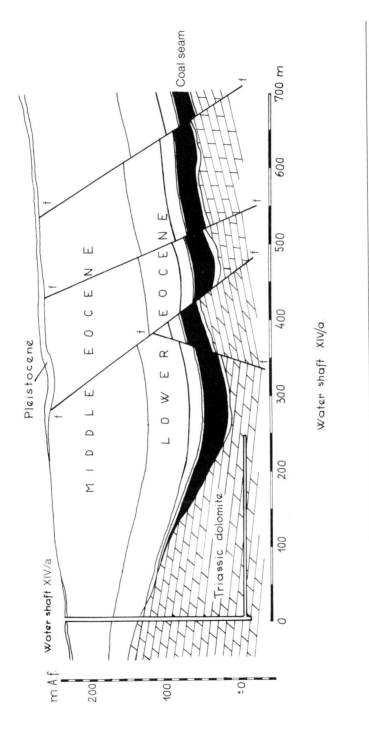

Figure 17.4. Geological cross section through the dewatering tunnel.

Instant combined water protection method

It was necessary to develop another method which resulted in safer mining and reduced environmental harm to a minimum. The so-called instant combined water protection method was found to best fulfill these requirements. It involves the driving of drainage tunnels and the drilling of boreholes under the seam which collect the water released by microstructural changes in the rock.

The expected discharge of water from the drainage system does not exceed the "spontaneously rising" discharge, and the expected zone of depression can be reduced by pumping.

To dewater new mines producing several million tons of coal per year, it is expected that a total quantity of water amounting to 2,523 to 3,364 liters per second (39,995 to 53,326 U.S. gallons per minute) will have to be pumped. This will be utilized in regional water supply systems, in order to supply a larger area of the country with good quality drinking water. In addition, it will also compensate for possible damage caused by dewatering.

A Subsidence Investigation of a Site Above Old Anthracite Workings in Pennsylvania

18

Dermot Ross-Brown, Senior Engineer,
Dames & Moore, Denver, Colorado, USA

Introduction

This chapter describes an investigation that was carried out in 1976 for the Perkins and Will Partnership to evaluate the subsidence potential of a proposed hospital extension in Scranton, Pennsylvania.

The site is located on a gentle slope and is underlain by at least eight anthracite coal seams, seven of which have been mined beneath all or portions of the site. The coal seams range in depth from about 12 to 104 meters (40 to 340 feet) below the surface. From shallowest to deepest, the seams are: Big Vein, Bottom New County, Clark, Dunmore No. 1 Top Split, Dunmore No. 1, Dunmore No. 2, and Dunmore No. 3. An idealized cross section giving the depth and thickness of the seams as well as the idealized extraction ratios of each is shown in Figure 18.1.

The relative locations of the hospital and the proposed extensions are shown in Figure 18.2, together with the locations and dimensions of the pillars in one of the seams. Note that no mining has supposedly occurred beneath the existing hospital, and none of the seams in other locations have been mined during the past 40 years.

The scope of the study was to examine the data already obtained by others, collect any new data which might be required, perform an independent analysis, and assess the subsidence potential at the site.

Design Procedure

Assessment of the extent of the problem

The first consideration was to try to evaluate the extent of the subsidence problem caused by the mining of the seven coal seams beneath the site. This is summarized in Table 18.1.

Mr. Ross-Brown is currently with Science Applications, Inc., Fort Collins, Colorado.

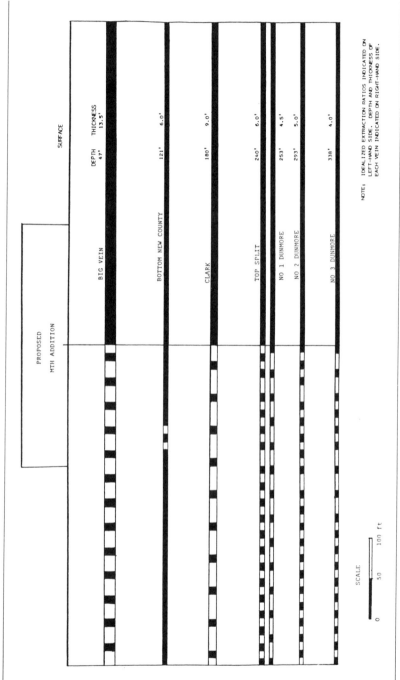

Figure 18.1. Idealized cross section through mined coal seams beneath the hospital extension.

Clearly, if a preliminary analysis indicates low stresses in the pillars in relation to the ultimate strength of the pillars, so that a high safety factor is indicated for all the pillars in the mine, then no problem exists (Condition A). Consequently, sophisticated analyses are not required and the hospital extension can be designed by largely ignoring the effects of mining.

On the other hand, if it is likely that some of the pillars in some of the seams will have a low safety factor, it is necessary to decide whether sufficient data is

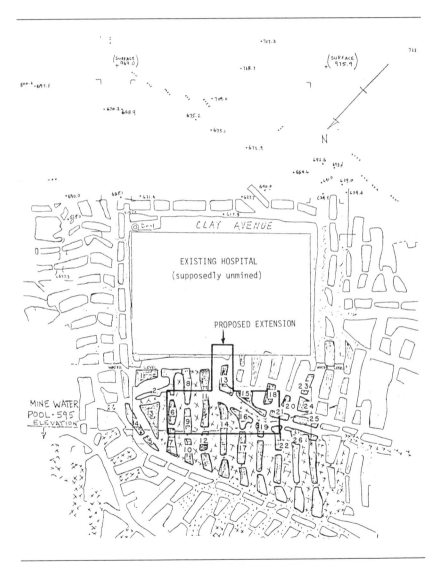

Figure 18.2. Pillar locations in the No. 2 Dunmore Seam.

Table 18.1. Extent of Problem

Condition	Type of problem	Remedy
A. Relatively low stresses. High safety factor.	No problem	No analysis required. Largely ignore mining effects.
B. Relatively high stresses. Low safety factors. Reliable data and realistic analytical techniques.	Tractable problem	Collect data, analyze and isolate problem areas. Remedial measures may then be selectively applied.
C. Relatively high stresses. Low safety factors. Poor data or inadequate analytical techniques.	Intractable problem	Analysis is not justified. Use a conservative design, e.g., grout all voids to provide competent support.

available to solve the problem and whether a suitable method of analysis is available. If the answer to both these questions is positive (Condition B), the problem is tractable and a rational design method can be applied. In this way the problem areas can be delimited and remedial measures selectively applied.

However, if the answer to either of these two questions is negative (that is, insufficient data or inadequate method of analysis), Condition C exists. Since an analysis cannot be justified, a conservative approach will have to be taken. For instance, this might entail grouting or flushing all the voids in all the seams.

The preliminary analysis indicated that most of the seams fell into the Condition B category, which justified the collection of further data and subsequent analyses. Two seams were not analyzed, however. In one case, because of the low extraction ratio in the Bottom New County, the pillars were considered to be stable (Condition A). In the other case, the Big Vein was not examined due to extensive mining and the lack of reliable data (Condition C). Consequently, a conservative design was suggested, which involved locating the bottom of the caissons below this seam level.

Procedure
In tackling problems of this type, it is first necessary to assemble some basic information on the problem. The geology is particularly important. For instance, it is necessary to know the locations and thicknesses of the various geologic materials and also the nature and extent of any discontinuities, such as faults and joints, which are present. Prior to analysis, the strength properties of these materials should be determined. Knowledge of the topography and geometric layout of each seam is also important.

Although there are some uncertainties concerning the geology and the accuracy of the mine plans, it was judged that the data was sufficiently accurate to justify a more sophisticated analysis than had been carried out before. In addition, several proven methods of analysis were available and two of these seemed suitable for analyzing the problem. The mine simulation analog at the University of

Minnesota was used to analyze the pillar stresses in each seam, and a displacement discontinuity computer program was used to study the vertical interaction effects between the various seams and the amount of surface subsidence. It was judged that pillar strengths could also be reasonably determined from Van Heerden's formula (1975) using actual pillar dimensions and strength values derived from a laboratory testing program.

Given the above circumstances, it appeared that a rational analytical approach was required. The procedure employed is summarized as follows:

A. *Basic data collection*
   - Geology
   - Geometry (elevations and mine plans)
   - Hydrology
   - Basic materials properties
B. *Problem definition*
   - Determination of likely modes of failure and behavior models. ("Punching" of floor or roof by pillars, roof collapse, pillar failure, etc.)
C. *Stress analysis using applicable model(s)*
   - Stress analysis for each seam allowing for horizontal pillar interaction
   - Check effects of vertical interaction between seams
   - Three-dimensional analysis, if necessary
D. *Pillar strength determination*
   - Laboratory testing
   - Field observations of pillar support and/or collapse
   - Application of suitable formulas, using specific pillar dimensions
E. *Subsidence calculations*
   - Calculate safety factors for pillars
   - Assess the acceptability of safety factor
   - Define problem area(s)
   - Identify and apply remedial measures

Interviews and Site Visits

Previous reports were studied and discussions were held with representatives of Joseph S. Ward and Associates, the U.S. Bureau of Mines, the Pennsylvania Department of Environmental Resources, and the Moses Taylor Hospital.

Initial site visits were carried out to examine the cores obtained from a previous boring program, to inspect the exterior and interior of the existing hospital and its environs for indications of subsidence damage, to evaluate the underground conditions, and to observe underground flushing operations.

Core examination
Cores were examined from six holes drilled as part of an earlier investigation. Nos. 1 through 4 went to depths of 23 to 27 meters (75 to 88 feet) terminating below the Big Vein Seam, and Nos. 5 and 6 went to depths of 85 to 88 meters (280 to 290 feet) terminating below the No. 1 Dunmore Seam. A lack of uniformity was

observed in the Big Vein Seam and its immediate roof and floor strata in the southwest corner of the site. Apart from this, bedding was uniform and normal with a slight dip. The cores comprised typical coal measure strata, shales, silt-stones, and predominantly competent massive sandstones. Three major sandstone units, each from 7.6 to 10.7 meters (25 to 35 feet) thick, were present in holes 5 and 6. At the Clark Seam horizon, 3 meters (10 feet) of core was missing in hole 5 and 3.4 meters (11 feet) in hole 6. At the No. 1 Dunmore Seam horizon, 3.4 meters (11 feet) were missing in hole 5 and 2.4 meters (8 feet) in hole 6. It was not possible to determine the lateral extent of any voids which may have been associ-ated with these missing core locations.

The thickening and thinning of the seam, and the presence of bone coal and dirt partings in the seam sections, indicated that mining layouts, and the thick-ness of extraction, might be erratic as much from force of circumstances as from design. Even where the seam was thick, the full height of coal was totally extract-ed. A typical layout of pillars in the No. 2 Dunmore Seam beneath the site is shown in Figure 18.2.

Ten samples were selected as being representative of the various rock and coal types that were present in the strata section and these were tested in the laboratory.

Hospital inspection
The existing hospital premises were inspected with a view to identifying possible evidence of subsidence damage, such as hairline cracks in walls, sticking windows and doors, sloping or uneven floors, evidence of access of the weather into the structure, and problems from service pipes and offgrade drains. Particu-lar attention was paid to the oldest and tallest buildings. Discussions were held with the chief maintenance engineer who had been employed at the hospital for 23 years, to determine whether any structural repairs which had been required could be attributed to subsidence damage. Surrounding property was also ex-amined from the exterior. However, the poor state of repair of many buildings, and the steeply sloping streets with uneven sidewalks made it difficult to differentiate between damage which may have been due to subsidence and that due to natural causes. It was not possible to examine some of the buildings, which had been re-ported in the press over the previous 25 years as being affected by sudden surface collapse, since they had been demolished. It should be noted, however, that they were not located in the immediate site area.

Examination of mine workings
A visit was made to the old mine workings in order to examine the underground conditions, to attempt to evaluate the pillar support competency, and to observe and assess the flushing operations. Access was gained to the old mines through a small shaft situated about 0.8 kilometer (0.5 mile) southeast of the site. At this point, due to seam inclination, the seams were present under approximately 46 meters (150 feet) of vertical cover. Conditions were examined in the No. 2 Dun-more, No. 1 Dunmore, Top Split of the No. 1 Dunmore, and Clark seams. Obser-vations were principally directed to the No. 2 Dunmore since this seam had been identified in the preliminary analysis as being of major concern.

The No. 2 Dunmore Seam had been more extensively extracted in the area of observation than it had been under the site as indicated by the mine plans. In an extraction averaging 1.8 meters (6 feet) in height, entry widths were commonly in the order of 9 to 11 meters (30 to 35 feet) as shown on the plans, in the area of observation. Due to waste material being hand packed around the pillars, it was not always possible to directly measure pillar and entry dimensions. The seam appeared to be of a fairly uniform character and extremely competent. Pillars were standing well and both roof and floor conditions were excellent. Support was provided spasmodically by wooden props, many of which had rotted away. Particular attention was paid to an old caving line beyond which the pillars had been robbed and which was inaccessible due to the roof collapse of massive, coarse-grained sandstone. Adjacent to the caving, some pillars were in an advanced state of stress and spalling of the pillar sides was occurring. Block samples of the seam were taken for analysis and laboratory testing.

The No. 1 Dunmore Seam was being control-flushed at a point some 366 meters (1,200 feet) from the shaft access. Mining refuse, which had been previously crushed to minus 12.7 millimeters (0.5 inch), was used as the fill. It was placed into position through 152-millimeter- (6-inch-) diameter pipes, so as to fill the void up to the roof as far as practicable. Because the seam was only 0.9 to 1.2 meters (3 to 4 feet) thick, extensive hand packs had been built along the sides of the main entries, using the floor stone which had been taken up to provide sufficient headroom for transport.

The Top Split of the No. 1 Dunmore and the Clark seams were not investigated beyond a small distance from the access shaft. From these limited observations the Clark Seam appeared to be more irregular in thickness than the other seams, with a more irregular roof and less well-defined partings.

Laboratory Testing

Core samples taken from boreholes Nos. 5 and 6 were sampled and tested in the laboratory. Tests were carried out on each of the three main rock types—sandstone/siltstone, shale, and anthracite—to determine their elastic properties and their uniaxial compressive strengths.

When it became apparent that the strength of the pillars in the No. 2 Dunmore might be critical, bulk samples of anthracite were collected from this seam during the mine visit. These samples were cored and tests were carried out to determine the elastic moduli and strength properties of this seam specifically.

The details of the individual test results are summarized in Table 18.2, which shows the competent nature of the strata beneath the hospital extension. The uniaxial compressive strengths of the sandstone, siltstone, and shale are relatively high.

Even the anthracite is strong. The No. 2 Dunmore anthracite appears to be stronger than anthracite from the upper seams based on four tests at a diameter/height ratio of 0.5. This general tendency, however, is confirmed by a total of 21 test samples at different diameter/height ratios. The individual results of strength versus diameter/height ratio for tests on the No. 2 Dunmore anthracite are shown in Figure 18.3.

Table 18.2. Summary of Rock Properties from Laboratory Testing

	Sandstone/ siltstone	Shale	Clark & upper seams	No. 2 Dunmore
Density (lb/cu ft)	166.4 (3)	170.4 (3)	96.8 (18)	99.0 (13)
Uniaxial strength (psi)	22,650 (3)	15,475 (3)	4,280 (2)	6,225 (2)
Young's Modulus (10^6 psi)	6.27 (3)	3.77 (3)	0.68 (6)	0.78 (3)
Poisson's Ratio	0.11 (3)	0.25 (3)	0.28 (6)	0.33 (3)

Note: Number of tests carried out is shown in parentheses in the table.

Stress Analysis

As a result of the site visits and discussions and after reviewing the laboratory data, it was concluded that pillar failure (as opposed to failure of the floor or roof) was the only likely mode of failure that needed to be considered.

Consequently, it was necessary to determine the stresses in the pillars. An elastic analysis was considered appropriate, since the condition of the pillars indicated that they were on the elastic portion of the load-deformation curve, and this was performed using an analog type of solution.

The mine simulation analog

An electrical resistance network analog consists of a large orthogonal array of electrical resistors simulating a "semi-infinite" prism of conducting material. One face of the network represents the seam level, which is a plane of symmetry in the physical problem, and the other faces represent infinite boundaries of the mass. Each junction of resistors in the network corresponds to an element of conducting material, and the potentials at the junctions satisfy the finite difference form of Laplace's equation (Jaeger and Cook, 1969).

The design of the University of Minnesota electrical resistance network analog, which was used in this study, is based on that of the original South African analog (Cook et al., 1966) but was revised to simulate a considerably larger prism of conducting material by Crouch et al., (1974). In both cases, the face of the network representing the seam contains 30 × 50 uniformly spaced "measuring nodes," and the physical extent of the network is increased by varying the sizes of the elements (and the values of the resistors) in the array. The University of Minnesota network is effectively 94 × 114 elements wide by 128 elements deep (Figure 18.4).

Each node in the central area can be connected, via a phone plug, to the ground plate, and groups of peripheral nodes likewise can be grounded by closing a switch. Thus, a seam can be modeled as being linearly elastic by connecting fixed resistances between the network nodes (including the periphery) and the grounding plate. In this case it is necessary to know the elastic properties of the seam and host material and the seam thickness. Then a suitable plug resistance can be chosen to model the problem.

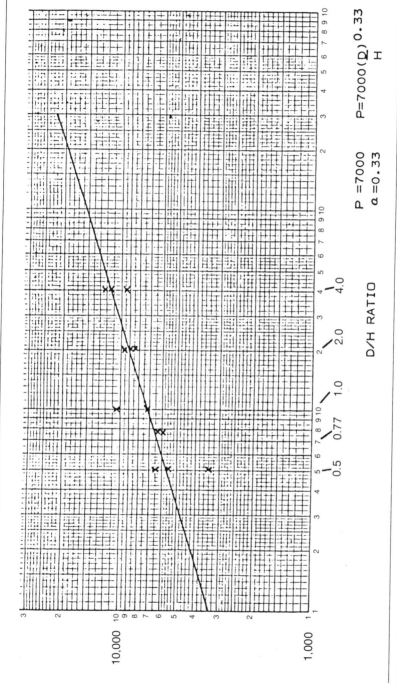

$$P = 7000 \left(\frac{D}{H}\right)^{0.33}$$

$P = 7000$

$\alpha = 0.33$

D/H RATIO

Figure 18.3. Strength versus D/H ratio for No. 2 Dunmore Seam.

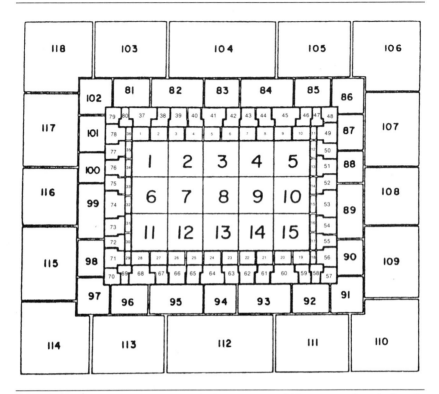

Figure 18.4. Diagram of front panel of University of Minnesota electrical resistance network analog.

The analog can also be used to model nonlinear seam behavior in order to take into account practically any type of inelastic seam behavior (e.g., failure of coal pillars or time-dependent deformation) or artificial support characteristics. This feature was not used, however, in the present study.

The analog is connected up to a RAMTEK color television screen. The stresses in the unmined portions of the seam may be displayed in such a way that different stress levels are represented by different colors. Sixteen stress level ranges can be represented in this manner. In addition, the precise stress level at each node may be found by direct measurement.

Preparation of data

A scale of one plug equivalent to an area 3.05 meters (10 feet) square was chosen to analyze all the seams. This enabled a detailed analysis to be performed on an area 152 × 91 meters (500 × 300 feet) beneath the hospital extension. To take account of the boundary effects an area 344 × 244 meters (1,130 × 800 feet) was actually modeled using the 118 peripheral switches. Using this system the 1,500 centrally located nodes were unaffected by conditions at the network boundaries.

Prior to setting up the underground geometry on the analog board, the hospital addition was superimposed on the pillar plan on each level and the plan was enlarged to the correct scale. The pillars were then traced off at this scale. The grid shown in Figure 18.5 was superimposed on this tracing using a light table, and the grid squares representing pillars were filled in with a pen. In most cases it was obvious whether a grid square represented a pillar or a void, but where the pillar boundary went through the middle of a square, the whole square was assigned as either a pillar or a void according to the relative areas of each in the square. The resulting grid map for the Clark Seam is shown in Figure 18.5. The grid map also shows the relative position of the hospital addition in relation to the pillars. The hospital addition was represented at the same scale and relative position on the grid map in the analysis of all the seams.

The vertical sections on either side of each seam were studied, and appropriate properties for the surrounding host material were chosen. For instance, if the roof and floor of the seam were predominantly sandstone or siltstone, a Young's Modulus approaching 3.5×10^5 kilograms per square centimeter (5×10^6 psi) was chosen for the host material. Where shale dominated, a value closer to 2.1×10^5 kilograms per square centimeter (3×10^6 psi) was used, and where coal seams or filling material were present in the vicinity, values closer to 7×10^4 kilograms per square centimeter (1×10^6 psi) were selected for the host material. A Poisson's Ratio of 0.11 was generally used for the host material.

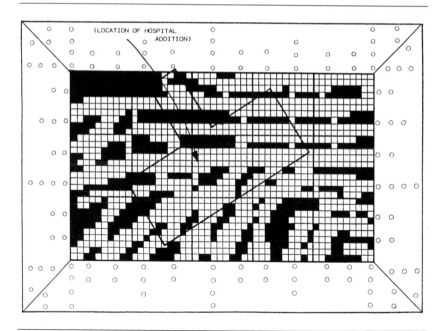

Figure 18.5. Grid map of initial pillar configuration in the Clark Seam. The relative position of the hospital extension is shown.

Based on the average laboratory results for the coal the elastic properties for the coal seams were always taken as:

- Young's Modulus $= 4.8 \times 10^4$ kilograms per square centimeter $(0.68 \times 10^6$ psi)
- Poisson's Ratio $= 0.30$

Given the above information and knowing the average height of each seam and the depth below surface, the resistance of the phone plugs was determined by calculation from the analog formula. Generally, either 1,000- or 5,000-ohm plugs were chosen to model each problem.

All the input data was entered onto a set of data sheets prior to setting up and running each problem.

When the pillar geometry was set up on the analog board and the overburden stress applied, the stresses acting on each coal area could be seen, color-coded on the TV screen. In all the analyses the extreme red was set to indicate a stress of 70 kilograms per square centimeter (1,000 psi) or greater, and each of the 16 colors was set to represent a stress range of 4.7 kilograms per square centimeter (66.6 psi); hence, a band of three colors represented a range of 14.1 kilograms per square centimeter (200 psi). (Since Figures 18.6 and 18.7 are black and white reproductions of the color-coded stresses on the TV screen, stress levels cannot be read directly from these figures.)

Figure 18.6. Photograph of pillar configuration and stress pattern in the Clark Seam as shown on TV screen.

In order to check the effect of pillar removal (either due to inaccurate survey, pillar robbing, or pillar collapse) on the stress distribution, pillars having an area less than 2.8 square meters (30 square feet) were removed. The resulting pillar configuration and stress pattern were also recorded digitally and on a colored photograph of the TV screen. Generally, very little increase in pillar stresses was indicated.

As a result of these simulations it was shown, for example, that the relative level of stresses in the Clark Seam is low (Figure 18.6) and that a reasonable amount of pillar removal would not have much effect on these stress levels.

A similar type of analysis was carried out for the Top Split, No. 1 Dunmore, No. 2 Dunmore, and No. 3 Dunmore seams.

The results of analyzing the base case of the No. 2 Dunmore are shown in Figure 18.7. Due to the greater depth, which is equivalent to an overburden stress level of 22.3 kilograms per square centimeter (317 psi), and the smaller pillar sizes, the pillar stresses are generally above 70 kilograms per square centimeter (1,000 psi). The maximum stress, averaged over an area of 3.05 × 3.05 meters (10 × 10 feet), was found to be 80.6 kilograms per square centimeter (1,147 psi).

The analog was also used to obtain closures. Closure may be defined as the convergence between the roof and the floor as a result of the excavation. The closures obtained were generally very small and reached a maximum of about 12.7 millimeters (0.5 inches).

Figure 18.7. Photograph of pillar configuration and stress pattern in the No. 2 Dunmore Seam as shown on TV screen.

Area of pillar collapse No. 2 Dunmore

During the underground visit, an inspection was made of an area southeast of the existing hospital, where a considerable number of pillars had been removed and caving had been permitted. One of the pillars showed signs of distress, but this may have been in part due to the pillar robbing operation. Since the pillars were smaller at this location than under the hospital (as shown by the mine plans) a decision was made to model this area on the analog.

From this exercise, a pillar strength of at least 120 kilograms per square centimeter (1,700 psi) was indicated from the intact pillars that were showing no signs of distress for any reason.

Analysis of vertical interaction effects

In order to check the vertical interaction effects between the seams, a displacement discontinuity analysis was run, and a check was made on the elastic stresses and displacements obtained from the analog solution. The vertical idealized section that was used is shown on the left-hand side of Figure 18.1.

The analysis showed that the vertical interaction effects between the seams could be largely ignored. The calculated stresses and displacements agreed with those from the two-dimensional analysis of horizontal seams obtained from the analog solution.

Design stresses in each seam

Taking into account the vertical and horizontal interaction effects between pillars, the maximum pillar stresses encountered in each seam, averaged out over an area of 28 square meters (300 square feet) were approximately:

Clark	46 kg/cm² (650 psi)
Top Split	56 kg/cm² (800 psi)
No. 1 Dunmore	67 kg/cm² (950 psi)
No. 2 Dunmore	79 kg/cm² (1,125 psi)
No. 3 Dunmore	86 kg/cm² (1,225 psi)

Because of the large surrounding abutments of unmined coal, the stresses in the No. 3 Dunmore pillars were not considered to be as serious as the stresses which were encountered in the No. 2 Dunmore, No. 1 Dunmore, and the Top Split pillars.

As mentioned previously, pillars were removed to simulate pillar collapse or previous mining in order to see how the pillar stresses might be affected. Removal of a moderate number of pillars, however, did not make a large difference to the level of stresses determined from the mine plan configurations.

Pillar Strengths

In order for coal pillars to be effective supports, the strength of each pillar must be greater than the stress to which it is subjected. While the stress in a pillar can be determined reasonably easily, the estimation of pillar strength is more difficult.

The variability of coal, both laterally and vertically, is one of the problems. For instance, coal usually contains three sets of discontinuities or weakness planes approximately normal to each other; these are usually referred to as face cleavage, butt cleavage, and bedding planes. In addition, the strength of specimens varies greatly depending on their size and shape. Because of these factors, among others, it is necessary to test many specimens and to resort to statistical procedures to determine pillar strengths.

The effects of specimen size and shape

Due to the effects of the discontinuities (such as cleavage and joints), laboratory tests are not usually representative of the behavior of a mine pillar. The reason is that a small specimen is usually a relatively intact continuous structure, and the smaller the specimen the fewer the discontinuities and, hence, the greater the strength. Conversely, as the specimen size increases, the strength of the specimen decreases, since the chances of encountering a critically oriented discontinuity increases.

Denkhaus, 1962, observed that the strength of metals is constant from a specimen size of 20 millimeters (0.8 inch) and upwards, so that large engineering structures, such as bridges and ships may be designed on the basis of strength tests on small laboratory samples. He predicted that the critical sample size for coal would be much larger. It was later shown by Bieniawski, 1968, that the critical specimen size for coal is about 1.5 meters (5 feet). Bieniawski's curve of strength against specimen size is shown in Figure 18.8. It therefore follows that there

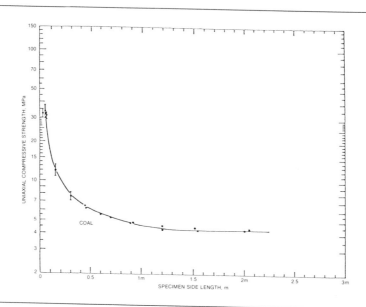

Figure 18.8. The phenomenon of the strength approaching asymptotically a constant value (after Bieniawski).

cannot be a reliable method of predicting pillar strength from tests on small laboratory specimens and that in situ tests on large specimens are required. In situ testing is time-consuming and expensive, however, and was not within the scope of the present study. Consequently estimates from laboratory tests and underground observations had to be made.

The ratio of the diameter or width (W) of the specimen to the height (H) of the specimen also affects strength. Low, fat pillars or specimens (high W/H ratios) are stronger than tall, slender pillars (low W/H ratios) of the same material. Most of the empirical formulas for estimating pillar strengths are based on W/H ratios of 1, that is, where the diameter or width is equal to the height of the specimen. The actual pillar dimensions can then be substituted into the formulas to correctly account for the shape effects.

Since the working height of the Dunmore seams is typically around 1.5 meters (5 feet), the size and shape effects can be summarized as follows: the strength of coal specimens tested at W/H ratios of 1 can be expected to decrease as the sizes of the specimens increase up to the size of a 1.5-meter (5-foot) cube. Above this size no decrease in the strength is anticipated. However, the strength of coal pillars in the mine can be expected to increase from this minimum as the pillar widths increase; this is because the pillar height is constant while the width is increasing (that is, W/H is increasing).

Pillar strength formulas
A large number of investigators have conducted laboratory tests on small coal specimens of various shapes. These tests were carefully reviewed, as were the pillar strength formulas derived by various workers.

Considerable weight was given to the work performed by Van Heerden, 1975, in South Africa, since he has carried out tests on in situ coal specimens similar to those of his colleague, Bieniawski, and appears to have appreciated the size effects better than most workers. Recently, he also made a comparison of the various pillar strength formulas. In doing this he found that the data from each colliery could be reasonably represented by a straight line and that when all the data obtained from the collieries was plotted in a dimensionless form, a single straight line could be fitted. The equation of this line is:

$$\frac{P_p}{P_s} = \left[0.64 + 0.36 \; \frac{W}{H}\right] \tag{1}$$

where P_p =   strength of pillar
$\quad\;\; P_s$ =   strength of specimen (assuming the dimensions of a 1.5-m [5-ft] cube)
$\quad\;\; W$ =   pillar width
$\quad\;\; H$ =   pillar height

Calculation of pillar strengths
Since large-scale in situ testing was not within the scope of the present study, it was necessary to estimate the strength of a 1.5-meter (5-foot) cube from the laboratory tests. While this is far from ideal, it was the best that could be done

under the circumstances. The calculation was performed using a procedure outlined by Holland and Gaddy, 1957, in which the relationship between the strength of a cubical specimen and its size is expressed by the following equation:

$$S_S = \frac{K}{\sqrt{D}} \qquad (2)$$

where S_S = strength of a cubical specimen in kg/cm^2 (psi)
D = edge dimension of specimen in cm (in.)
K = coefficient depending upon the characteristics of the coal tested —units are in kg/cm$^{3/2}$ power (lb./in.$^{3/2}$)

Note that the equation has a similar shape to Bieniawski's empirical curve shown in Figure 18.8.

Applying Equation 2 to the test results of 492 kilograms per square centimeter (7,000 psi) for a 5.08-centimeter (2-inch) cube in the case of the No. 2 Dunmore Seam, $K = S_S\sqrt{D} = 492\sqrt{5.08} = 1,108$ kilograms per centimeter$^{3/2}$ (9,900 pounds per inch$^{3/2}$). When this K value was substituted into Equation 2, for a 1.52-meter (5-foot) cube the strength was calculated as follows:

$$S_S = \frac{K}{\sqrt{D}} = \frac{1108}{\sqrt{152}} = 89.9 \text{ kilograms per square centimeter } (1,280 \text{ psi}).$$

In a similar manner, a strength of 77 kilograms per square centimeter (1,095 psi) was calculated for the Clark and upper seams.

These strength values were used in the Van Heerden formulas using actual pillar dimensions for 60 pillars beneath the site, and the strength of each pillar was independently calculated. (Independent checks were also made using various other approaches and other pillar formulas).

Evaluation of Pillar Safety

Safety analysis

The ratio of actual stresses in the pillar to failure stress (or strength) is known as the factor of safety for that pillar, i.e.,

$$\frac{\text{stress required to cause pillar failure (strength)}}{\text{stress acting on pillar}} = \text{Factor of safety } (F)$$

Sixty pillars in the three most critical seams beneath the site were analyzed in detail. The average and minimum factors of safety obtained were:

Seam	No. of pillars analyzed	Average F	Minimum F
Clark	18	2.35	1.79
No. 1 Dunmore	18	3.85	2.84
No. 2 Dunmore	24	2.71	1.90

Note that the factor of safety chosen as being acceptable in a particular situation depends on the relative importance of the problem and the consequences should failure occur. The factor of safety is also to some extent a factor of "ignorance" where the available data base is unreliable or inadequate. When ignorance is high a higher factor of safety must be used to produce a safe design.

Note that there is also a tendency for the pillar strength to decrease with time due to deterioration and weathering, so that the factor of safety generally decreases as time passes. Sometimes the stresses also change with time, especially if mining operations are being carried out in the immediate vicinity.

Discussion and Conclusions

The computer modeling and stress analysis performed were based on the dimensions and configurations of pillars as shown on available mine plans. The underground visit to confirm the accuracy of the plans was conducted about 0.8 kilometer (0.5 mile) southeast of the site. At that location, the relative positions and dimensions of the pillars as shown on the plans were found to be nominally correct. It was assumed, therefore, that the underground conditions beneath the site were similar to those encountered during the underground inspection visit.

Consequently, based on the information that was available, it was concluded that there was no immediate danger of pillar collapse beneath the hospital extension. An adequate factor of safety for the pillars under conditions of static loading was indicated for all the seams. Therefore, no major surface subsidence was anticipated.

However, it must be stated that it is impossible to estimate the reduction in pillar strength resulting from time deterioration and possible deterioration by water.

Although the analyses, as well as the historical record, did not indicate that a subsidence problem existed, "flushing" was recommended to guard against possible future deterioration of the pillars at the site. It was considered that filling the voids of the mined-out areas, would be more beneficial in protecting the existing pillars from spalling than in providing meaningful support of the roof rock. It would tend to minimize rather than prevent roof failure as the flush material could not actively provide support until some compaction had taken place.

Since the investigation was carried out, an extensive flushing operation has been completed at the site.

Acknowledgments

The study was performed for John Kenney of the Perkins and Will Partnership, whose guidance and support are appreciated.

Two of the author's colleagues at Dames & Moore who played a major role in this study were Derek Steele, who contributed his mining and subsidence experience, and Bernard Archer, who contributed to the engineering geology and provided day-to-day management of the project.

At the University of Minnesota, where the stress analysis was performed, contributions to the study were provided by Dr. M. Hardy, M. Christianson, and M. Voegele.

The author would also like to acknowledge the help given by a number of people in the Scranton area, in particular the U.S. Bureau of Mines personnel at Wilkes-Barre, especially C. Kuebler, I. Williams, and C. Sauer.

References

Bieniawski, Z. T., 1968, "In Situ Strength and Deformation Characteristics of Coal." Engng Geol., vol. 2.

Cook, N. G. W., Hoek, E., Pretorius, J. P. G., Ortlepp, W. D., and Salamon, M. D. G., 1966, "Rock Mechanics Applied to the Study of Rockbursts." J. S. Afr. Inst. Min. Metall., 66.

Crouch, S. L., Hardy, M. P., Sinha, K. P. and Fairhurst, C., 1974, "A Computer-controlled Electrical Resistance Network Analog for Rational Mine Design." USBM Grant-Report (GO122107), Univ. of Minnesota.

Denkhaus, H. G., 1962, "A Critical Review of the Present State of Scientific Knowledge Related to the Strength of Mine Pillars." J. S. Afr. Inst. Min. Metall., vol. 63.

Holland, C. T. and Gaddy, F. L., 1957, "Some Aspects of Permanent Support of Overburden on Coal Beds." Proceedings, West Virginia Coal Mining Institute.

Jaeger, J. C. and Cook, N. G. W., 1969, "Fundamentals of Rock Mechanics," Methuen and Co. Ltd., London.

Van Heerden, W. L., 1975, "In Situ Complete Stress-Strain Characteristics of Large Coal Specimens." J. S. Afr. Inst. Min. Metall.

Strata Control by North American Coal Corporation in the Pittsburgh No. 8 Coal Seam, Ohio

19

Michael S. Roscoe, Division Geologist,
The North American Coal Corporation,
Powhatan Point, Ohio, USA

Introduction

North American Coal Corporation's Central Division is currently achieving maximum extraction in room and pillar mining of the Pittsburgh No. 8 Coal seam while maintaining the needed strata stability for safe and continuous operation. Since the regional geologic conditions do not permit total recovery of pillars in room and pillar panels, the percentage of extraction in these areas must be watched closely. In the past, increasing the extraction percentage past certain limits has, at times, caused extremely dangerous and costly squeezes.

It is the purpose of this chapter to present a general overview of the Central Division's operations and mining methods and the way in which the above problem in strata stabilization was approached and solved. Two studies by C. T. Holland (1957, 1973) concerning this problem are presented along with their practical application. Roof control problems relating to geology as well as methods employed to obtain long-term stability of roof, floor, and ribs are also discussed. Future plans to bring together existing data from different fields of research so that hazard zones can be better delineated are outlined. A brief outline of a longwall subsidence study is also included. It is hoped that the information presented here can be of help to those facing similar problems in similar operations.

Geography and Statistics

North American Coal Corporation's Central Division in southeast Ohio is comprised of two subsidiary companies: Quarto Mining Company and The Nacco Mining Company, and three parent company mines. The majority of the division's properties (six of seven) are located in Belmont and Monroe Counties along the Ohio River (Figure 19.1). Situated along the western edge of the unglaciated Allegheny Plateau, the local terrain is quite steep with a relief of 122 to 153 meters (400 to 500 feet) being common. Regional elevations range from 186 meters (610 feet) on the Ohio River to over 427 meters (1,400 feet) to the west. In this setting, North American Coal Corporation either owns or has under lease 48,564 hectares (120,000 acres) of Pittsburgh No. 8 Coal. Approximately 32,376 hectares (80,000 acres) are in remaining reserves, representing some 360 million tons of high-quality steam coal.

Production from these reserves comes from six mines (five shafts and one drift), ranging in operational age from 5 to 70 years. Tonnages are produced from two longwall units, 58 continuous miner units, and four conventional sections and total output is in excess of 6 million tons per year.

All coal produced within the division must be cleaned prior to rail or barge shipment to electric utilities. Seven preparation facilities range in age from 3 to 70 years and incorporate gravity separation and mechanical drying methods, with two of the preparation plants having additional thermal drying units. The cleaned product contains 10 to 18 percent ash, 5.5 to 8 percent moisture, 2.8 to 3.9 percent sulfur, and has an 11,000 to 12,100 Btu content.

Geologic Data

The Pittsburgh No. 8 Coal seam is Upper Pennsylvanian in age and defines the lower boundary of the Monongahela Group. The coal developed along the

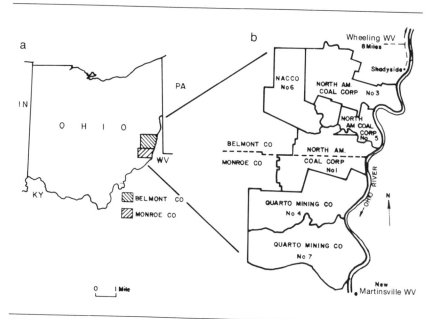

Figure 19.1. Map showing (a) location of Central Division's main property and (b) further breakdown into individual mines and subsidiary companies.

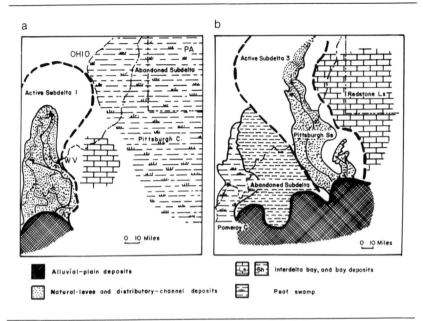

a   b

Figure 19.2. Paleogeographic maps of early Monongahela times. (a) Approximate regional boundaries of Pittsburgh Coal deposition, and (b) generalized regional geology of overlying strata (Donaldson, 1973).

fringes of the Dunkard Basin and extends over much of southwestern Pennsylvania, northern West Virginia and southeastern Ohio (Figure 19.2). North American Coal Corporation's properties are situated along the southwestern edge of this basin, and coal thicknesses in this area range from 0.91 to 3.05 meters (3 to 10 feet) averaging approximately 1.68 meters (5.5 feet) (including partings). Regional dip is to the southeast, amounting to between 3.05 and 12.2 meters (10 and 40 feet) per mile and is generally consistent with very few structural irregularities. Elevation of the coal to the south is approximately 107 meters (350 feet) rising to 244 meters (800 feet) along the northern edge of the properties where it begins to outcrop.

The seam is made up of a main bench coal 1.2 to 1.5 meters (4 to 5 feet) thick, containing several thin and variable partings, and a rider coal 152 to 305 millimeters (6 to 12 inches) thick which is separated from the main bench coal by a parting 25 to 203 millimeters (1 to 8 inches) thick that is consistent both in occurrence and position throughout the area. In general both the rider and main bench coals thin and become more variable to the south, reaching 0.61 meter (2 feet) or less immediately to the south and southwest of North American's present holdings. The rider coal grades laterally into a coaly to carbonaceous shale in the southern portion of the properties. The only interruption to these regional trends is the occurrence of clay veins and faults which are restricted to certain portions of the properties.

The strata immediately above the rider coal varies from a bedded to non-bedded calcareous mudstone-claystone—2.4 to 3.7 meters (8 to 12 feet) thick—in the north to the Pittsburgh Sandstone—3.05 to 12.2 meters (10 to 40 feet) thick—in the southwest (Figure 19.3). Overlying this unit is the Redstone Limestone 3.05 to 5.5 meters (10 to 18 feet) thick, a bedded argillaceous limestone which thins and rises toward the area of sandstone deposition. Above the

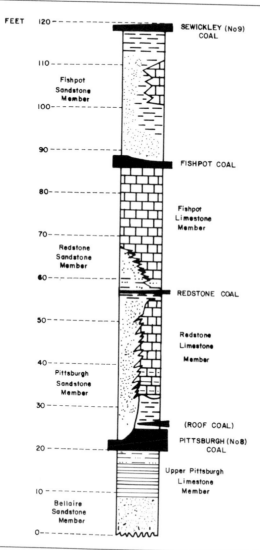

Figure 19.3. Generalized stratigraphic column of a portion of the Monongahela Group.

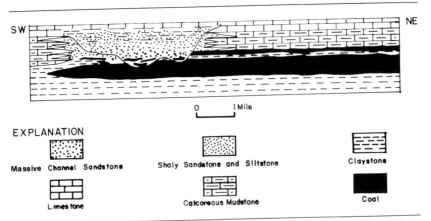

EXPLANATION

Massive Channel Sandstone

Shaly Sandstone and Siltstone

Claystone

Limestone

Calcareous Mudstone

Coal

Figure 19.4. Geologic cross section, showing lateral gradation from bay to deltaic deposition.

Redstone Limestone, this sequence (cyclothem) is repeated with the Redstone Coal and associated strata. Successively higher strata up to 244 meters (800 feet) thick which comprise the remainder of the Monongahela Group and the lowermost portion of the Dunkard Group, do not differ much from this generalization. The floor, a compact silty mudstone 0.3 to 0.91 meter (1 to 3 feet) thick, is consistent over the entire property and is underlain by a hard calcareous siltstone.

The variation in "roof" lithology cited above represents a change in the depositional environment from bay to deltaic channel deposition (Figure 19.4). General channel direction is northwest to southeast with transport to the northwest (Donaldson, 1969). Complete erosional cutouts associated with channeling are rare, but partial scouring of the coal increases in the extreme southwest corner of the properties. Roof control problems relating to this lithologic transition are discussed under the general heading of roof control.

Mine Design

Controlling factors

Basic mine design and development in the Pittsburgh No. 8 Coal seam in southeast Ohio are controlled in varying degrees by the following existing conditions:

1. Types of overlying strata.
2. Geographic locations.
3. Cleat direction.
4. Regional structure.

Since the immediate roof generally consists of bedded to nonbedded, occasionally highly slickensided mudstone-claystone, 2.4 to 3.7 meters (8 to 12 feet) thick with very low supportive qualities, approximately 152.4 millimeters (6 inches) of top coal (roof coal) must be left to protect this unit from moisture and deterioration. In areas where the rider is present, this simply means mining up to the bottom of the rider. But in areas where the rider is absent or dirty, the protective coal becomes the topmost portion of the main bench. In this situation, the determination of the exact thickness of roof coal being left during mining is difficult, and extremes in either direction are costly. The strength of the overlying Redstone Limestone, 3.05 to 5.5 meters (10 to 18 feet) thick, further dictates that complete extraction using secondary recovery methods in room and pillar panels should not be attempted. Pressure overrides, due to the poor caving characteristics of this limestone, prohibit this method of maximum reserve extraction. Both factors, roof coal and partial extraction, effectively reduce available reserves of the division while continually hampering rates of production as well as percentages of extraction.

Geographic location, cleat direction, and regional structure, play minor roles in mine design but impose the following restrictions. Proximity to the Ohio River (transportation access) limits initial development in three directions (north, south, and west), while cleat directions (face: north 75° ± west and butt: north 15° ± east) control to a minor degree the ease of extraction and the direction of room and pillar panel development. Regional dip comes into play if and when pumping of mine water is required and also during transportation considerations. Otherwise, local variations in the general structure (steepening dip) have only rarely controlled direction and/or type of development.

General layout

The general layout of the Central Division's mines incorporates a system of mains and submains, the design of which has changed with increasing experience and knowledge. The mains of the Division's newer mines reflect this change. In these mines, the mains (east to west) are made up of three independent groups of four entries separated by two abutment barrier pillars measuring 61 by 762 meters (200 by 2,500 feet) of pressure arch yield pillar design. The outside two groups are generally used for intake and return air courses while the middle group serves as the service entries (Figure 19.5). Service entries need to be developed and protected so that they will remain open during the life of the mine. Air courses, on the other hand, have to function less than 10 years since air shafts will be sunk at regular intervals during main development and mine advancement. One-way track haulage (two separate lines) is planned for Quarto's No. 7 mine.

Submains (north to south) intersect the mains at intervals of either 762 meters (2,500 feet) as in room and pillar panel development, or 1,524 to 1,829 meters (5,000 to 6,000 feet) as in longwall development. They consist of one group of seven to nine entries, three of which are used for intake, three for return; the remaining entries are used for service. This submain system is designed assuming dead loading with centers based directly on amounts of overload with a safety factor of 2.

Room and pillar panel entries (east to west) intersect the submains at intervals of approximately 305 meters (1,000 feet) and are generally four or five in number. Longwall entries intersect at intervals of 183 meters (600 feet) and

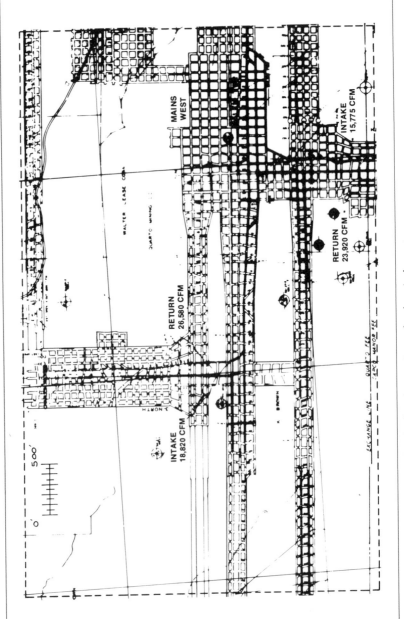

Figure 19.5. Mine plan of Quarto No. 4, showing main and submain layout.

there are currently three in number. Roof bolting in room and pillar work and in main and submain development is carried out on standard 1.22-meter (4-foot) centers. The types (resin and mechanical) and lengths (1.52 to 3.05 meters [5 to 10 feet]) of bolts that are used vary from mine to mine with selection depending upon the local geologic and geographic conditions.

Since the basic mining occurs in room and pillar panels, the economics of extraction depend upon the approach and design in these areas. Because of this, the extraction and design of room and pillar panels is continually changing to meet variations in local conditions and advances in theory relating to pillar, floor, and roof strengths. A discussion of these adjustments follows.

Squeezes and Pillar Design

The position and thickness of the overlying Redstone Limestone have prevented total extraction in room and pillar panel areas and, because of this, there is a tendency to leave insufficient amounts of coal for protection. Dimensions of pillars and entries (percentages of extraction) that will maintain stability of the roof and floor must therefore be watched very closely. Further complicating the situation is the fact that amounts of cover can vary by as much as 152 meters (500 feet) in as little as 305 meters (1,000 feet). Because of this variation, stable conditions in one area of a panel might not exist for the remainder of that panel or even into adjacent identically developed panels. Squeezes that were caused by this variation have occurred in sections that were thought to have sufficient coal left. The following discussions on two of these squeezes outline the history of North American's efforts at their prevention.

Powhatan No. 1 mine squeeze

The first of these squeezes occurred at the Powhatan No. 1 mine in April 1957 and has been selected for discussion because the study which followed represents North American's first attempt at designing a pillar and entry system that would balance stability requirements with maximum extraction. The squeeze occurred in a section undergoing full retreat with an overburden cover of 140 to 195 meters (460 to 640 feet) near an area that had experienced a minor squeeze the previous year. Pillar dimensions in the affected area measured 12.2 × 4.6 meters (40 × 15 feet) with 5.2-meter (17-foot) entries. Larger pillars were left at varying intervals as a method of protection (Figure 19.6). The extraction percentage of this configuration was 66 percent omitting the large pillars and 61 percent including them. A subsequent study by C. T. Holland showed that the squeeze resulted from the leaving of insufficient coal and that the larger pillars were not large enough to prevent or slow the progress of the squeeze once it had started. Spacing of these larger pillars might even have shared in the squeeze.

Parameters for various pillar and entry systems were then established relating coal height and compressive strength to amount of cover. Compressive strength of the Pittsburgh No. 8 Coal, a moderately strong coal, was determined in the laboratory using 76.2-millimeter (3-inch) cubes and then extrapolating to larger sizes. Some of these specifications can be seen in Tables 19.1 and 19.2. In these specifications, the maximum amount of cover for a panel is assumed to extend

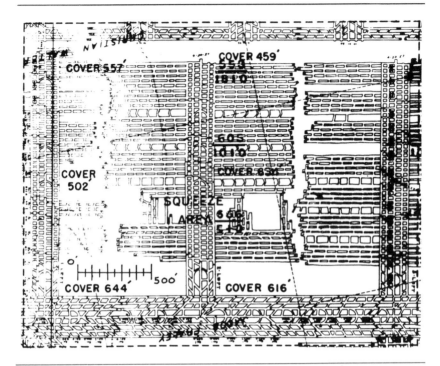

Figure 19.6. Mine plan of Powhatan No. 1 mine section 2 Left off F South, showing pillar size and configuration, and area affected by squeeze.

Table 19.1. Estimated Strength of Coal Pillars

Pillar thickness (ft)	Pillar least width (ft)	Room width (ft)	Estimated pillar strength (psi)
7½	23½	19½	1,500
7½	21	17½	1,420
7½	26½	19½	1,600
7½	24	17½	1,520
7½	29½	19½	1,710
7½	27	17½	1,620
7½	32½	19½	1,770
7½	30	17½	1,700
7½	37	19½	1,890
7½	34½	17½	1,825

Source: C.T. Holland, 1957.

Table 19.2. Average Unit Load on Pillars at Various Depths and Percentages of Extraction

Percent of coal recovered	64	63	62	61	60	59	58	57	56	55	54	53	52	51	50	49	48
Thickness of cover (ft)							*Average load on pillars (psi)*										
300	830	810	790	770	750												
400				1,030	1,000	975	950										
500					1,250	1,220	1,190	1,160	1,130								
600								1,400	1,360	1,330	1,300	1,280					
700											1,520	1,490	1,460	1,400			
800													1,660	1,630	1,600	1,570	1,520

Source: C. T. Holland, 1957.

over the whole panel thereby creating a margin of safety. Each future room and pillar panel was to be designed using these specifications, and subsequent mining generally followed these guidelines with only minor changes occurring in the basic pattern of the room and pillar panel. It should be noted, however, that Holland's specifications were not strictly followed, and the tendency to overmine an area still persisted.

Squeezes that occurred after Holland's 1957 investigation happened where anomalous situations existed or where the new specifications had not been followed. A more detailed evaluation of these squeezes is currently being carried out by members of the engineering staff.

Quarto's No. 4 mine squeeze

The second stage (1973) of pillar design was initiated during the planning of Quarto's No. 4 and No. 7 mines in response to a squeeze that occurred in a room and pillar panel in Quarto's No. 4 mine. C. T. Holland was again called in to determine the cause and establish new parameters that would allow for maximum safe extraction.

The squeeze under discussion occurred in a proposed 244 by 610 meters (800 by 2,000 feet) room and pillar panel 1.5-meter + (5-foot +) thick coal under cover

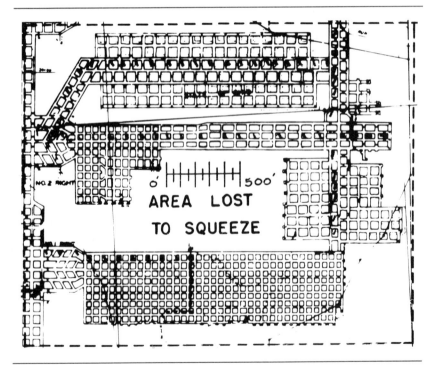

Figure 19.7. Mine plan of Quarto No. 4 mine section 1 Right off 1 North, showing area lost to squeeze, and configuration and size of remnant pillars.

of 122 to 230 meters (400 to 750 feet) of overburden (Figure 19.7). Mining procedures during this time involved driving four entries and developing rooms on the return side during advancement. The remaining rooms were then developed on the intake side during retreat. All entries were 5 meters (16 feet) wide, breakthroughs at 5.5 meters (18 feet) 15 × 15-meter (50 × 50-foot) centers. The resulting squeeze caused the loss of approximately half of the panel.

It was later determined that, instead of the 43 percent extraction that the above dimensions would indicate, a much higher percentage of from 57 to 65 percent had occurred. This was caused by driving breakthroughs wider than designed and rounding off corners for easier movement and transport. With this reduction in the resulting pillar size, loading increased to 141 to 282 kilograms per square centimeter (2,000 to 4,000 psi) while actual pillar strength was about 141 kilograms per square centimeter (2,000 psi).* Using past experience and an improved confidence in the "pillar formula," Holland developed a new set of specifications and proposed a different sequence of room and pillar development. One table of these specifications has been reproduced (Table 19.3). It should be noted that a high safety factor[†] is required since secondary recovery (slabbing) will reduce this number to the 1-2 range.

The proposed sequence involves driving four entries a distance of 762 meters (2,500 feet) and developing rooms in retreat. Slabbing, the mining of a single pass on two sides of a pillar, takes place in retreat from each room. The time required in room development with this method should not exceed four and a half to six months. Since changing over to this method, major squeezes have been averted and threats of their occurrence have been avoided by temporary adjustments in pillar size (Figure 19.8).

The engineering staff is presently reevaluating the history of mining conditions in relation to actual pillar and entry design in some of the division's older mines. It is hoped that the trends that are found to exist can be applied to a refinement of the existing design data. This will in turn be utilized in all future room and pillar panel design work. Preliminary data indicate that squeeze occurrences are related to method of panel development with full retreat appearing to be more susceptible than others.

The engineering staff is also looking into the possibilities of total extraction in room and pillar panels. It is believed that with the correct approach coupled with the proper designs and methods, total pillar extraction is possible. Initial efforts will most likely occur in those areas where the Redstone Limestone rises and thins.

Roof Support

Even when pillar and entry systems have been designed according to specification, problems in roof control continue to arise because of the three factors mentioned

---

* Pillar strength is determined using the Holland Gaddy Formula $S = (k\sqrt{L})/T$ where S = pillar strength, L = least lateral dimension of pillar in inches, T = thickness in inches and k = strength obtained in testing cubical specimens.
† The safety factor is determined by dividing the strength of the pillar by the load on the pillar with both quantities being in pounds per square inch.

Table 19.3. Revised Pillar and Entry Designs

Thickness of mined section (ft)	Room width (ft)	Before driving rooms & breakthrough (ft)	After driving rooms & breakthrough (ft)	Coal left unmined (%)	Load on pillars (psi)	Strength of pillars (in²)	Safety factor of pillars
				300 ft cover			
6	16	53 × 53	37 × 37	49	615	2,527	4.1
6	18	53 × 53	35 × 35	44	688	2,394	3.5
6	20	53 × 53	33 × 33	39	774	2,254	2.9
				400 ft cover			
6	16	58 × 58	42 × 42	52	783	2,603	3.3
6	18	58 × 58	40 × 40	48	841	2,501	3.0
6	20	58 × 58	38 × 38	43	932	2,367	2.5
				500 ft cover			
6	16	62 × 62	46 × 46	55	908	2,677	2.9
6	18	62 × 62	44 × 44	50	993	2,552	2.6
6	20	62 × 62	42 × 42	46	1,090		
				600 ft cover			
6	16	66 × 66	50 × 50	57	1,045	2,725	2.6
6	18	66 × 66	48 × 48	53	1,134	2,628	2.3
6	20	66 × 66	46 × 46	49	1,235	2,527	2.1
				700 ft cover			
6	16	70 × 70	54 × 54	60	1,176	2,552	2.6
6	18	70 × 70	52 × 52	55	1,268	2,603	2.1
6	20	70 × 70	50 × 50	51	1,372	2,552	1.9*

6	16	74 × 74	58 × 58	800 ft cover	61	1,302	2,819	2.2
6	18	74 × 74	56 × 56		57	1,396	1,715	2.0
6	20	74 × 74	54 × 54		53	1,502	2,652	1.8*
6	16	79 × 79	63 × 63	900 ft cover	64	1,415	2,888	2.0
6	18	79 × 79	61 × 61		60	1,509	2,796	1.9*
6	20	79 × 79	59 × 59		56	1,614	2,701	1.7*
6	16	86 × 86	70 × 70	1,000 ft cover	66	1,509	2,932	2.0
6	18	86 × 86	68 × 68		63	1,599	2,865	1.8*
6	20	86 × 86	66 × 66		61	1,698	2,819	1.8*

Source: C. T. Holland, 1973.
* Not recommended for use.

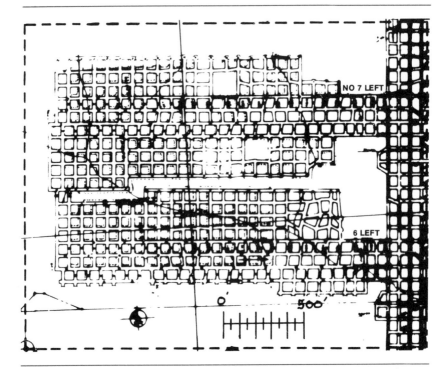

Figure 19.8. Mine plan of Quarto No. 4 mine section 7 Left off 1 North, showing minor pillar adjustment.

in the following list:

1. Changes in lithology.
2. Existence of fractures and/or joint sets.
3. Proximity to drainage.

Although not thoroughly understood, all three of these factors have caused and will continue to cause roof control problems within the corporations's Central Division.

As was stated earlier, lateral changes in the lithologic makeup of the roof occur to the southwest (mudstone-sandstone) (Figure 19.2*b*). In areas of immediate change from sandstones to silty mudstones, pressure differences due to differential compaction will cause instability in the roof and ribs. Poor cohesion between different types of strata will also cause problems when advancement reaches these areas. Effects of this type are common knowledge and to be expected in these areas, but North American Coal is presently experiencing roof control problems in areas of more subtle changes in roof geology.

Lateral gradation from the calcareous to the silty phase of the mudstone-claystone facies has created areas of lateral and vertical interfingering of these two similar rock types (Figure 19.9). In these areas, Quarto No. 7 mine has experienced a rash of recent roof falls which extend up to the more competent Redstone Limestone. Since intersections are more susceptible to these falls (longer spans), bolt pattern and lengths have been changed in intersections to compensate for this somewhat weaker strata. In this case, a complete switch to 3.05-meter (10-foot) resin bolts was initiated, to penetrate the limestone. Although this solution would appear obvious, many combinations were attempted before going over to complete 3.05-meter (10-foot) coverage. Preliminary results are encouraging, but more time is needed to fully evaluate this adjustment. The experience gained from the study of this geologic trend can and will be utilized as other areas of the mine advance into this zone.

The presence of fracture and joint sets is also an important factor causing unstable roof conditions. Very little work has been done to delineate these features, and their importance has only recently been realized. To improve this condition,

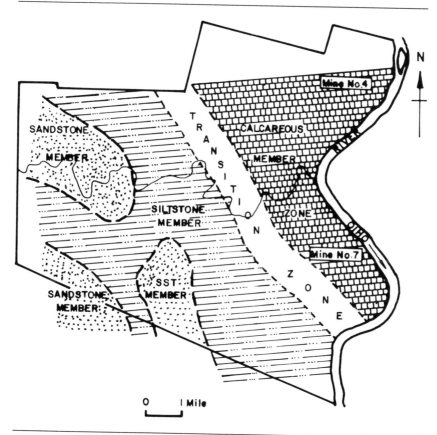

Figure 19.9. Lithofacies map of Quarto Mining Company's properties, showing variation in the geology of the immediate roof.

North American Coal, in cooperation with the U.S. Bureau of Mines (USBM) Technical Support Groups in Denver and Pittsburgh, is currently involved in the mapping of these linear fractures, using high-altitude photos and radar imagery. Preliminary indications are that this method holds great promise for delineating potentially hazardous zones. Although only specific areas have been looked at, using this technique, future plans include property wide coverage and evaluation.

Probably the most common areas of roof failure are those that are directly associated with drainage proximity or an extremely low amount of cover. In these areas, water, weathering, and complicated stresses due to unloading, combine to create poor roof conditions in strata that would have otherwise been stable. Experience has shown that the critical amount of cover needed to protect a certain type of roof varies drastically when other factors are present. In other words, any combination of the factors discussed above increases or magnifies their importance. A federally funded study is presently being conducted in the tri-state area relating all these factors to the occurrence of roof falls, and it is hoped that a computer analysis of the assembled data will provide information concerning existing but nonrecognizable trends.

Floor

Problems in floor stabilization within the Central Division's mines are rare, but have been encountered where the following two factors occur:

1. Excessive cover in conjunction with poor pillar and entry design.
2. Areas near large bodies of water (Ohio River abandoned mines).

In areas of higher cover (up to 244 meters [800 feet]), minor amounts (0 to 305 millimeters [0 to 1 foot]) of heaving do occur even with optimum room and pillar design. These amounts are generally not large enough to restrict movement and tend to stabilize with time. It should be noted that certain amounts of bottom are removed along the service entries of newer mines, increasing the effective height of the opening as well as removing the uppermost portion of bottom that tends to deteriorate under load. This process partially alleviates a potential problem along the more important entries. Higher amounts of cover in conjunction with poor room and pillar design have caused difficulties in the past (see section on squeezes), but current design parameters have virtually eliminated this problem.

Water-induced stability problems are usually encountered in areas that are in close proximity to secondary drainage and are usually local in nature, short lived and controllable. These situations are exceptions to the rule since the Pittsburgh No. 8 Coal seam, under normal conditions, is dry in this region. Major problems do occur, however, when advancement reaches the Ohio River. In one case, the presence of an abandoned waterfilled mine with a 61-meter (200-foot) hydrostatic head in combination with the Ohio River has apparently "pushed" water to greater distances than either condition would have done by itself. The pumping of the accumulated water during the mining of these sections did little to slow the deterioration of the "premoistened" floor, and abandonment prior to full advance was required (Figure 19.10). No special efforts at prepumping to change

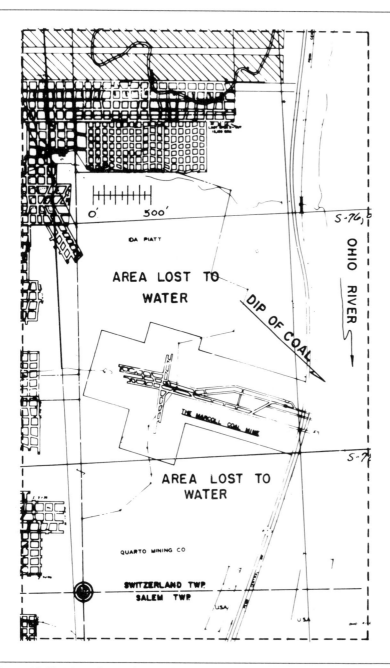

Figure 19.10. Mine plan and map of Quarto No. 4 mine, showing area lost due to wet floor conditions.

the existing conditions were attempted prior to mining in these areas. The resulting block of coal is now considered unminable. It should be noted that conditions improved going up dip and full advancement to the Ohio River is being made.

Mains

Another stabilization problem, that of deterioration, occurs along main and submain entries which are used for haulage and personnel movement. In newer mines, these entries will be used for 30 years or more making the stability of the roof and ribs a basic requirement.

The standard and common practice of installing wood planks or steel straps has been and will continue to be carried out along main line haulageways in the older mines of the division as well as submains in the newer mines. This method does much to protect and improve the roof but does nothing to eliminate or prevent minor amounts of spalling of roof and rib materials. To solve this problem, Quarto's No. 4 and No. 7, and Nacco No. 6 mines as well as other mines in the Ohio Valley are completely covering the roof and ribs along main line service entries. This is accomplished by bolting 51 × 51-millimeter (2 × 2-inch) 14 to 16 gauge galvanized wire mesh against the roof and ribs and then covering it with Mandoseal, a vermiculite-based concrete-like substance. The resulting surface not only seals out moisture but provides minor amounts of support. Straps or planks are still installed prior to application when conditions require it. Guniting has also been used in the past to protect entries at other mines within the division.

Although this method is expensive and labor intensive, long-term savings are realized both in main line cleanup and maintenance and in producing safer haulage conditions. The testing of another form of total covering—total encasement with steel—is on the drawing boards, and if tested, side by side comparison of the two systems will be possible.

Longwall Mining

North American Coal currently has two longwall panels in operation and plans to add two more prior to 1980. With an increased number of operating longwall panels, there will be a corresponding increase in the amount of disturbed ground. Even though an attempt has been made to restrict longwall reserves to certain areas, problems relating to surface subsidence and groundwater will still occur.

Surface subsidence over the first extracted panel (Quarto No. 4) is presently being monitored on a biweekly basis during advancement at over 150 surface locations. Another surface study is planned over a panel which is overlain by a slightly different section of strata. It is hoped that the information obtained from these two panels can be used to derive accurate predictive methods which are applicable to both the local terrain and geology. Preliminary data indicate that the angle of draw and amount of subsidence differ from values that are arrived at using the formulas established by the National Coal Board in England (NCB, 1975).

Changes in the existing water table over the extracted panels may also prove troublesome both for the short and long terms. Although water levels and production capabilities are presently being monitored at only one location, additional water wells are to be drilled and monitored over future panels. Hydrological data obtained from these wells will provide an insight into what actually happens to the water table in disturbed ground.

Conclusions

Regional geologic conditions have placed certain restrictions on the economic development of the Pittsburgh No. 8 Coal seam in southeast Ohio. North American Coal Corporation's Central Division has shown that even with restrictions the balance between underground stability requirements and economically attractive percentages of extraction can be realized. Full utilization and application of existing expertise and theory has helped obtain this balance. Future and current studies concerning pillar design and roof control will further refine the existing knowledge thereby permitting better economics of operation.

References

Adler, L., and Sun, M. C., 1968, "Ground Control in Bedded Formations." Research Division Virginia Polytechnic Institute, Bulletin 28.

Berryhill, H. L. Jr., 1963, "Geology and Coal Resources of Belmont County, Ohio." U.S. Geological Survey Professional Paper 380.

National Coal Board Mining Department, 1972, "Design of Mine Layouts."

Donaldson, A. C., 1969, "Ancient Deltaic Sedimentation (Pennsylvanian) and Its Control on the Distribution, Thickness and Quality of Coals" *in* Some Appalachian Coals and Carbonates: Models of Ancient Shallow Water Deposition.

"Feasibility of Longwall Mining at Powhatan Mines No. 4, 6, and 7—Phase 1." 1976, PD-NCB Consultants Limited, Unpublished Consultants Report.

Given, I. A., 1973, "Room and Pillar Methods" *in* SME Mining Engineers Handbook, Vol. 1, Chapter 12.4.

Holland, C. T., 1973, "Mine Pillar Design" *in* SME Mining Engineers Handbook. Vol. 1., Chapter 13.8.

Holland, C. T., 1973, "Pillar and Entry System Designed for Quarto Mines No. 4 and 7, North American Coal Company, near Powhatan Point, Ohio." Unpublished Consultants Report.

Holland, C. T., 1957, "Report on a Proposed Plan of Mining to Eliminate Squeezes at the Powhatan No. 1 and No. 3 Mines of The North American Coal Corporation." Unpublished Consultants Report.

McCulloch, C. M., W. P. Diamond, B. M. Bench, and M. Deul, 1975, "Selected Geologic Factors Affecting Mining of the Pittsburgh Coalbed." Bureau of Report of Investigation, RI 8093.

Moebs, N., 1977, "Roof Support Structures and Related Roof Support Problems in the Pittsburgh Coalbed of Southwestern Pennsylvania." RI 8230.

Parker, J., and J. Maher, 1976, "Rock Mechanics for Profit—A Practical Approach." Unpublished Seminar Proceedings.

"Subsidence Engineers' Handbook." 1975, National Coal Board Mining Department.

Application of Effectual Strata Control for Better Coal Reserve Recovery in Deep Coal Mines

20

Niles E. Grosvenor, Vice President-Western Operations,
and Richard L. Scott, Mining Engineer,
Gates Engineering Company,
Denver, Colorado, USA

Introduction

Statement of problem

There are two compelling reasons for a dedicated effort to improve the percentage recovery of the known minable coal reserves in the world:

1. Conservation of the limited and assured coal reserves is imperative because consumption of the reserves for ongoing benefits is definable by annual tonnage rates and geographical location.
2. Capital resources available for new ventures are usually disbursed on a priority basis, and are measured by the benefits returned to the owner in relation to the time rate of capital recovery on a project, rather than the total capital return over the life of the project.

Solution to the problem

Effectual strata control entails studying the overburden of coal seams from all geological aspects in order to anticipate the mode of subsidence for the design of a deep coal mine. A mine is designed to control the location and dispersion of the anticipated overburden weight as subsidence occurs. Retreat mining prevents loss of caving of life-of-the-mine airways, haulageways, and underground service openings and avoids the loss of large areas of recoverable coal from overruns by subsidence stress concentrations.

Method of problem solving

The sources of information used in preparing this paper are as follows: Gates Engineering Company in-house knowledge and files; a computer literature search by the Colorado School of Mines Arthur Lakes library; a doctoral thesis and a master's thesis by graduate-degree candidates of the Colorado School of mines; conversations with friends in the industry and related mine researchers.

The more identifiable sources of information are listed in the reference section at the end of this chapter.

Investigation

Coal reserves

In his inaugural lecture at the University of New South Wales in November 1976, Professor Roxborough presented a table of the principal coal mining nations, showing the number of years remaining in the lives of known recoverable in-place coal reserves at 1974 rates of coal production and percentage recovery. For example, Professor Roxborough projected the years of available coal life in the United States—at an annual extraction growth rate of 8.5 percent, achieving 1.1 billion tons per year by 1985—to be 43 years. He also listed the years of life for growth rates of 3 percent and 5 percent to be 80 years and 58 years, respectively. Even though Professor Roxborough stated that deep coal production in the United States will amount to only 30 percent of the total by 1985, increases in percentage recovery or improved conservation of coal reserves are, therefore, a national goal for everyone. The same imperative applies to all coal producing nations.

Construction of a new deep coal mine requires considerable capital expenditure before any coal is produced or any capital recovery begins. Front-end capital requirements are for such items as land and water acquisition, surface transportation roadways, mine site preparation, storage area and structures and surface facility construction, utility services, mining equipment, and initial supplies. In the United States, a significant item of front-end expense is incurred to comply with Environmental Protection Agency laws, regulations, and rules. The front-end capital cost per ton of extracted coal is reduced by an amount related to the increased percentage recovery of coal reserves.

Engineers, therefore, should use improved percentage recovery of coal reserves as the basic premise for mine design. First, a mining layout must be designed that avoids the risks of losing any coal reserves during mine development through errors in strata control. Next, a mining layout must be designed that avoids the risks of prematurely losing any permanent mine openings during room and pillar coal extraction through judgment errors in strata control. This design method is referred to as the engineer's approach to mine design, and the objective of the procedure is to develop the mine on the advance and to extract coal on the retreat.

In the real world, however, the basic premise does not change from the goal of improved percentage recovery of coal reserves. Risks of losing coal reserves are taken by attempting to maximize coal production in two ways: from development headings by increasing their widths; and from outbye room and pillar coal extraction near to the shaft, thus bringing the mine up to desired mature production in a shorter time frame. The latter design method is referred to as the stockholder's (capital investor's) approach to mine design, and the objective is to mine both on the advance and on the retreat. The typical engineer actually designs a mine on an intelligent compromise between losing strata control and obtaining immediate coal output.

Professor Roxborough further stated, "Indeed the future exploitation of coal resources will be determined not so much by what is economically recoverable, but rather what is technically recoverable." Retreat mining, the engineer's approach, is the basis used herein. Retreat mining is defined as driving development headings to the boundaries of the in situ recoverable coal reserves and extracting the coal from room and pillar areas in a preplanned, orderly manner for maximum percentage recovery. Boundaries probably would be determined by property ownership lines and geological phenomena to determine actual mining regions or section boundaries.

Retreat mining is a conservative coal mine design because the mine development for planned room and pillar coal extraction may be accomplished with little risk of strata control loss. Moreover, the loss of strata control when using the room and pillar method, causing possible loss of a fringe mine section, is not as devastating as the loss of a future outbye section of remaining recoverable coal reserves. The chief advantage of retreat mining is that percentage extraction is enhanced, because the necessity to safeguard access to a future mining area does not exist. Another advantage of the method is that a predetermined direction of strata breakline is created and kept moving to prevent build-up of abutment stress on the coal seam to the magnitude where the pillars and unmined faces crush, terminating operations in that section of the mine.

Computer literature search
The computer literature search pertaining to strata control in deep coal mining provided a 154-listing printout of international bibliographies with abstracts published since 1969. In general, the publications deal with strata control based on underground stress-strain measurements after mines were designed, oriented, laid out and developed to an extent. Only one paper deals with the actual design, orientation, and layout of a deep coal mine by effectual strata control, i.e., by studying the overburden of coal seams from all geological aspects (Campbell et al, 1975).

Simco Peabody mine
The Simco Peabody mine, Peabody Coal Company, Coshocton County, Ohio, is an operating deep coal mine with an annual production capability of approximately 600,000 to 700,000 tons of coal. Campbell et al, 1975, used known geological investigation tools to identify the signs that predict the location of different roof conditions. The tools were satellite and high-altitude aerial photographs, detailed geological and rock mechanics field surveys, exploration drill hole cores, and geophysical electric logs. A geologic model of the Simco property was constructed from the data. Peabody engineers were able to design, orient, and lay out the deep coal mine using predicted strata conditions. The accuracies of the predictions have been verified by the subsequent coal mine development. The study's technical competence and significance is described by Campbell et al:

Nothing new or unique is involved (in the strata stability predictions), except perhaps the coordination of the work between mining engineer and geologist; the mining engineer relates prior experience of where bad top has been encountered, while the geologist defines regions that may be hazardous in the future.

There is, however, another unique, although subtle, aspect of the Simco project. The geologists are associated with the Kennecott Copper Corporation, Ledgemont Laboratory, Lexington, Massachusetts—a corporate basic research organization completely divorced from the Peabody Coal Company which was a Kennecott subsidiary company at the time. Thus, the Simco project was a Ledgemont undertaking and, therefore, was not under the time and budget restraints which an operating company usually places on the exploration department. Historically, an exploration program in the coal industry is provided limited time and barely-sufficient funds to make coal reserve and quality estimates.

Block caving

It was noticed, when reviewing the abstracts, that none of the deep coal mine strata control investigators had cross-referenced studies with the metal mining industry; more specifically, with the block caving mining method. Block caving mining engineers work in the realm of letting down the overburden strata with control and safety.

Conversations were held with metal mining friends with current block cave mine design experience in molybdenum and copper mining. From these conversations, it was deduced that the expenses of mine design, orientation, layout, development, and surface plant facilities for a block cave mine are utter risks, except for the work performed by the geological and rock mechanics engineers during the orebody evaluation period.

The persistent risk in the exploration for a block cave mine is whether or not the exploration drilling program actually yields a representative sample of the geologic conditions in a nonhomogeneous rock mass. The new Henderson mine of the Climax Molybdenum Company, just coming on stream, had a front-end capital investment of well over 500 million U.S. dollars before the first block of ore was undercut and management learned that the orebody would cave, and that the mine design was satisfactory. Therein, Climax exhibits the philosophies of patience and the value of ample exploration funding not generally evident in the coal industry.

Block caving mine designers attempt to learn the directions in which the kinetic forces of strata subsidence will act for use in mine design. The Simco Peabody coal mine appears to have been designed on such an effectual strata control basis. Block caving mine designers plan for effectual strata control on the basis of continuing motion of the mass and for deliberate instigation of instability.

Coal mine engineers usually design to resist strata stresses and strains, or to stop motion and maintain stability. The literature suggests that coal mine rock mechanics and strata control research appear to be directed toward how much load a pillar can take before crushing, or how wide a roadway may be driven before the roof falls, so that men and equipment may be kept safe and in order to mine the maximum percentage of coal on the advance without incurring the danger of roof failure.

More cross-reference and information exchange should occur between coal mining and metal mining. The disciplines are aloof from one another; for example, mine supply manufacturers understand the existence of the two-fold nature of supplying "rock bolts" to metal mines and "roof bolts" to coal mines. In the United States, both industries are now under the same health and safety law, which will induce knowledge exchange.

A general supposition is that engineering tools exist to provide improved percentage coal extraction (particularly in room and pillar mining) through effectual strata control when fewer constraints are imposed on engineers.

Geological and rock mechanics engineering

Geological engineers can study satellite and high-altitude aerial photographs with an optical resolution of a quality to permit the identification and plotting of faulting and lineament phenomena not obvious during surface geological reconnaissance. Interpretations of geological information gained from exploration drill hole chips, cores, and geophysical logs are sophisticated to the point that geological and rock mechanics engineers are able to draw substantive conclusions to build reliable geological models of subsidence stresses. (Geophysical logging of coal exploration drill holes is an extrapolation of an exploration tool originally developed for uranium exploration in the western United States.)

Rock mechanics in the United States, as a distinct engineering discipline, slowly is being introduced into mining from underground construction excavation research. Rock mechanics, generally, is relegated to a staff member in the mine engineering department and not budgeted as a separate technical service. Rock mechanics engineering expertise usually is sought after the fact: for example, after a surface mine pit wall has collapsed; a tunnel portal is engulfed by a landslide; a coal mine section is abandoned due to roof failure; a mine service shaft loses alignment; and so on. Engineering expertise usually comes from universities or private consultants. (The United States Bureau of Mines is the important research center in the USA.) A good example is that of the Simco Peabody coal mine design where the applied geological and rock mechanics expertise was imported from the Ledgement Research Laboratory of the parent mining corporation.

A little-used source of what should be important basic information in coal exploration drilling is the Daily Drilling Rig Tour Report prepared by the driller. The tour reports are used to develop drilling expense allocations; but the information is seldom, if ever, reproduced on the geological stratigraphic drill hole logs or included in the engineering report developed from the drilling program. Such items as "lost circulation," "bit stuck-hole caving," "inrush of water," "set casing," "hole abandoned—cement-plugged-and-skidded rig" must have some message for geological and rock mechanics engineers.

Effectual strata control efforts begin when a property is declared to be a coal mining prospect. The initial parameters to be established are the existence and extent of the recoverable coal reserves and the quality and minability of the coal. Estimates of the extent or quantity and minability of the coal reserves involve percentage-of-coal-extraction estimates which, at the termination of mining, are related to strata control.

The geologist constructs a field map bearing satellite and high-altitude aerial photograph interpretations for use by both himself and the rock mechanics engineer. Both engineers do detailed fieldwork. The geologist lays out the exploration drilling program, and the rock mechanics engineer measures angles and directions of jointing to estimate the amount and direction of stresses the strata have, or will have, upon being undermined. The rock mechanics engineer also plots the estimated locations and directions of subsequent subsidence arch-abutment stresses. The necessary coal mine feasibility (mining costs) estimate is dependent upon the work of the geologist rather than the rock mechanics engineer. The percentage-of-coal-extraction estimate over the life of the mine is based upon prior or nearby coal mining experience.

Mine Design

Mine design and orientation (mining direction) is based on the judgment of a rock mechanics engineer. The application, or concept of principles, is the same for both deep mining and surface mining (consider a large disseminated deep copper mining open pit).

A geological mine model is a model of the surface, the underlying strata, and the geological data therein. Mine opening locations have not been determined. Rock mechanics and geological engineers jointly determine subsidence stress abutment lines on the model in relation to fault and lineament locations. The rock mechanics engineer applies the nature and effects of forces to the model (applied mechanics). The surface location of fault and lineament planes are projected to the elevation of the coal seam to guide location of mining sections.

Mine design, orientation, and layout are confusing names because the tasks overlap. The engineering work associated with each task by the joint efforts of the rock mechanics and mining engineers is concurrent.

Mine orientation

Orientation of the mine is described by the rock mechanics engineer applying effective strata control to determine the line of subsidence abutment when the strata are undermined. Rock mechanics and mine engineers work out compromises to lay out the directions that the mine region, mine section roadways, and ventilation access entries should follow. The mining engineer cannot follow theoretical routes because of imposed constraints, such as the degree and direction of coal seam dip and other practical matters, drawn from mining experience essential to constructing a mine. Location of the main entry to the surface should be subjective to the orientation of the coal mine, rather than subjective to the convenience of surface facilities and transportation, within practical limits.

Layout of mine sections

The layout of mine sections should have configurations dictated by the location of significant faults and lineaments rather than symmetrical configurations laid out on a drafting table from preselected mine access. Applied effectual strata control uses fault and lineament planes for subsidence to allow the undermined area to collapse in order to become self-supporting. Thus, the load forces on chain pillars are lessened by destroying the length of the arch. Configurations and sizes of the mine sections suggest whether room and pillar, longwall, or shortwall methods—or a combination—should be contemplated. Layout permits the mining engineer to study the nature of attitude of the coal seam, measure lineal dimensions of mining sections, and estimate the life of the equipment utilized in order to ensure that the capital equipment investment is recovered through depreciation.

The mine layout should be further refined by organizing several compatible mine sections into a mining region and the total mine into different regions. Two reasons for designing a mine in this manner are as follows:

1. Totally distinct and separate ventilation systems may be engineered to permit the isolation of a mining region, if not a section, from the remainder of the mine in the event of a fire or other catastrophic mishap.
2. Mine planning to develop different regions concurrently brings the new mine coal production rate to the desired mature rate in a shorter calendar time frame; this results in a quicker return on capital investment.

The optimum mine layout involves setting up a retreat mining program with the line of retreat parallel to the line of the subsidence arch-abutment line and perpendicular to the mine section access entry. The direction of movement of the abutment stress line is determined during the mine operation. This determination is made in order to lay out the mine so that the stress line is never coincident with an access entry heading, thereby avoiding the creation of an unsupported roof beam of any length. Crosscuts between entry headings are sacrificed at the risk of roof failure, but this risk may be reduced by chain pillars not being lined up side by side.

Mine entry headings should be looked upon as mine development work and should be designed as narrow as traffic safety clearance permits. Roadway parting areas, intersections, should be located in competent strata where the heading width can be increased over a short distance rather than being located by drafting room design. Ventilation headings may be designed wider than roadways, as a local roof failure in a ventilation roadway does not necessarily interrupt the business of mining coal. A better ventilation design, however, is to provide an adequate ventilation cross-sectional area by the addition of headings; say, from a five-heading entry design to a seven-heading entry. The best available technology may not be applied to entry-heading design as long as the coal industry looks upon driving headings or entries as another coal producing section rather than as a separate function identified as mine development.

Mine planning
Mine planning is the location from a point in time to the mine development (time being either days from the mine beginning or to a future calendar date). Room and pillar mine planning practice has been to develop a coal reserve as rapidly as possible to provide working places for a sufficient number of miner units to achieve a desired coal production rate. Coal mining on the advance and on the retreat are looked upon as the same, except that the miner unit daily production rate is acknowledged to be greater on the retreat; there are no mine extensions, no ventilation construction, and lesser room support costs on the retreat. Improved percentage coal extraction is recognized as a worthwhile goal, subject to the ability of controlling the immediate roof for the sake of safety.

Improved percentage extraction through effectual strata control forces mine planning to develop the mine on the advance and to produce coal on the retreat. The procedure is not new, as longwall miners utilize this method. The mine plan must continue to provide working places for a sufficient number of miner units to achieve a desired coal output rate. The mine planning difficulty will be to substitute the term "mine development on the advance" for the historical term "mining on the advance."

Longwall mining on the advance is practiced. However, the advantage of bringing a longwall panel into production without the time lag of completely developing the panel for retreat mining is offset by the risk of losing the access entries, the longwall equipment, and the remainder of the unmined coal reserve in the panel through roof failure. The subsidence arch, also, has lateral abutments whose stress concentrations would be parallel to, and possibly coincident with, the line of panel access entries, causing roof failure and crushing of the pillars when mining on the advance.

If improved percentage coal reserve extraction or gross total coal sales over the life of a coal mine is the agreed upon long-range goal of a project, the mine engineer must use the best available technology based on effectual strata control in the mine orientation and layout. Immediate constraints that may conflict with the best available technology are as follows:

1. A mine plan must be approved by appropriate governmental agencies to obtain a permit to mine (privately owned land included). The plan must be in compliance with all laws, rules, and regulations.
2. Investment money is provided on a priority basis on the comparative rapid recovery of the investment rather than on the basis of total rate of return over the life of the mine.

A suggested practical partial solution to improving percentage coal extraction through effectual strata control is to begin the mine design in the accounting department. Mine development should be a cost center separate from coal production. The resulting coal production from mine development work should be added to the other mined coal as "development tons" and credited to the cost of development. Performance of the development unit would be measured by meters of advance (as in metal mining) rather than tons produced.

Further fringe benefits which might result from setting up mine development on the advance as a cost center are as follows:

1. Separate development labor crews from other mine unit competition to permit better work control.
2. Utilize different work rules.
3. Measure work accomplishment on an identifiable performance.
4. Investigate contracting the major portion of mine development.
5. Induce manufacturers to field-test prototype equipment in a more sophisticated mode of operation and maintenance.
6. Prove that percentage coal reserve extraction may be enhanced by applied strata control and mining on the retreat.

Conclusions

The life of the known recoverable coal resources in the world are finite and limited in utilization. The concern of every coal operation must be to improve the percent extraction of recoverable coal through application of the best available technology.

Technology exists for application of effectual strata control to coal mine design, orientation, layout, and planning in order to achieve a greater percentage extraction of minable coal reserves.

Except for the very deep coals, coal mine industry philosophy must change to accept the proposition that a coal mine operation is developed on the advance and coal is extracted on the retreat for optimum percentage extraction of coal reserves.

References

Abel, J. F., Jr., Ph.D., 1978, Colorado School of Mines, Golden, Colorado; personal conversation and files, March.

Agapito, J. F. T., 1972, "Pillar Design in Competent Bedded Formations." Doctoral thesis, Colorado School of Mines, Golden, Colorado.

Campbell, J. A. L.; Petrovic, L. J.; Mallio, W. J.; Schulties, C. W.; 1975, "How to Predict Coal Mine Roof Conditions Before Mining." Mining Engineering, October.

Eatough, M. G., 1978, Gardner-Denver Company, Denver, Colorado; personal conversation, March.

"1977 Keystone Coal Industry Manual," 1978, McGraw-Hill, New York.

McDonald, B., 1978, Arthur Lakes Library, Colorado School of Mines, Golden, Colorado: computer literature search, March.

Morris, C. L., 1978, Climax Molybdenum Company, Golden, Colorado; personal conversation, March.

Panek, Dr., L. A., 1978, United States Bureau of Mines, Denver, Colorado; personal conversation, March.

Reeder, R. T., 1978, Colorado School of Mines, Golden, Colorado; personal conversation and files, March.

Roxborough, F. F., 1976, "World Coal—Resources, Reserves, and Recovery." Inaugural lecture as Professor and Head, School of Mining Engineering, University of New South Wales, St. Ives, New South Wales, Australia, November.

Stewart, C. L., 1977, "Rock Mass Response to Longwall Mining of a Thick Coal Seam Utilizing Shield-Type Supports." Master's thesis, Colorado School of Mines, Golden, Colorado.

Stewart, R. M., 1978, Wyoming Mineral Corp., Denver, Colorado; personal conversation, March.

Underground Mine Hazard Analysis Technique

21

Richard D. Ellison, Executive Vice President,
D'Appolonia Consulting Engineers, Inc.
Pittsburgh, Pennsylvania, USA

Introduction

Stability conditions in underground mines are highly dependent upon geologic conditions as well as the mine geometry and method. This chapter introduces a systematic procedure for identifying important geologic conditions and evaluating them singularly and in appropriate combinations to predict entry behavior or to design the mine so that undesirable factors are minimized. The method is termed the *Geologic Hazard Analysis* technique for its application to avoidance of hazardous conditions. The method could also be termed:

1. *Geologic Problem Analyses* technique if the condition being analyzed is not hazardous, but still represents the evaluation of conditions in order to minimize difficulties.
2. *Geologic Profitability Analyses* technique if the method is used to optimize mining so that maximum production and recovery are realized for the selected method, while safety and contingency items are minimized.

The basic example presented deals with establishing requirements for entry geometries and support systems to provide long-term stability for a thick western U.S. coal seam. Secondary examples relate to: the evaluation of probable caving conditions behind longwall supports; the anticipated magnitude of water inflow for different areas of the mine; and the potential for spontaneous combustion in the relatively young subbituminous coal.

The example used is patterned after a project performed for the Dravo Corporation and the Rocky Mountain Energy Company in Wyoming. However, the actual geologic conditions illustrated, including the overall shape of the property, have been totally modified in order to make the example general and unrelated to final decisions made with regard to mining for that project.

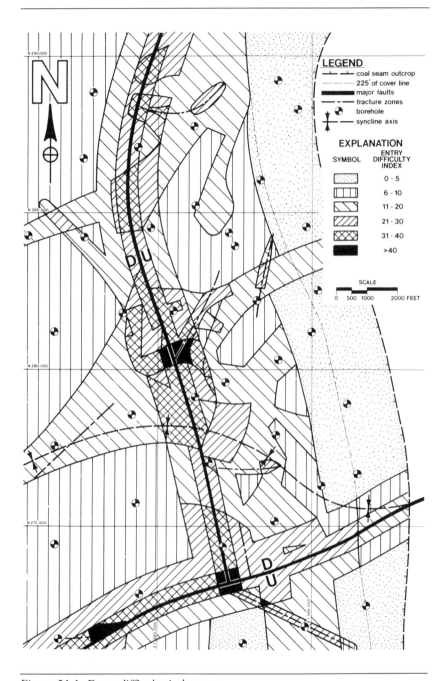

Figure 21.1. Entry difficulty index map.

General Capabilities of the Method

The end result of the evaluation is a map of the entire proposed mining area showing the relative importance of the characteristic being considered. The example shown in Figure 21.1 is for the difficulty of developing entries. The "entry-difficulty index" varies from zero to greater than 25. The lowest numbers are associated with relatively ideal conditions where only minimal or normal development, support, and maintenance conditions are anticipated. The highest index numbers represent zones where the greatest amount of difficulty with entry development, support, and maintenance will occur. The value of this type of map for planning, design, and in-mine observations is evident.

The designation of any index must necessarily be somewhat subjective because it is designated before mining is accomplished and must, therefore, be based on an appropriate level of analysis of the available data. If this evaluation is accomplished following a logical process, however, the results should be very reliable and accurate, particularly in relation to intuitive judgments determined on the basis of less organized processes.

The results can be used for many different purposes, such as those summarized in Table 21.1. The degree to which the method should be applied at any given mine is dependent upon:

1. The level of difficulty anticipated.
2. The importance of having very accurate predictions.
3. The amount of mining that has occurred in adjacent areas which will provide direct empirical indications of mining requirements and costs.

In some cases, the method should not be used because it will not contribute to improved safety, costs, or recovery. In many cases, however, the method can be the most cost-effective step in the exploration and design programs.

Example factors that can be considered are summarized in Table 21.2. Major attractions of the method are: it is very easy to choose only those factors of importance to the condition being investigated; and the factors can be easily

---

Table 21.1. Typical Applications

General minability
Exploration planning
Method applicability
Recoverability
Development costs
Maintenance costs
Feasibility
Prediction and partial prevention of surprises
Overall mine plan development
    – geometry
    – sequence

Table 21.2. Example Factors

The mechanical and physical rock properties
Lithology and structure
Intensity and characteristics of bedding and jointing
Fracture zones or faults
Aquifer locations
Water pressures and effects on rock properties
Gas pressures
Seam thickness
Dip
Thickness and condition of draw slate or rider coals
In situ stresses
Cleat orientation
Washout potentials

evaluated based on their combined effects. Often, the combination of two or three conditions can have a major impact, while the existence of only one of the conditions would have an effect.

Development of Hazard Map for Entry Development

A geologic section through the mine used for this example is shown in Figure 21.2. The seam to be mined is located near the bottom of the section at a depth varying from 250 to more than 300 meters (800 to 1,000 feet) deep. The seam thickness varies from 6 to 9 meters (20 to 30 feet). The sequence of overburden rock varies across the site. Of importance to this example, the rock immediately above the coal varies from thick, competent sandstone near the central portion of the site to thick shales in the northern portions of the mine. A number of significant fracture zones could be mapped in aerial photographs and satellite imagery and verified in the field. Also, two large, nearly vertical faults occur as illustrated in Figures 21.1 and 21.2.

An initial step in any site-specific program is to perform geologic evaluations using all available information, including:

1. Interpretation of aerial photographs and satellite imagery.
2. Review of any published data for the area, nearby wells, or adjacent mines.
3. Visitations to nearby mines, if possible.
4. Detailed examination of any preliminary boring data.

This initial study will provide a characterization of the range of anticipated geologic conditions and their general distribution throughout the site. The next step is to drill additional boreholes located to generate necessary information in areas of greatest importance. This step is most economical when these boreholes are combined with those normally required for exploration. The only added cost then is that some of the borings are used for special sampling, laboratory testing,

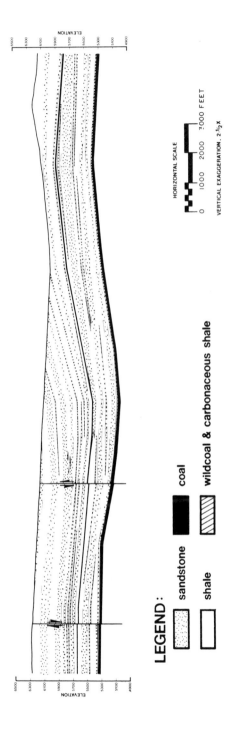

LEGEND:

sandstone

coal

shale

wildcoal & carbonaceous shale

HORIZONTAL SCALE

0 1000 2000 3000 FEET

VERTICAL EXAGGERATION: 2½X

ELEVATION

6500 6300 6100 5900 5700 5500 5300 5100 4900

Figure 21.2. Geologic section looking west.

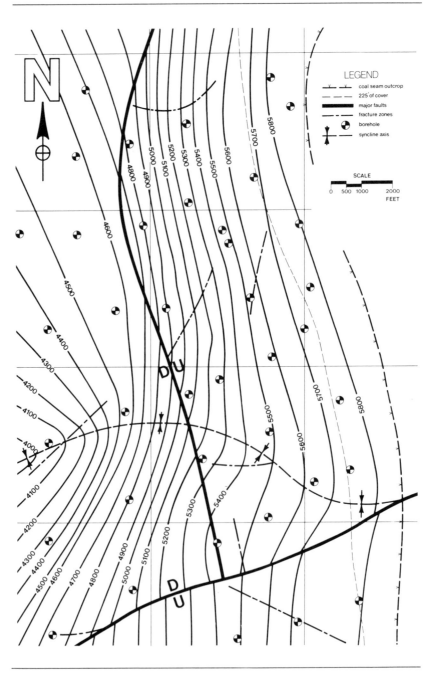

Figure 21.3. Base map.

and in situ testing in order to develop improved geologic, rock mechanics, hydrologic, and other required parameters. For the example presented in this chapter, these tests included:

1. Complete coring of six boreholes.
2. Partial coring at selected zones in all boreholes.
3. Detailed geophysical logging of all boreholes.
4. Drilling of oriented core in several boreholes to assure proper interpretation of cleat direction.
5. Conduct of drill stem permeability tests at selected locations in several boreholes.
6. Setting of piezometers at selected aquifers in several boreholes.
7. Conduct of laboratory strength tests on representative core samples of the roof and floor rocks and the coal.
8. Geochemical and mineralogical testing of selected rock and coal samples.

From these data, the base map illustrated in Figure 21.3 is developed to include the elevation of the coal seam and the major structural features of the area. Also, a series of cross sections similar to that shown in Figure 21.2 is drawn and zones of representative conditions are selected for preliminary design of the particular mining item being considered. In this case, Figures 21.4 and 21.5 show examples of preliminary design alternatives performed to determine approximate entry size, spacing and geometry, and support systems required to suit the different types of geologic conditions. For example, requirements are different where shale or sandstone above the roof is very thick, or where thin shale lies beneath thick sandstone and high groundwater pressures, or where conditions are highly variable as a result of the differential compaction of rock types that are rapidly changing.

Based on the preliminary designs, evaluations of coal recovery, development, support and maintenance costs, and any operating impacts are estimated for inclusion into the hazard analyses. The importance of each pertinent geologic factor is identified and tables are developed which relate the different geologic factors (including combinations of factors) to the mining requirements.

The next step is to prepare a series of overlay maps for each of the important geologic factors. Figures 21.6 and 21.7 illustrate such maps overlaid on the base map for the thickness of sandstone above the coal seam, and the thickness of coal left in the longwall gob line. Additional overlay maps used for this example include:

1. Thickness of wild coal and carbonaceous shale at the roof.
2. Thickness of shale to the first sandstone unit.
3. Groundwater pressure above the coal.
4. Sulfur isograd for the upper two feet of coal.

All of these individual maps, or any two or more of these maps, can then be overlaid in order to provide a composite "picture" of all of the factors which control mine entry development behavior. This process is illustrated in Figure 21.8.

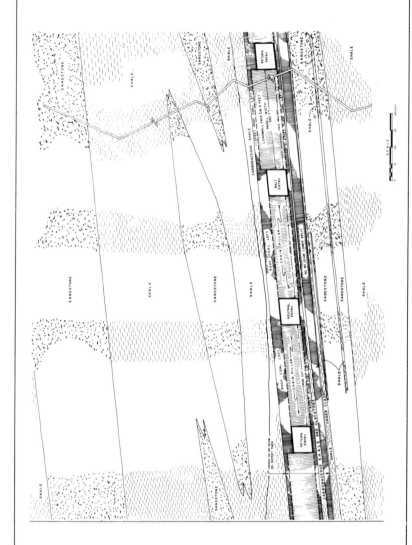

Figure 21.4. Example entry arrangement for one of several roof conditions.

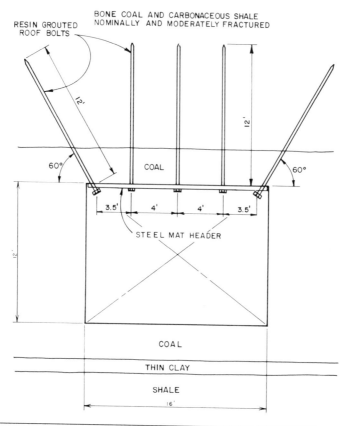

Figure 21.5. Example of one of several types of roof support systems.

As discussed below, the procedure can be accomplished either manually or using a compute:. Both procedures have their advantages and disadvantages. For the initial study of any given mine, the manual procedure is preferred because it requires maximum "hands-on" evaluation to assure that important characteristics and conclusions are not missed. The computer methodology has the obvious advantages of being very quick and efficient allowing the evaluation of any combination of conditions.

The best results will always be obtained when the importance of each overlay or each combination of overlays is determined by a "round-table" group consisting of individuals with varying backgrounds, experiences, and responsibilities to the project. For example, the meeting can include the mining engineer, the structural geologist, the rock mechanist, the hydrogeologist, the safety engineer, the economic analyst, and others. This assures that input from all disciplines is included and permits areas of insufficient data to be quickly identified so that those data gaps can be efficiently closed.

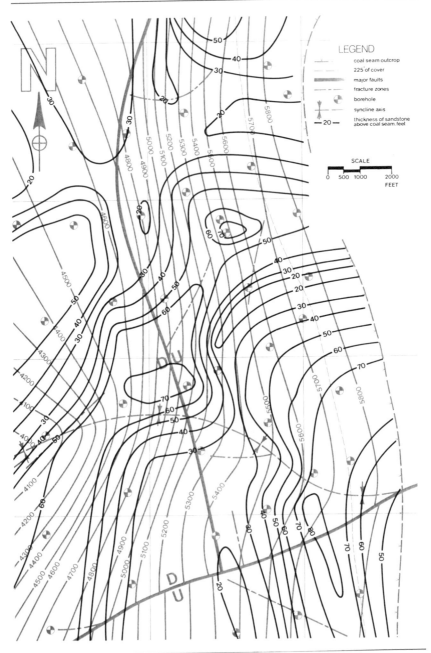

Figure 21.6. Example of composite showing sandstone thickness above the coal seam and base map.

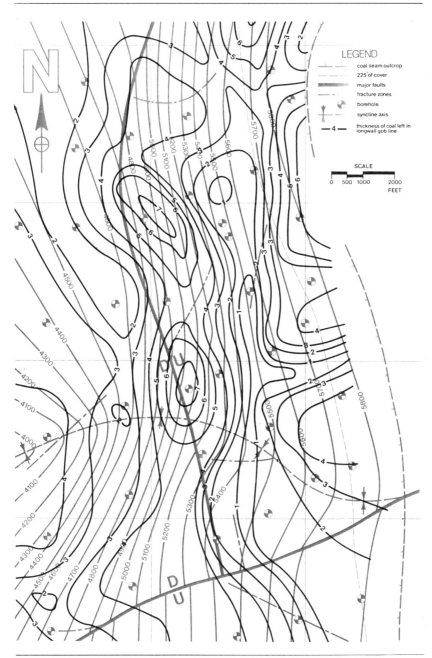

Figure 21.7. Example of composite showing thickness of coal left in the gob and base map.

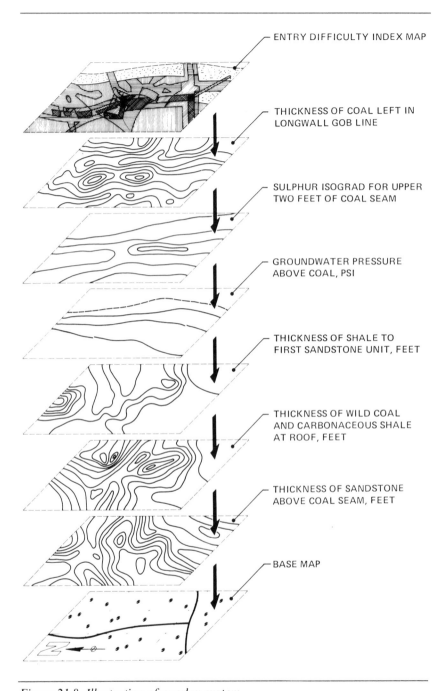

ENTRY DIFFICULTY INDEX MAP

THICKNESS OF COAL LEFT IN
LONGWALL GOB LINE

SULPHUR ISOGRAD FOR UPPER
TWO FEET OF COAL SEAM

GROUNDWATER PRESSURE
ABOVE COAL, PSI

THICKNESS OF SHALE TO
FIRST SANDSTONE UNIT, FEET

THICKNESS OF WILD COAL
AND CARBONACEOUS SHALE
AT ROOF, FEET

THICKNESS OF SANDSTONE
ABOVE COAL SEAM, FEET

BASE MAP

Figure 21.8. Illustration of overlay system.

Table 21.3. Entry Hazards *1/3*

Geologic structure	Hydrostatic pressure (psi)		
	0-50	50-200	>200
0-500 feet from fault	10	16	20
500 feet to 1/4 mile from fault	6	9	12
Differential compaction	6	9	12
0-500 feet from fractured zone	3	6	8
Near synclinal axis	1	3	5
Topographic unloading	1	3	5

The assignment of quantitative hazard numbers is accomplished by considering the various characteristics and their overall impact. Tables 21.3 and 21.4 are two examples of the tables used in the subject example. Table 21.3 shows where entry hazards were indexed considering geologic structure features at three levels of hydrostatic pressure. The worst condition, of course, is near a major fault where rock fracturing will be high and where groundwater pressure is high. The resulting water inflow will create rapid weathering of any rock materials and potential instability due to pressures acting on rocks surrounding the entries. The relative best conditions occur where only minor structural features exist and the groundwater pressure is low. A ranking of zero, not shown on the table, would be designated when there are no particular structural features and when groundwater pressures are low.

Table 21.4 shows the example of entry hazards related to roof lithology. When there is a thick layer of carbonaceous strata and high groundwater pressures, very significant difficulties with opening of bedding planes and rapid weathering of the material can be expected. On the other hand, when there is a reasonably thick layer of competent shale and a low groundwater pressure, the difficulties with both bedding separation and weathering leading to roof falls will be relatively small.

A series of these types of "subjective" indices derived from quantitative predictions from preliminary design analyses are then formulated into a single desig-

Table 21.4. Entry Hazards *2/3*

Roof lithology	Hydrostatic pressure (psi)		
	0-50	50-200	>200
More than 10 feet carbonaceous strata	5	7	8
0-10 feet carbonaceous strata under 50 feet of sandstone	4	5	6
0-10 feet shale	1	5	8
10-30 feet shale under 50 feet of sandstone	0	2	4

Table 21.5. Entry Hazards *3/3*

Hazard index	Area (acres)	Percent of mine area	Estimated cost multiplier
0-5	280	7	1.1
6-10	1,440	36	1.25
11-20	1,360	34	1.5
21-30	680	17	2.5
31-40	200	5	7.0
>40	40	1	10.0
Total = 4,000		**Average = 1.9**	

nation of the hazard index for all zones within the mining area. They are summarized on the entry difficulty index map, as illustrated in Figure 21.1.

The relative impacts of each hazard identification are summarized in Table 21.5. This table clearly indicates that any of the zones (six percent of the mining area) with hazard indices between 31 and 40 or greater than 40 probably should not be planned for mining. Also, the 17 percent of the mine within the 21 to 30 hazard index range must be considered very carefully because the cost of entry development is still 2.5 times the cost that would be realized in relatively ideal conditions. The remaining 77 percent of the mine falls within estimated cost multipliers of 1.1 to 1.5 and is very appropriate for inclusion in the basic mine plan.

Referring to Figure 21.1, it becomes apparent that if it is desired to minimize unnecessary difficulties during the first several years of operation, zones with an index of less than ten would be desirable for initial development. This could be particularly important for the example mine because its thickness dictates that a high mining system be used, which in and of itself will require substantial start-up efforts.

It is emphasized that the individual boundaries between hazard indices shown in Figure 21.1 are not exact since they are based on interpretations of several geologic factors, each of which is interpolated between borings. Nevertheless, the general pattern of conditions illustrated in Figure 21.1 is appropriate and will be realized when mining is commenced.

Additional Example Uses of the Method

Without discussing the detailed methods of evaluation, three additional uses of the method are illustrated in Figures 21.9, 21.10, and 21.11 for the example mine. Figure 21.9 shows the anticipated difficulty of realizing desired caving behind the longwall supports. In the 0-5 and 6-10 index areas shown in Figure 21.9, caving will occur readily and will assure that high stresses in the supports are not generally realized. In the areas designated with indices greater than 20, caving and occasional bridging should be expected with resulting potential difficulties associated with stresses in the supports and excessive spalling of the face. Again, the

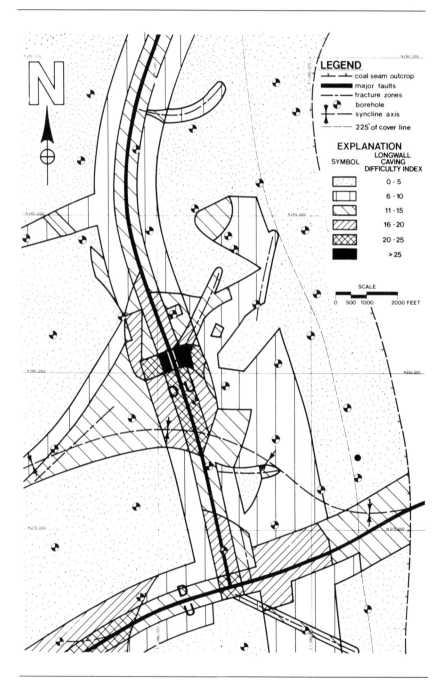

Figure 21.9. Longwall caving difficulty index map.

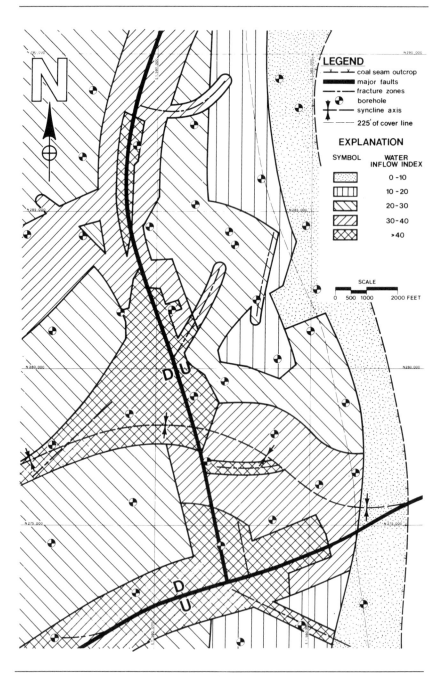

Figure 21.10. Water inflow index map.

desirability of starting the operation in the "good" zones, so that the equipment and mining system can be debugged with minimal geologic problems, is evident.

Figure 21.10 shows the prediction of groundwater flow that is anticipated into the mine after longwall caving is completed. The difference in estimated inflow varies from as little as 4.5 liters (1 gallon) per minute per acre for the area with an index less than ten, to as much as 45 liters (10 gallons) per minute per acre for the areas with an index greater than 40. Because significant portions of groundwater inflow must be handled for the entire life of the mine, the desirability of avoiding the greatest inflow areas during the early stages of mining is evident. A secondary use of the data shown in Figure 21.10 is the demonstration of the real probable impact of any mining on regional or local groundwater conditions so that permitting for the mine can be obtained in the shortest possible time.

Figure 21.11 illustrates a hazard evaluation of spontaneous combustion potential throughout the example mining area. To the knowledge of the author, this is the first evaluation of its type accomplished for a major mining operation in the United States. As illustrated, the entire mine area is divided into three areas of spontaneous combustion potential, varying from high potential to increased potential to most severe potential. Like many western coals, this relatively young subbituminous coal does have the potential for spontaneous combustion throughout the proposed mining area. The mine plan and design must minimize this potential and provide for all safety and equipment recovery requirements in the event that any combustion should be encountered. Obviously, as the potential increases, and certainly in the very severe area of potential combustion, extreme care will be required to avoid undesirable losses of coal reserves or the trapping of equipment in areas that cannot be easily extinguished. The parameters that were used in these unusual analyses included:

1. International rank classification of the coal.
2. Pyritic sulfur content of the coal.
3. Pyritic sulfur content of the roof rock.
4. Rate of heating in hydrogen peroxide (I index).
5. Rate and magnitude of oxygen absorption (S index).
6. Amount of coal left in the gob line following longwall mining.
7. Thickness of highly carbonaceous strata and wild coals directly overlying the coal seam.

Computerized Application

The greatest advantage of using the manual procedure of evaluating the individual overlays to develop any "hazard" map is that the hands-on procedure assures that experienced judgment decisions are made at each step within the study program. It is acknowledged, however, that utilization of the computer can greatly speed up the procedure, reducing costs, and also permit many complex combinations of geologic factors to be easily analyzed.

Figure 21.12 again shows the entry difficulty hazard map with a highlighted circled zone used to illustrate the output of the computerized system in Figure 21.13. The numbers shown in Figure 21.13 are the summarization of hazard index

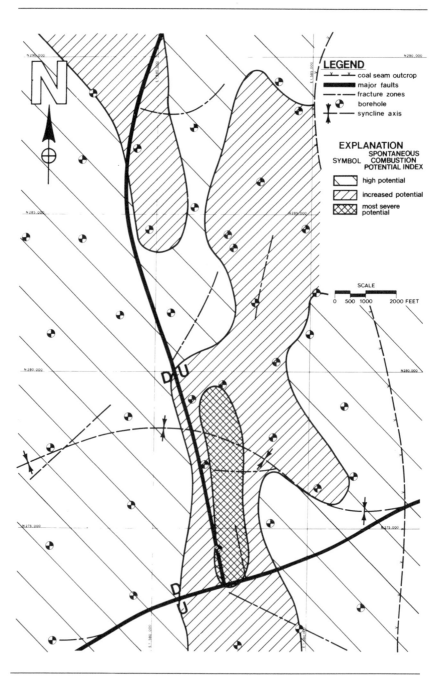

Figure 21.11. Spontaneous combustion potential index map.

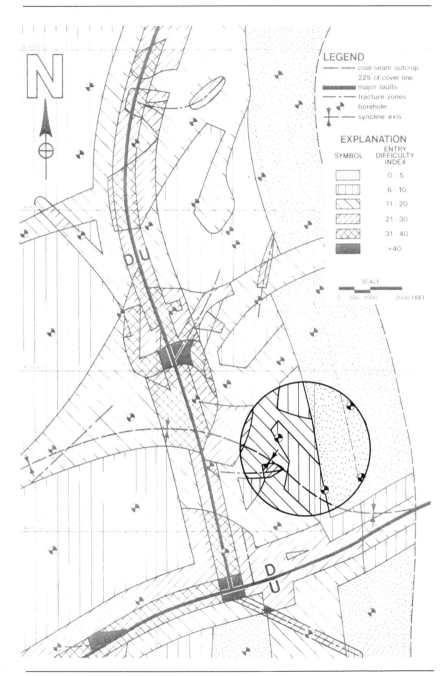

Figure 21.12. Location of detail for computer output shown on Figure 13.

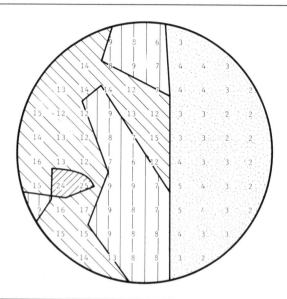

Figure 21.13. Example of computer output.

at any given location based upon the designated combination of general geologic structure and roof lithologic properties appropriate to mine entry development. Similar maps can be quickly prepared for any item of interest simply by designating the geologic factors that should be combined and the relative hazard index (or other type of designated index) appropriate to that particular condition.

Preparation Plant Refuse and Waste Disposal

Stabilization of Coal Waste Deposits

22

Robert I. Fujimoto, Chief, Mine Waste Branch,
Mine Safety and Health Administration, U.S. Department of Labor,
Denver, Colorado, USA

Background

In the early 1920s, when mechanical loading was first introduced in underground mines, only 0.3 percent of bituminous and lignite production was automatically loaded. This percentage increased to 12.2 percent by 1934, and by 1970 had reached 97 percent. The less exacting machine-mining, which removes substantial amounts of material above and below the seam, has increased the proportion of coal waste to coal in the material transported to the preparation plant. Prior to 1940, the ratio was approximately 1 to 10, or 100 kilograms of waste per metric ton of coal (200 pounds of waste per short ton of coal). By 1969 this ratio was about 1 to 5. It is now estimated that between 1965 and 1980, the total amount of coal waste produced by mining operations will increase from 100 million tons to a minimum of 200 million tons per year.

Before coal mines were mechanized, only the thicker and better seams of coal were developed on a large scale. The coal was mined, picked, and loaded by hand. Consequently, only marketable coal was transported to the surface and the waste material remained underground. With new and improved mining equipment, coal seams with a larger percentage of impurities could be mined, and the mixture of coal and rock could be economically transported to the surface and cleaned of impurities before marketing. Because this material had little or no economic value, it was disposed of as economically as possible and in such a manner that the disposal did not interfere with the production operation. The removal of waste rock from marketable coal is a continuing operation throughout the life of most coal mines. The largest volume of solid waste is produced at the preparation plant where the coal is crushed, sized, washed, and separated from rock and other impurities to achieve a grade of coal meeting the market demand. Coal wastes are ordinarily separated into two classes: coarse (greater than 1 millimeter), and fine (less than 1 millimeter).

Coarse refuse is typically transported from the preparation plants to the disposal area by conveyor belts, tramways, or trucks. Fine refuse is usually transported hydraulically through pipes to settling ponds as a slurry.

Until about 15 years ago most of the fine refuse was transported hydraulically through pipes and disposed of in natural drainage areas or underground mines. Since that time, regulations have forced the mine companies to cease pumping directly into natural drainages. Most companies have followed the practice of pumping refuse into settling ponds before releasing the clearer water into natural drainage.

The pumping of slurry into settling ponds was initiated during the late 1950s in order to comply with water pollution laws. Although the pumping of slurry behind coarse coal waste embankments clears the water of coal fines, it saturates the embankments and their foundations, creating an unstable condition and tremendously increasing the hazard potential. In the beginning there was little hazard potential since the ponds were small. The magnitude of the problems grew as the facilities increased in size until the Buffalo Creek failure. There had been many reported, but unrecorded, failures of small structures. However, little concern was displayed since there were no severe losses of life or property. This situation dramatically changed with the Buffalo Creek failure. It became woefully and belatedly apparent that if these facilities are not designed and constructed properly, the possibility of failure exists. The results of such a failure could have severe consequences equal to or exceeding the Buffalo Creek disaster. Unfortunately, we seem to react only when there is a tragedy which results in several deaths. Laws are usually written in response to, and at the expense of, deaths such as the ones that occurred at Buffalo Creek.

The following statement by W. A. Wahler, of W. A. Wahler & Associates, accompanied an article by William E. Davies, a member of the Interior Task Force Committee ("Buffalo Creek Dam Disaster: Why It Happened," *Civil Engineering, ASCE*, July 1973):

The dam's failure is an example of the results of hastily conceived legislation . . . requiring instant conformance without technological assurance that its requirements can be safely met. Faced with fines and jail sentences for polluting the streams after the date of compulsory compliance, most of the coal mine operators in Appalachia approached the problem pragmatically; they learned that by running the [slurry] "black water" containing coal dust through a waste pile, the water emerged clean. Alas, a solution to the dilemma of what to do was found. It was quick and cheap.

Prior to the Buffalo Creek disaster there was essentially no enforcement of regulations on coal waste deposits. Since the 1966 disaster at Aberfan, South Wales, the problem of stability of waste embankments had been recognized and there was an overall awareness that similar problems existed in the Appalachia region; however, little or no emphasis was placed on correcting the deficient waste embankments.

It was only after the Buffalo Creek coal waste dam failure that regulations regarding refuse dams and retaining dams were initiated to insure safe coal waste embankments. The intent of the regulations is to prohibit the construction and use of unsafe coal waste facilities for either coarse or fine refuse.

A post-failure investigation at the Buffalo Creek site and many other facilities suggested that any one of numerous other structures could have failed catastrophically, given the proper set of circumstances. The absence of unusual storm activity immediately prior to the failure called attention to the many structural inadequacies of the facility. Several studies identified the reasons why failure of such a structure could occur. They are:

1. Inadequate foundation preparation.
2. Lack of zoning and compaction in the embankment.
3. Lack of adequate water-control facilities, such as spillways, decants, and diversion ditches.
4. Lack of seepage collars along embankment conduits, allowing water to move along the outside of the pipe deep within the embankment.
5. No control of water within the embankment.

In light of what is known today about the Buffalo Creek flood, it is apparent that this unfortunate disaster could have been prevented through proper design and construction of the refuse facility, better utilization of available materials, and periodic inspection. Operators are fortunate in one respect in that an almost endless supply of material is available to construct safe refuse impounding structures, whereas a typical dam designer must limit quantities to control costs.

The following short article is reprinted from *Engineering News-Record**. A similar article appeared in *Coal Age* magazine.

The already close association between the mining and construction industries should become even stronger as a result of this month's meeting of the International Commission on Large Dams (ICOLD).

The more than 2,000 delegates from 56 countries met in Mexico City to address, among other questions, the problems involved with special types of fill dams. The mining industry, it turns out, not only builds dams with the tailings or waste produced in ore processing, but only recently found that one of its big tailings dams is at the top of the list of the world's largest dams.

The 274-million-cu-yd New Cornelia tailings dam in Arizona displaced Pakistan's front-runner, Tarbela, which has a volume of 191 million cu yd.

Because of the Arizona structure and the failure of the non-engineered Buffalo Creek tailings dam in West Virginia (ENR 3/2/72 p. 10), ICOLD delegates decided civil engineers must participate in designing such dams.

According to several ICOLD papers, the huge embankments built by mining companies, "may become potentially hazardous structures, particularly in earthquake areas, by virtue of the nature of mine residues and the method of disposal, generally by hydraulic placement of the materials."

According to an Australian delegate, "Much of the design of tailings dams to date is quite empirical and has been the responsibility of the mining profession. It is now evident that civil engineers and dam designers . . . should participate in the design of these dams." The delegates concluded that the study of the liquefaction potential of tailings materials "under normal and earthquake conditions is the most vitally important field of inquiry.

"The possible failure of these structures represents a considerable public hazard." one delegate said, "and it is clearly necessary for the dam experts to enter into dialogue with our mining colleagues on the design of tailings dams and use our joint efforts to improve the security to the public."

Design Criteria

The state-of-the-art in the design and construction of earth dams can be applied almost across the board to the design and construction of coal waste structures, except in the areas of combustibility, potential rapid material degradation, and extended construction periods.

The fields of geotechnical, hydrologic, and hydraulic engineering are well covered and as long as the latest technology regarding earth dam engineering is utilized, most safety requirements will be met. Because federal regulations were general and engineers may not have known what the technical requirements were, many of the initial plans submitted were not acceptable. It was for this reason that guidelines were developed so that a common ground of understanding could be established. The "Design Guidelines for Coal Waste Structures" were developed utilizing basic criteria used by the major dam design agencies in the United States. The guidelines are continually updated as the state-of-the-art for the safe and orderly disposition of coal waste is advanced. Generally, the following criteria, taken from the guidelines, are the most recent technical considerations that need be implemented into the design of coal mine waste structures in order to meet federal requirements.

Current, prudent engineering practices require a conservative approach to provide maximum flood protection for water-retension structures located where failure may cause loss of life or serious property damage. Therefore, designs of water, sediment, or slurry impoundments should be based on the probable maximum precipitation of 6-hour duration. A 20 percent reduction in the probable maximum precipitation (PMP) is allowed for impoundments east of the 105th meridian which have drainage areas less than 10 square miles. For areas west of the 105th meridian, inflow design floods should be prepared using both the probable maximum thunderstorm 1-hour rainfall and the probable maximum 6-hour general-type storm rainfall. The more critical of the two inflow design floods should be used in the design of the structure. If it can be shown that the failure of an impounding structure would not cause loss of life or serious property damage, then a lesser design criteria may be used if information substantiating such a decision is submitted by the coal companies. A 100-year frequency storm of 6-hour duration (one-percent probability) is the minimum storm permitted in the design of any impoundment.

The design freeboard distance between the low point on an impounding structure and the maximum water elevation for the anticipated design capacity should be at least three feet. However, in situations where sufficient documentation is provided indicating that adequate freeboard is assured so that there is no possibility of the embankment being overtopped, a lesser freeboard may be acceptable. Many factors are involved in the determination of freeboard requirements. Items that should be considered include: duration of high water level in ponds, effective wind fetch, water depth, potential wave runup on embankment slope, and the ability of the embankment to resist erosion. The crest should slope to force all drainage to the upstream side of the embankment.

The design freeboard distance between the top of bank of any spillway or diversion channel and the maximum water surface in the channel should be at least $1.0' + .025v(d)^{.33}$ where v = velocity in ft/sec and d = depth in feet.

Under normal conditions, diversion ditches around an impoundment should be designed in accordance with the appropriate state regulations. Diversion ditches around embankments that cannot impound water are generally required to pass the runoff from a 6-hour duration, 100-year frequency storm.

When any emergency outlet structure for an impounding facility is being checked, any diversion ditches should normally be neglected as part of the outlet structure. If a diversion ditch is being considered to pass runoff water in lieu of a spillway around an impoundment, the ditch should be designed and constructed under the same design specifications as a spillway.

The "SCS Handbook," Notice 4-102, August 1972, is an acceptable reference for hydrology design considerations for coal waste structures.

Another suitable reference is the Bureau of Reclamation's publication entitled "Design of Small Dams," Revised Reprint, 1974.

Impoundments in which part of all of the inflow from the design storm is to be stored, shall be subject to a drawdown criteria. The drawdown criteria is met if 90 percent of the volume of water stored during the design storm can be evacuated within 10 days from the facility.

All pipes and conduits through embankments should be provided with anti-seep collars. The length of the line of seepage along the line of contact between the embankment and the barrel and the anti-seep collars should be about 20 percent longer than the length of pipe or conduit lying within the zone of saturation. Pipes and conduits should be constructed with provisions to prevent clogging. A suitable reference on prevention of clogging is "Debris-Control Structures," Bureau of Public Roads, Hydraulic Engineering Circular No. 9, February 1964.

Surfaces of channels and diversion ditches should be capable of withstanding the expected maximum velocity of the design flow without undue erosion or scour. A good reference for the sizing of riprap for open channels is "Use of Riprap for Bank Protection," Bureau of Public Roads, Hydraulic Engineering Circular No. 11, June 1967.

The stability of an impounding structure should have minimum, static and dynamic factors of safety of 1.5 and 1.2 respectively, under full anticipated design capacity.

For dry refuse piles that do not and cannot impound water, slurry, and/or silt, the coal refuse should be placed with side slopes no steeper than 27 degrees (between benches) and spread in a maximum of 2-foot-thick layers, or designed to minimum, static and dynamic safety factors of 1.5 and 1.2 respectively, or more, with steeper slopes and/or thicker layers.

Foundations for refuse piles and impoundments must be properly prepared by removing all vegetation and undesirable material in order to achieve a firm foundation.

Filters, drainage blankets, etc., that are so thin that contamination may occur during construction, are not considered adequate. Normally, a blanket of well-graded material five feet thick is preferred; three feet is the minimum and will require special construction considerations to be acceptable. If, in general, the proposed construction requires close field control to assure that the facility is properly constructed, then careful consideration must be given to all elements of the design. A good reference on filter design requirements is "Design of Small Dams."

When a coal company has requested approval to raise the height of an impoundment by upstream construction over slurry sediment, the following is recommended:

> The coal company performs suitable tests on the slurry (subsurface investigation) to prove that the slurry has sufficient strength for stability and support of the added material and the construction of the dam addition must be engineer-controlled and suitably compacted in layers.

> The dumping of material over the freeboard area of the dam crest to extend and raise the embankment is not allowed unless it is in accordance with an approved plan.

Closed-circuit coal waste is generally considered a poor structural material, and its use for embankment construction should be viewed with caution. Normally, the material contains a considerable amount of water and therefore should be regarded as a fluid with no structural integrity. If closed-circuit coal waste is to be used as a structural component in an embankment, the embankment should be designed and constructed to the previously established safety factors, using soil properties determined for the particular coal waste material.

Regulations Overview

Once regulations are enacted it can normally be expected that they will not disappear and that very few changes will take place as long as the general problem persists. It is the profession's responsibility, and the agencies', to see that the requirements are reasonable and we should remain on top of the problem at all times. When laws are written, which is usually a result of a disaster, reasonable standards do not often result and changes need to be enacted. If we don't act, then we may be stuck with unreasonable requirements.

Coal waste disposal regulations are directed toward two different problems. One, the problems associated with degradation of the environment, and two, the problems concerning safety of the miners and the public. Many state regulations are more concerned with environment and reclamation requirements than with the public safety.

Initially the federal regulations ignored the environmental problems and limited their concern to safety of miners. The enactment of the Surface Mining Reclamation and Enforcement Act of 1977, and possibly of the Environmental Protection Agency, has a tremendous environmental impact.

The requirements to meet environmental and safety regulations will result in an increased cost of coal production, but it has been an area of neglect and one that can be satisfied reasonably if it is approached on a technical basis.

There need to be some adjustments to avoid the overlapping responsibilities of the different regulatory agencies. Hopefully, in time this problem can be overcome.

Conclusions

Unless present regulations on coal waste structures are followed and other failures because of poor construction prevented there will be unreasonable requirements added to the present regulations. To prevent over-regulation government and industry must work together to ensure that prudent engineering practices in accordance with present requirements are applied.

However, if the failures had not occurred at Aberfan, South Wales and Buffalo Creek, West Virginia, and if the coal industry was building its structures in accordance with good engineering practices, there would not be regulations in the United States today regarding the safety of coal waste structures.

Regulations by no means will eliminate the hazards created by coal waste deposits unless a genuine effort is made by industry to construct its structures properly. A regulatory agency will never be able to totally enforce its requirements if companies intend not to comply. So many violations can be disguised that compliance cannot be insured unless the entire construction of all coal waste deposits is continually inspected; and that would be an unfeasible task.

In order for an effective program to work, the requirements must be sensible and those involved, in addition to having a positive attitude, must be convinced that they are applicable to the program. In the case of mine waste disposal, the use of stringent earth dam criteria will usually provide a safe mine waste structure. The criteria need to be applied by engineers who are knowledgeable and have experience in the geotechnical, hydraulic, and hydrologic fields.

Due to the rise in environmental and safety restrictions regarding the disposal of coal wastes, it is advisable that companies approach the technical problems of hiring in-house engineers with this background or get a reliable consultant who can provide them with the required services. This will restrict other failures from occurring and hopefully will hinder the need for further regulations in the disposal of coal mine wastes.

References

1978, "Design Guidelines for Coal Waste Structures." U.S. Department of Interior, Mining Enforcement and Safety Administration, Denver and Pittsburgh Technical Support Centers.

Rose, Jerry G., 1976, Proceedings of the Second Kentucky Coal Refuse Disposal and Utilization Seminar, College of Engineering, University of Kentucky.

Schlick, Donald P., 1975, "Federal Interest in Coal Mine Waste Disposal." U.S. Department of Interior, Mining Enforcement and Safety Administration, IR 1023.

Engineering Properties of Coal Mine Refuse

23

C. Y. Chen, Engineering Manager,
Michael Baker, Jr., Inc., Beaver, Pennsylvania, USA

Introduction

Increasing production of coal increases the generation of coal refuse. There are two major types of coal refuse: coarse and fine refuse. In general, coarse refuse consists of run-of-mine material arising from driving the roadways or drifts and the coarse material separated in the washing plant. Fine refuse consists mainly of slurry and tailings. Slurry is the fine material remaining in suspension in the water after the washing process. Tailings are the fine rejected material from the froth flotation process used for cleaning fine coal.

Early disposal techniques, which are still quite popular, involved simply piling coal refuse on dumps. The dumps thereafter became a significant source of air, water, and visual pollution in coal mining regions. Another common disposal technique involves the construction of embankments out of coarse refuse to confine fine refuse and water generated from preparation plants. Most of these embankments are poorly engineered and constructed and have also become safety hazards to the outskirts of embankments and/or downstream areas.

The problem generated by coal refuse disposal was highlighted by a tragic event in the United Kingdom in 1966 where a refuse bank slide at Aberfan killed and injured many school children and caused a great deal of property loss. The aftermath of the Aberfan disaster has generated many studies, under the leadership of the British National Coal Board, to search for a better solution in dealing with this disposal problem. In the United States, it was not until the Buffalo Creek disaster in 1972, that national attention began to focus on this problem. The disaster which killed 125 people, injured hundreds, and left more than a thousand people homeless, was the result of the collapse of a coal waste dam in the valley of Buffalo Creek, West Virginia. Following the Buffalo Creek disaster, a series of investigations were conducted to determine the cause of the disaster. Those studies and investigations, conducted both in the United States and abroad, have come to the conclusion that in order to ameliorate the coal refuse disposal problem, it is essential to understand the material properties necessary to utilize coal refuse material in large-scale engineering construction and to properly plan and design disposal areas to become safe and useful reclaimable land. Only by doing this can a disposal system that is economic and environmentally acceptable be planned and implemented.

This study, consequently, focuses the investigation on the engineering properties of coal refuse.

In addition to its inherent earthy properties, coal mine refuse also possesses three unique properties: degradation, combustion, and acid generation potential.

Based on the analyses of the data collected from the United Kingdom and the United States, the engineering properties affecting those three unique properties are discussed. The findings and conclusions resulting from this study should provide a better understanding of the properties of coal refuse and should also be useful to the engineering profession and the mining industry.

List of Symbols

The symbols used throughout this chapter, and their definitions, are listed below:

A = percent ash content
ANOVA = analysis of variance
B = Btu per pound
C = percent fixed carbon
c = unit cohesion
$C.I.$ = confidence interval
$d.f.$ = degrees of freedom
D = degree of compaction
D_{10} = effective grain size or the size at which 10 percent is finer
D_{50} = average grain size or the size at which 50 percent is finer
e = void ratio
F = fine fraction or percent finer than No. 200 U.S. standard sieve by weight
F ratio = variance ratio
G = specific gravity
K = number of regressor
k = permeability
$\log$ = common logarithms
$\ln$ = natural logarithms
MD = maximum dry density
n = number of data
OMC = optimum moisture content
p = corresponding probability (for one-sided) of a larger absolute value of t
r = number of means compared
R = sample coefficient of correlation
R^2 = sample coefficient of determination
$\overline{R^2}$ = corrected (for degree of freedom) R^2
s = sample standard deviation

S_{xy} = standard error of estimate
Sp^2 = pooled variance of samples
S'_{20} = drained shear strength at 1.41 kg/cm² (20 psi) normal stress
S'_{50} = drained shear strength at 3.52 kg/cm² (50 psi) normal stress
t = t statistics assuming null hypothesis is true, or student's t variable
V_a = percent air void, expressed as a percentage of total volume
$V\underline{C}$ = percent volatile matter and fixed carbon
$\overline{X}$ = sample mean of X
Y_i = measured value
$\ddot{Y}_i$ = calculated value
μ = population means
σ = population standard deviation
φ = angle of shearing resistance

Degradation and Shear Strength

The studies conducted in Great Britain by Taylor and Spears, 1972, and Wimpey Laboratories Ltd, 1972, for the British National Coal Board (NCB) have indicated that the mechanical breakdown during placement of material is likely to be the predominant cause of degradation and that the long-term physicochemical changes, after the refuse bank has been constructed, are only significant on the surface of the bank. Accordingly, data from various sources including the National Coal Board were obtained to determine the extent of degradation caused by a given compactive effort (Standard Proctor) and the effect of such degradation on shear strength of coarse refuse.

Laboratory test results

A total of 26 laboratory compaction test results were obtained from Doyle et al., 1975, Moulton et al., 1974, and Wimpey Laboratories Ltd., 1972. Seventy-six sets of coarse refuse drained strength results with gradation curves were obtained from Doyle et al., 1975; Taylor and Spears, 1972; Wimpey Laboratories Ltd., 1972; Busch et al., 1974; Poellot and Almes, 1975; and W. A. Wahler and Associates, 1973. A total of 33 sets of fine refuse drained strength results with gradation curves were obtained from Doyle et al., 1975; Wimpey Laboratories Ltd., 1972; Busch et al., 1975; Poellot and Almes, 1975; and W. A. Wahler and Associates, 1973. Among the 33 results, only 20 sets of strength results have corresponding specific gravity results. The shear strengths used in the analyses were determined from two different normal stress levels from the effective Mohr strength envelope obtained from Consolidated-Undrained or Consolidated-Drained Tests. Also used in this study were 24 fixed carbon content with corresponding specific gravity results obtained from Doyle et al., 1975.

The t distribution was used to compare the fine fraction or the portion finer than a No. 200 sieve before and after laboratory compaction by the Standard Proctor method. From 26 sets of data, the paired value or the difference in fine fraction for each refuse sample before and after compaction was calculated. This new set of data consisted of a new statistical sample for use in the comparison of two populations; namely, the fine fractions before and after compaction. The average difference between two populations at 95 percent confidence level, as calculated from *t* distribution, was 4.01 ± 2.84, and at 99 percent confidence level was 4.01 ± 3.85. These intervals all exclude the values of zero difference. Consequently, the null hypothesis that there is no difference can be rejected with a very high level of confidence. Figure 23.1 displays the various test results along the line of equality and indicates that the fine fraction does increase significantly after laboratory compaction by the Standard Proctor method.

Regression and correlation analysis
In order to determine whether or not the shear strength of coal refuse is affected by the fine fraction, an attempt was made to establish the relationship between the shear strength and the fine fraction by regression and correlation analysis. A total of 76 sets of coarse refuse shear strength and fine fraction results were

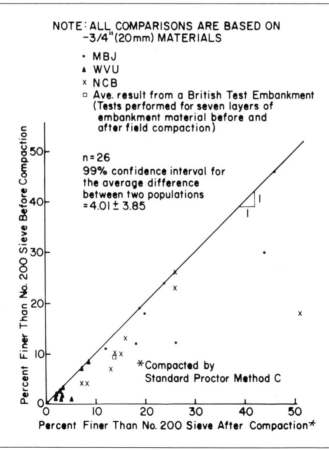

Figure 23.1. Graph displaying the test results after laboratory compaction of coarse refuse.

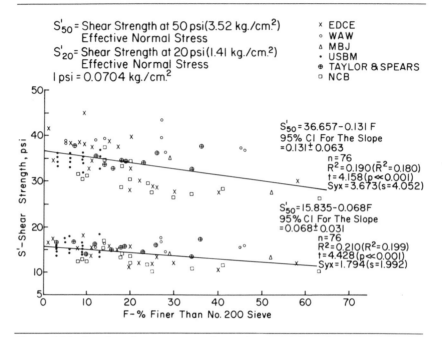

S'_{50}= Shear Strength at 50 psi(3.52 kg./cm².)
Effective Normal Stress
S'_{20}= Shear Strength at 20 psi(1.41 kg./cm².)
Effective Normal Stress
1 psi = 0.0704 kg./cm².

x EDCE
o WAW
Δ MBJ
• USBM
⊕ TAYLOR & SPEARS
◻ NCB

S'_{50}= 36.657−0.131 F
95% CI For The Slope
=0.131 ± 0.063
n=76
R^2=0.190($\bar{R}^2$=0.180)
t=4.158(p≪0.001)
Syx=3.673(s=4.052)

S'_{50}=15.835−0.068F
95% CI For The Slope
=0.068 ± 0.031
n=76
R^2=0.210(R^2=0.199)
t=4.428(p≪0.001)
Syx=1.794(s=1.992)

Figure 23.2. Results of regression and correlation analysis for coarse refuse.

obtained for study from six different sources, as previously described. Also obtained for analysis were 33 sets of fine refuse shear strength and fine fraction results from five different sources. For each strength test result, two shear strengths were determined from the effective Mohr strength envelope at 1.41 and 3.52 kilograms per square centimeter (20 and 50 psi) normal stress respectively. These two sets of shear strengths were then compared and regressed, separately, with the set of fine fractions obtained from the gradation curves representing the refuse samples from which the respective strength tests were performed.

Coarse refuse. Figure 23.2 shows the results of regression and correlation analysis for coarse refuse. The results indicate that the fine fraction does affect the shear strength quite significantly with a confidence level of at least 95 percent. As shown in Figure 23.2, the 95 percent confidence intervals for the slopes of both equations all exclude the zero values of the null hypothesis. Also the t values calculated for both equations were 4.158 and 4.428 for high and low normal stress levels, respectively, whereas the critical t value at 95 percent confidence level was only about 2.0. Consequently, the null hypothesis that there is no effect or no relation between the shear strength and fine fraction of coarse refuse can be rejected with very high levels of confidence.

Regression analysis does show the existence of the relationship between shear strength and fine fraction with very high levels of confidence. However, as shown by the calculated sample coefficients of determination, R^2, which are 0.19 and 0.21 for high and low normal stress levels, respectively, this relationship was not very strong. This is reasonable and expected because of the wide variety of data sources and the fact that many other factors such as void ratio, density, and moisture content, etc., which also affect the shear strength, were not entered into the analysis.

Fine refuse. Regression and correlation analyses were also performed for 33 sets of fine refuse data. The results of the analyses are shown in Figure 23.3. The t values calculated were 3.294 and 2.247 for the equations with high and low normal stress levels, respectively. The critical t value, however, was found to be 2.042 for the 95 percent confidence level. This means that the effect of fine fractions on shear strength is significant at the 95 percent confidence level at both normal stress levels. This conclusion is also confirmed by the 95 percent confidence intervals calculated for the slopes of both equations. The relatively low strength of correlation for the low normal stress level was found to be more pronounced than that of the coarse refuse.

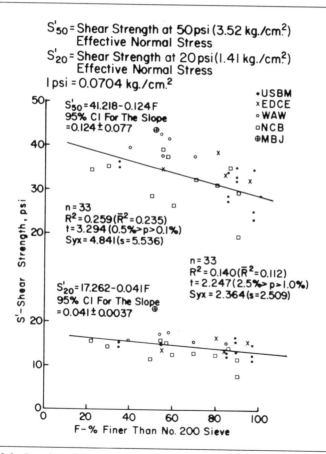

S'_{50} = Shear Strength at 50 psi (3.52 kg./cm.²) Effective Normal Stress

S'_{20} = Shear Strength at 20 psi (1.41 kg./cm.²) Effective Normal Stress

1 psi = 0.0704 kg./cm.²

• USBM
× EDCE
○ WAW
□ NCB
⊕ MBJ

$S'_{50} = 41.218 - 0.124 F$
95% CI For The Slope $= 0.124 \pm 0.077$

$n = 33$
$R^2 = 0.259 (\bar{R}^2 = 0.235)$
$t = 3.294 (0.5\% > p > 0.1\%)$
Syx $= 4.841 (s = 5.536)$

$n = 33$
$R^2 = 0.140 (\bar{R}^2 = 0.112)$
$t = 2.247 (2.5\% > p > 1.0\%)$
Syx $= 2.364 (s = 2.509)$

$S'_{20} = 17.262 - 0.041 F$
95% CI For The Slope $= 0.041 \pm 0.0037$

S' – Shear Strength, psi

F – % Finer Than No. 200 Sieve

Figure 23.3. Results of regression and correlation analysis for fine refuse.

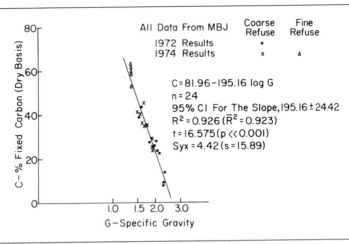

Figure 23.4. Graph showing the relationship between specific gravity and fixed carbon (coarse and fine refuse).

Table 23.1. Summary of Regression Analysis between Shear Strength and Fine and/or Specific Gravity

		Models	
Statistics	$S'_{20} = 21.083$ $- 30.316$ (Log G)	$S'_{20} = 18.661$ $- 0.0495F$	$S'_{20} = 21.953$ $- 24.643$ (Log G) $- 0.028F$
R^2	0.354	0.219	0.411
$\bar{R}^2$	0.318	0.176	0.342
t			
F		2.245 $(2.5\% > p > 1\%)$	1.286 $(25\% > p > 10\%)$
Log G	3.137 $(0.5\% > p > 0.1\%)$		2.354 $(2.5\% > p > 1\%)$
95% C.I.			
F		±0.0463	±0.0458
Log G	±20.304		±22.086
S_{yx}	2.052	2.255	2.015
s	2.483	2.483	2.483

Note: These results were obtained by analysis of fine refuse. S'_{20} = drained shear strength at 1.41 kg/cm² normal stress, and $n = 20$ is the number of data.

It was thought that a better regression model for shear strength might be obtained by including coal content as another independent variable. Unfortunately, the data did not allow the inclusion of coal content directly, but an indirect assessment was possible. There is a strong relationship between carbon content and specific gravity as shown in Figure 23.4, which indicates a very high degree of confidence level and coefficient of determination. Carbon content, therefore, was introduced indirectly by adding specific gravity as another regressor in the regression analysis.

The introduction of specific gravity as another regressor in the regression analysis for coarse refuse was not successful, because the available data obtained were found to be too scattered and insufficient to establish any meaningful relationship with the shear strength of coarse refuse. On the other hand, the use of specific gravity as another regressor in the multiple regression analysis was found to be very promising in the analysis for the shear strength of fine refuse.

A total of 20 sets of fine refuse shear strength data, with a corresponding fine fraction and specific gravity for each shear strength result, were obtained from various sources as previously described. Regression analyses, both simple and multiple, were performed. The results of these analyses summarized in Tables 23.1 and 23.2, indicate that the multiple regression models with both fine fraction

Table 23.2. Summary of Regression Analysis between Shear Strength and Fine and/or Specific Gravity

	Models		
Statistics	$S'_{50} = 50.766$ $- 82.512 \,(\text{Log } G)$	$S'_{50} = 45.918$ $- 0.159F$	$S'_{50} = 54.067$ $- 60.977 \,(\text{Log } G)$ $- 0.106F$
R^2	0.518	0.449	0.681
$\bar{R}^2$	0.491	0.418	0.643
t			
F		3.825 $(p < 0.1\%)$	2.949 $(0.5\% > p > 0.1\%)$
Log G	4.398 $(p \ll 0.1\%)$		3.523 $(0.5\% > p > .01\%)$
95% C.I.			
F		±0.0875	±0.0758
Log G	±39.414		±36.530
S_{yx}	3.983	4.261	3.333
s	5.584	5.584	5.584

Note: These results were obtained by analysis of fine refuse. S'_{50} = drained shear strength at 3.52 kg/cm^2 normal stress, and $n = 20$ is the number of data.

and specific gravity as regressors are the better models. The multiple regression models, while weakening the statistical significance of each independent variable somewhat, as they should have, increased the strength of correlation and reduced the standard error of estimate quite substantially as shown in Tables 23.1 and 23.2.

Combustion, Air Content and Related Properties

Coal mine refuse may be ignited accidentally by external agencies, such as the tipping of hot ashes or the lighting of open fires or braziers. The most common cause of burning, however, is spontaneous combustion of carbonaceous materials, often aggravated by the presence of pyrite. This is an oxidation process in which the material combines with oxygen from the air with the evolution of heat. Oxidation usually proceeds very slowly at ambient temperatures, but increases rapidly and progressively as the temperature rises. Neither oxidation, spontaneous combustion, nor burning will occur in the absence of air. For burning to occur with visible flame it is normally necessary for the material to be combustible, to reach its specific ignition temperature, and for sufficient oxygen to be supplied from the air.

Factors influencing spontaneous combustion

The factors which influence spontaneous combustion are temperature, coal rank, pyrite content, moisture, air voids, and specific surface. The details of how those factors qualitatively affect spontaneous combustion have been discussed by Thomson et al., 1970.

It is very difficult, if not impossible, to determine and quantify every factor which affects spontaneous combustion. However, for a given type of coal refuse, air voids, and to a certain extent the moisture content, can be easily and economically determined and controlled both in the laboratory while undergoing tests and in the field.

Air content/voids. The effectiveness of controlling the air voids to reduce the combustion potential has been documented by Isaac and Troughton, 1971, and Isaac, 1972, for a British instrumented demonstration embankment constructed of coal refuse at the Cortonwood colliery in the South Yorkshire Area. Consequently, air content was selected as a variable to be related with various degrees of compaction at the optimum moisture content determined by the Standard Proctor method. The extent of the decrease in air content as the degree of compaction increases was also statistically analyzed to confirm that the population means of the air voids differ quite significantly with the change in the degree of compaction. In addition, the 95 percent simultaneous confidence intervals for the difference in population means of air voids between various degrees of compaction were also statistically determined. This allowed the estimation of the population parameters of the change in air voids as the degree of compaction varied. The aforementioned analyses (Chen, 1976) were performed respectively on 15 representative U.K. coal refuse results, on 16 U.S. Appalachian bituminous coal refuse results, and on the combined U.K. and U.S. results. This chapter presents only the result of analysis on U.S. refuse.

Since the combustible matter and some other compositions within the coal refuse are also important for potential combustion, those variables, such as carbon content, volatile matter, and caloric value, were consequently used to statistically relate with the specific gravity of coal refuse. The maximum dry density, determined from the Standard Proctor method, is important in determining the degree of compaction in the field and the air contents of refuse mass. This maximum dry density was, therefore, used as a dependent variable statistically correlated with specific gravity of coal refuse.

Standard Proctor Results. The U.S. data used in this study were obtained from Doyle, et al., 1975. The U.K. data were obtained both from Wimpey Laboratories Ltd., 1972, and Isaac and Troughton, 1971. All U.S. data were from the Appalachian bituminous coalfield. As for the U.K. data, seven were from the demonstration embankment at Cortonwood colliery and another eight were from eight different representative collieries throughout the country. All U.K. data also represent bituminous coal refuse except in one case from a semianthracite region.

Using the Proctor results obtained, percent air voids were calculated for various degrees of compaction. The percent air void is defined as the volume of the air void expressed in terms of the percentage of the total volume of a given coal refuse mass. The results and statistics of average air void versus various degrees of compaction are shown in Figure 23.5. The F ratio calculated during the

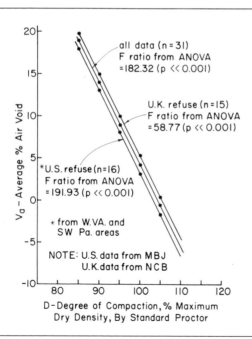

Figure 23.5. Results of average air void versus degree of compaction (coarse refuse).

analysis of variance (ANOVA) was intended to test the hypothesis that population means of air void at various degrees of compaction are all equal. This F ratio
is defined as the ratio of explained variance between the various degrees of compaction to the unexplained variance within various degrees of compaction. This
analysis was performed to statistically confirm that the population means of air
void between various degrees of compaction differ quite significantly with the
change in the degree of compaction. The calculated F ratios were 191.93, 58.77,
and 182.32 for the U.S., U.K., and combined data respectively. However, the corresponding critical F values at 99.9 percent confidence level for all cases were
only about 4.4. The apparent high calculated F ratio consequently rejects the null
hypothesis of equal population means with very high levels of confidence.

Having concluded that the compaction does significantly change the percent
air void, the amount of change or difference was calculated by the method of
multiple comparisons or simultaneous confidence intervals using the following
equation:

$$\Sigma C_i \mu_i = \Sigma C_i \bar{X}_i \pm \sqrt{F_{0.05}} \; Sp \; \sqrt{\frac{(r-1)}{n}(\Sigma C_i^2)} \tag{1}$$

Where $\Sigma C_i = 0$

$F_{0.05}$ is the critical value of F

Table 23.3 shows the 95 percent simultaneous confidence intervals for the difference in population means of air void between various degrees of compaction.
The results indicate that substantial reduction in air void can be achieved even
with a 5 percent incremental increase in degree of compaction.

Table 23.4 summarizes the results of regression analyses performed between
specific gravity and various compositions and maximum dry density determined

---

**Table 23.3. The 95% Simultaneous Confidence Intervals for the Difference
in Population Means of Air Voids between Various Degrees of Compaction
(Standard Proctor)**

	85%	90%	95%	100%	105%	110%
85%	0	4.85	9.69	14.54	19.30	24.24
90%	− 4.85	0	4.84	9.69	14.45	19.39
95%	− 9.69	− 4.84	0	4.85	9.61	14.55
100%	−14.54	− 9.69	− 4.85	0	4.76	9.70
105%	−19.30	−14.45	− 9.61	− 4.76	0	4.94
110%	−24.24	−19.39	−14.55	− 9.70	− 4.94	0

Note: The listed values above have a range of ±3.25. All differences are statistically
significant and the 95% simultaneous confidence intervals were calculated from equation
(1) with $r = 6$ and $n = 16$. Differences in population means were estimated from sample
means (coarse refuse, U.S. data).

Table 23.4. Summary of Regression Analysis between Specific Gravity, G, and Various Properties of Coarse Refuse

Dependent variable	B	C	VC	MD
Independent variable	G	$\text{Log } G$	$\text{Log } G$	G
No. of data (n)	19	19	19	31
Intercept	21593.72	73.95	100.65	14.70
Coefficient	-8031.30	-167.80	-189.50	40.85
95% C.I. for coef.	± 1148.05	± 32.35	± 47.92	± 3.84
t	11.703	10.95	8.344	21.732
R^2	0.889	0.876	0.805	0.943
$\bar{R}^2$	0.882	0.869	0.794	0.941
S_{yx}	0.77×10^3	3.98	5.90	3.68
s	2.25×10^3	10.98	12.94	15.03

Note: All data were from the Appalachian bituminous coal region except the one with $n = 31$, which included 15 U.K. results with one from the semianthracite region.
MD = Maximum Dry Density in pcf at Optimum Moisture Content (OMC), 95% CI for $OMC = 12.09 \pm 0.87$ ($s = 2.36$).

by Standard Proctor, respectively. The results indicate that the specific gravity affects those properties quite significantly and that the correlation for each analysis is excellent. This is evidenced by high t, R^2 and low S_{yx} values for each analysis.

Permeability

According to Bragg, 1973, the perpetual treatment of acid water from coal refuse dumps or from embankments constructed with coal refuse may be uneconomical. The only practical and economical method available today in controlling this pollution problem appears to be the isolation of acid generating ingredients such as pyrite from the other reagents—oxygen and water. Accordingly, the control of the permeability and air content of mine waste has become one of the most important measures in achieving this purpose.

This chapter focuses the study on the permeability of coal refuse by determining the relationship between the permeability determined in the laboratory and various index properties, and by assessing the effect of the reduction in void ratio caused by compaction or other means on the permeability of coarse and fine coal refuse.

Laboratory permeability results

A total of 47 sets of laboratory permeability results with void ratios and gradation curves were obtained from Busch et al., 1974, Doyle et al., 1975, and W. A. Wahler and Associates, 1973. Thirty-eight sets of laboratory permeability results with void ratio and gradation curves for fine refuse were also obtained from Busch et al., 1975, and W. A. Wahler and Associates, 1973. Those coal refuse samples were all obtained from the Appalachian bituminous coal region.

As a first step toward determining the effect of void ratio on permeability, the traditionally recognized straight line relationships between k and e^2, and log k and e, as presented in the soil mechanics textbooks such as those by Lambe and Whitman, 1969, Sowers and Sowers, 1970, and Tuma and Abdel-Hady, 1973, were examined for both coarse and fine refuse. For coarse refuse, the linear relationship for k versus e^2 was found to have a $R^2 = 0.066$ and, for log k versus e the $R^2 = 0.161$, indicating the lack of explanatory power of these two models. Other statistics of these analyses obtained further support this conclusion. As for fine refuse, the similar analyses for both models showed even poorer results. The equations obtained from regression analyses suggest two intuitively incorrect models, i.e., k decreases as e increases. Consequently, those models as suggested in the literature for soils do not apply to coal refuse, and particularly not to the fine coal refuse.

Model established

As described in various textbooks by Lambe and Whitman, 1969, Sowers and Sowers, 1970, and Tuma and Abdel-Hady, 1973, and in the paper by Barber and Sawyer, 1952, the void ratio and grain size of the soil do affect its permeability. Because of this prior knowledge and the variables used in establishing regression models for permeability versus void ratio and various grain sizes for mill tailings by Bates and Wayment, 1967, and the study conducted by Mittal and Morgenstern, 1975, on tailing sands, the effect on permeability, k, by void ratio, e, average size, D_{50}, fine fraction, F, and their various products were examined for coarse refuse. This was followed by the performance of a series of simple regression analyses with $\ln(k)$ as the dependent variable and $\ln(eD_{50})$, $\ln(e)$, $\ln(D_{50})$, and $\ln(e^2D_{50})$ as independent variables.

Coarse refuse. The results of those simple regression analyses revealed that the model with $\ln(e^2D_{50})$ as an independent variable was the best model among those four, as indicated by the highest t value, R^2, and the lowest standard error of estimates. Further analyses using multiple regression were performed. Discarding the models that showed the independent variables to be collinear, these analyses revealed a best regression model with two independent variables, $\ln(e)$ and $\ln(D_{50})$. This model confirms that the model with one independent variable of $\ln(e^2D_{50})$ as previously discussed is a very good model. This multiple regression model when expressed in a nonlinear form yielded the following model:

$$k = 393\, e^{3.631}\, D_{50}^{1.311} \tag{2}$$

This model is shown in Figure 23.6 which also shows the comparison between calculated and measured $\ln(k)$. Consequently, for coarse refuse, the model given by equation 2 is the best model obtained in this study. Although the R^2 is only 0.504 and $\overline{R}^2$ only 0.481, it is a good model considering wide variety and scatter of the permeability results (range $= 0.031 \times 10^{-6}$ to $1,600 \times 10^{-6}$ centimeters per second, average $= 116.46 \times 10^{-6}$ centimeters per second and standard deviation $= 307.00 \times 10^{-6}$ centimeters per second obtained for this analysis.

Fine refuse. Regression analyses were also performed for 38 sets of fine refuse data. A similar procedure used for coarse refuse was followed to select variables

$$\ln(k) = 5.975 + 3.631 \ln(e) + 1.311 \ln(D_{50})$$
$$(\text{or } k = 393 e^{3.631} D_{50}^{1.311})$$
$$R^2 = 0.504 \quad \bar{R}^2 = 0.481$$

$t = 5.07$ for coef. of $\ln(e)$ (95% C.I. $= \pm 1.445$)
$t = 5.11$ for coef. of $\ln(D_{50})$ (95% C.I. $= \pm 0.518$)
$n = 47$

$$Syx = \sqrt{\frac{\Sigma(Y_i - \hat{Y}_i)^2}{n - K - 1}} = 2.22 \,(s = 3.085)$$

$$(K = 2)$$

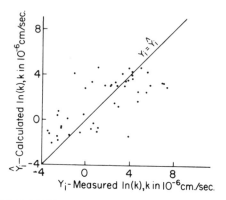

Figure 23.6. Graph showing the comparison between calculated and measured ln (k) for coarse refuse.

and models in performing the regression analyses. In addition to the parameters, e, D_{50} and F, as used in analyzing the coarse refuse, an additional parameter, D_{10} the effective size or the size at which 10 percent is finer, was also used (D_{10} was not used in analyzing the coarse refuse because of insufficient data).

A total of eight independent variables, $\ln(eD_{50})$, $\ln(e/F)$, $\ln(F)$, $\ln(eD_{10})$, $\ln(D_{10})$, $\ln(e)$, $\ln(D_{50})$, and $\ln(e^2 D_{50})$ were regressed on $\ln(k)$ individually. Among those eight bivariate regression models, the results indicate that the best model is the one with $\ln(eD_{10})$ as the independent variable. It has the highest t (8.929) values and R^2 (0.689), and the lowest S_{yx} (1.193). The regression equation is:

$$\ln(k) = 8.709 + 1.033 \ln(eD_{10})$$

$$\text{or } k = 6057 (eD_{10})^{1.033}$$

In order to search for a better regression model, multiple regression analyses were performed. By discarding the models whose independent variables are collinear, the model with two independent variables of $\ln(e)$ and $\ln(eD_{50})$ is considered to be the best model for fine refuse. The regression equation of the best model for fine refuse in nonlinear form is, therefore:

$$k = 2033 \, e^{3.486} \, D_{50}{}^{1.235}$$

This model is shown in Figure 23.7 which also displays the comparison between the calculated and measured $\ln(k)$. This model is considered to be a good one, given that the permeability results (range $= 0.3 \times 10^{-6}$ to 886×10^{-6} centimeters per second, average $= 63.406 \times 10^{-6}$ centimeters per second and standard deviation $= 160.152 \times 10^{-6}$ centimeters per second obtained and used in the analysis are quite scattered.

   Discussion. The model for predicting the permeability of mill tailings as developed by Bates and Wayment, 1967, was not pursued in this study because of the resultant multi-collinearities which developed when analyzing the refuse data. Besides, the Bates and Wayment model deals only with the mill tailings

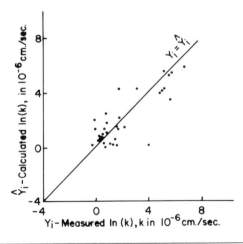

$\ln(k) = 7.617 + 3.486 \ln(e) + 1.235 \ln(D_{50})$
(or $k = 2033 e^{3.486} D_{50}{}^{1.235}$)
$R^2 = 0.695 \ \bar{R}^2 = 0.678$
$t = 3.266$ for coef. of $\ln(e)$ (95% C.I.$= \pm 2.169$)
$t = 8.860$ for coef. of $\ln(D_{50})$ (95% C.I.$= \pm 0.283$)
$n = 38$

$$S_{yx} = \sqrt{\frac{\Sigma(Y_i - \hat{Y}_i)^2}{n-K-1}} = 1.248 (s = 2.167)$$

$(K=2)$

Figure 23.7. Graph showing the relationship between calculated and measured ln (k) for fine refuse.

Table 23.5. Ranges of Material Properties Used in Regression Analysis

	Coarse refuse ($n = 47$)	Fine refuse ($n = 38$)
Permeability, k (10^{-6} cm/sec)	0.031-1600	0.3-886
Void ratio, e	0.202-1.14	0.495-1.256
Average size, D_{50} (mm)	0.14-7.0	0.0022-0.4
Effective size, D_{10} (mm)	–	0.00023-0.088
Fine fraction, F	0.03-0.46	0.10-0.97

which are cohesionless, free of clays and micas, and which have a well-defined grain size distribution.

The results of the preceding analyses, both for coarse and fine refuse, strongly indicate that, among all the index properties selected and used in the regression analyses, the void ratio appears to be the most important one.

The results of multiple regression analyses also indicate that the coefficients of determination for the case performed for fine refuse are generally higher than the case performed for coarse refuse. The reason for this could be attributed to the fact that the data for coarse refuse were obtained from three different sources whereas the data for fine refuse were obtained from only two different sources. Because of this, the stochastic errors were smaller for the fine refuse than for the coarse refuse, thereby increasing the coefficient of determination in the regression analysis.

Table 23.5 shows the ranges of material properties used in the regression analyses for permeability and other index properties.

Assumptions. Certain assumptions were made for the regression analyses where inferences were made for the parent populations from which samples were drawn. These assumptions are mainly the independence and normality of data. Because of the methods and procedures used in developing and obtaining data, and the fact that each sample was obtained from an infinite population, the probabilistic sampling procedure was assured. Accordingly, the most important assumption, the independence of the random variables, is considered to be valid. The normality assumption is met because of the sample size, the use of t distribution and the transformation of variables. Visual inspection of the data distributions confirmed the normality assumptions.

Summary of Various Material Properties

The 90 percent confidence intervals for the population means of various coal refuse properties for each classification made in accordance with the Unified Soil Classification system and for all data are summarized in Tables 23.6 and 23.7. The sources of data including methods of determination have been previously described. The conditions under which those materials were tested, were either samples obtained from "undisturbed" shelby tubes or materials fabricated or remolded to in situ or anticipated field conditions.

Table 23.6. The 90% Confidence Intervals for Average Coarse Refuse Properties Versus Unified Soil Classification (USC)

Classification (USC)	No. of data (n)	Permeability (K [x10^{-6} cm/sec])	Void ratio (e)	No. of data (n)	Shear strength	
					c' [psi]	(tan ϕ')
GM	—	—	—	7	2.07 ±1.62	0.66 ±0.03
GW-GM	4	171.275 ±173.970	0.297 ±0.054	12	2.37 ±0.97	0.67 ±0.03
SM	28	161.278 ±123.760	0.449 ±0.081	20	0.455 ±0.351	0.693 ±0.039
SW-SM	—	—	—	5	0	0.76 ±0.11
CL	12	0.673 ±0.534	0.247 ±0.015	—	—	—
All data*	57	183.697 ±72.601	0.354 ±0.045	81	1.36 ±0.341	0.652 ±0.016

*Includes those data without classification and those with classification where the number of data are fewer than are shown in the table.

Table 23.7. The 90% Confidence Intervals for Average Fine Refuse Properties Versus Unified Soil Classification (USC)

Classification (USC)	No. of data (n)	Permeability (K [×10⁻⁶ cm/sec])	Void ratio (e)	No. of data (n)	Shear strength	
					(c'[psi])	(tan φ')
ML	22	21.268 ±26.484	0.845 ±0.056	11	1.735 ±0.946	0.654 ±0.061
SM	10	233.683 ±149.25	0.775 ±0.098	5	2.620 ±3.782	0.704 ±0.026
MH	6	1.798 ±1.004	1.092 ±0.084	3	2.867 ±3.760	0.500 ±0.265
All data*	38	71.461 ±45.511	0.865 ±0.049	33	1.562 ±0.646	0.649 ±0.036

*Includes those data without classification and those with classification where the number of data are fewer than are shown in the table.

Conclusions

The following are the conclusions derived as a result of this study. These conclusions represent only the types of coal refuse investigated; namely, the Appalachian bituminous coal refuse and British coal refuse, as indicated in the sources of data. They are not the general conclusions to be applied to other types of coal refuse which were not investigated in this study.

1. The increase in fine fraction from degradation after the laboratory compaction by the Standard Proctor method is quite significant.
2. The amount of fine fraction significantly affects the shear strength.
3. The shear strength decreases with the increase of fine fraction for both coarse and fine refuse. The coefficients of determination for the established statistical relationships, however, are not very high.
4. The content of fixed carbon was found to be highly correlated with specific gravity of coal refuse.
5. The shear strength of fine refuse decreases with an increase of specific gravity. Since the coal content can be expressed as a function of specific gravity, this conclusion also suggests that the shear strength of fine refuse may increase with the increase of coal content.
6. The multiple regression equation is a better model in estimating the shear strength of fine refuse from the fine fraction and specific gravity.
7. The increase in compactive effort will substantially reduce the percent air void, and the population parameters of the reduction in air void as degree of compaction increases can be determined.
8. The specific gravity which was determined on the material finer than a No. 10 sieve was found to be highly correlated with the heating value and other compositions for Appalachian coal refuse. Consequently, the specific gravity can be used to estimate the heating value and other compositions of coal refuse from the equations developed in this study.
9. Based on the analysis of the data collected for U.S. Appalachian and various typical U.K. refuse, the maximum dry density determined by the Standard Proctor has an excellent correlation with the specific gravity determined on the material finer than a No. 10 sieve.
10. The laboratory permeability results which were obtained almost at full saturation and corrected to the constant temperature of 20 degrees Centigrade show considerable scatter.
11. The approximate linear relationships between $\log k$ versus e and k versus e^2 which are traditionally believed to be true for soil cannot be reasonably established for coal refuse.
12. The best equations which can be used to describe the empirical relationships between the permeability and both void ratio and grain size are the multiple regression equation with both e and D_{50} as independent variables. This conclusion is true for both coarse and fine refuse.
13. Among all those factors affecting permeability, the void ratio e is the most important one. This suggests that the reduction in void ratio, particularly by compaction, can substantially reduce the permeability of the material in accordance with the statistical relations developed in this study.

14. Within the ranges of various material properties shown in Table 23.5, the equations developed in this study can be used to describe and assess the permeability from various index properties.

Acknowledgments

The author wishes to acknowledge Michael Baker, Jr., Inc., for the use of computer facilities and for the assistance rendered in typing and drafting of the manuscript.

References

Barber, E. S. and Sawyer, C. L., 1952, "Highway Subdrainage." Proceedings of 31st Annual Meeting, Highway Research Board, Washington, D.C.

Bates, Robert C., and Wayment, William R., 1967, "Laboratory Study of Factors Influencing Waterflow in Mine Backfill-Classified Mill Tailings." R.I. 7034, Washington, D.C., U.S. Department of the Interior, Bureau of Mines.

Bragg, K., 1973, "Pollution Potentials and Interactions in Mine Waste Embankments in Canada—Their Costs and Prevention." Extraction Metallurgy Division, Report No. EMA 73-12, Department of Energy, Mines and Resources, Mines Branch, Ottawa, Canada.

Busch, R. A., Backer, R. R., and Atkins, L. A., 1974, "Physical Property Data on Coal Waste Embankment." R.I. 7964, U.S. Department of the Interior, Bureau of Mines.

Busch, R. A., Backer, R. R., and Atkins, L. A., 1975, "Physical Property Data on Fine Coal Refuse." R.I. 8062, U.S. Department of Interior, Bureau of Mines.

Chen, C. Y., 1976, "Investigation and Statistical Analysis of the Geotechnical Properties of Coal Mine Refuse." Ph.D. Thesis, University of Pittsburgh, Pennsylvania.

Doyle, F. J., Chen, C. Y., Malone, R. D., and Rapp, J. R., 1975, "Investigation of Mining Related Pollution Reduction Activities and Economic Incentives in the Monongahela River Basin." PB. 244352/AS. NTIS, Springfield, Virginia.

Isaac, A. S., and Troughton, S. J., 1971, "An Instrumental Demonstration Embankment at Cortonwood Colliery—Interim Report." Mineral Products Report No. M.P. 6, Coal Research Establishment, National Coal Board, Stoke Orchard.

Isaac, A. S., 1972, "An Instrumented Demonstration Embankment at Cortonwood Colliery—Final Report." Mineral Products Report No. M.P. 13, Coal Research Establishment, National Coal Board, Stoke Orchard.

Lambe, T. William and Whitman, Robert V., 1969, "Soil Mechanics." New York: John Wiley and Sons, Inc.

Mittal, Hari K., and Morgenstern, Norbert R., 1975, "Parameters for the Design of Tailings Dams." Canadian Geotechnical Journal, Vol. 12, No. 2.

Moulton, L. K., Anderson, D. A., Seals, R. K., and Hussain, S. M., 1974, "Coal Mine Refuse: An Engineering Material." First Symposium on Mine and Preparation Plant Refuse Disposal, Coal and the Environment Technical Conference, Louisville, Kentucky.

Poellot, J. H. and Almes, R. G., 1975, "Personal Communication." E. D'Appolonia Consulting Engineers, Inc., Pittsburgh, Pennsylvania.

Sowers, George F. and Sowers, George B., 1970, "Introductory Soil Mechanics and Foundations." 3rd edition; New York, The Macmillan Company.

Taylor, R. K., and Spears, D. A., 1972, "The Geotechnical Characteristics of a Spoil Heap at Yorkshire Main Colliery." The Quarterly Journal of Engineering Geology, Vol. 5, No. 3.

Thomson, G. M., Rodin, S., Eccles, D. E., and Webb, S. B., 1970, "Spoil Heaps and Lagoons." National Coal Board—Technical Handbook, London.

Tuma, Jan J., and Abdel-Hady, M., 1973, "Engineering Soil Mechanics." Englewood Cliffs, New Jersey, Prentice Hall.

W. A. Wahler and Associates, 1973, "Analysis of Coal Refuse Dam Failure, Middle Fork, Buffalo Creek, Saunders, West Virginia, Volume 2 of 2, Appendices." P.B. 215-143/9, NTIS, Springfield, Virginia, Appendix B.

Wimpey Laboratories Ltd., 1972, "Review of Research on Properties of Spoil Tip Materials." National Coal Board Research Project, Wimpey Laboratories Ltd., Hayes, Middlesex, England.

Engineering Aspects of the 1972 Buffalo Creek Dam Failure

24

Richard L. Volpe, Manager of Technical Services,
W. A. Wahler & Associates, Palo Alto, California, USA

Introduction

On February 26, 1972, a coal refuse dam near Saunders, West Virginia, owned and operated by the Buffalo Mining Company, failed catastrophically. The resultant flooding of the Buffalo Creek Valley caused the deaths of 125 persons, the loss of over 500 homes, and extensive flood damage to other property in the valley. Like the Aberfan disaster in South Wales (Bishop, 1966), the Buffalo Creek failure drew to public attention the problems of coal refuse disposal. Prior to these events, the growing problems of refuse disposal had been a matter of little, if any, public concern, either in the United Kingdom or in the United States. In the United States, however, the ramifications of the events in Buffalo Creek Valley included the passage by Congress of the National Dam Inspection Act and the Strip Mine Act of 1972. The failure at Buffalo Creek also had a significant influence in the formulation of two new regulatory agencies; namely, the Office of Surface Mining under the Department of Interior and the Mining Safety and Health Administration under the Department of Labor.

Immediately after the disaster several investigations were started, some under government auspices, and others initiated by groups of concerned citizens. The most exhaustive geotechnical study of the failure was performed under contract to the U.S. Bureau of Mines by a private consulting firm (W. A. Wahler & Associates, 1973). This yearlong study, which produced extensive documentation on the failure, included the first major effort in the United States to develop the engineering properties of coarse and fine coal refuse material. This chapter attempts to summarize that work and present the engineering aspects of the failure analysis, including the establishment of refuse materials properties.

Area, Topography, and Hydrology

Buffalo Creek is located in Logan County, the center of the southern West Virginia bituminous coal field. Logan, the county seat, is about 113 kilometers (70 miles) southwest of Charleston. The town of Man, about 19 road kilometers (12 miles) southeast of Logan, is sited at the confluence of Buffalo Creek and the Guyandotte River. Buffalo Creek Valley extends east-northeast from Man, a distance of about 24 kilometers (15 miles), and the creek has its headwaters at a point common to Logan, Boone, and Wyoming counties.

The topography of southwestern West Virginia, including the Buffalo Creek area, is characterized by rugged hills. The action of numerous small streams has completely dissected the Appalachian plateau and created a region of high ridges and sharp, V-shaped valleys. The heavily forested valley walls generally slope at about 2.0 horizontal to 1.0 vertical; a dense growth of bushes and small trees comprises most of the ground cover. The surrounding hills attain a maximum elevation of 869 meters (2,850 feet) above sea level, and their summits are roughly concordant with the old plateau surface. There is very little level or gently sloping upland topography, and valley floors are, in general, very narrow with minor alluvial terrace development.

The flood plain of Buffalo Creek averages 122 meters (400 feet) in width. In places, it is considerably narrower, and rarely exceeds 152 meters (500 feet) in width. Buffalo Creek flows southwesterly in a sinuous course, covering 27 stream kilometers (17 miles) between Saunders and Man, a straight distance of 19 kilometers (12 miles). The elevation at Saunders, where the flood wave caused by the dam failure entered Buffalo Creek, is about 457 meters (1,500 feet). During the next 6.4 stream kilometers (4 miles) the elevation drops 91 meters (300 feet) and then slopes less abruptly to Man, where it is 213 meters (700 feet).

Buffalo Creek, with a drainage area of nearly 117 square kilometers 45 square miles), is one of the larger tributaries within Logan County to the Guyandotte River, which flows northwesterly past Man and Logan to discharge into the Ohio River near Huntington. The coal refuse dams that failed flooding Buffalo Creek Valley, were actually located on the Middle Fork, a tributary to Buffalo Creek. The total drainage area of Middle Fork is 301 hectares (745 acres); the area above the failed dam (Dam No. 3) is 265 hectares (654 acres).

Description of the Refuse Dams

The refuse dams and the history of their development emerged from the records of hearings held in West Virginia (1972) and in the U.S. Congress (1972) to determine the sequence of events which led to the failure.

Prior to the catastrophic failure, there were four dams in Middle Fork Valley. They were referred to as Dam No. 1 through Dam No. 4, proceeding in a downstream to upstream direction. The pools behind these dams were referred to as Pools No. 1 through No. 4, respectively. In addition to the dams, there was also a very large coarse refuse dump located near the mouth of the valley.

Apparently these embankments had not been constructed on the basis of engineered plans, and no studies had been made of the engineering properties of the coal refuse from which they were built. Neither the mining company nor the men building these structures considered them to be dams in the classical sense. As was brought out in testimony at government hearings, the specific intent of these refuse structures was to clarify the "black water" from the coal preparation plant

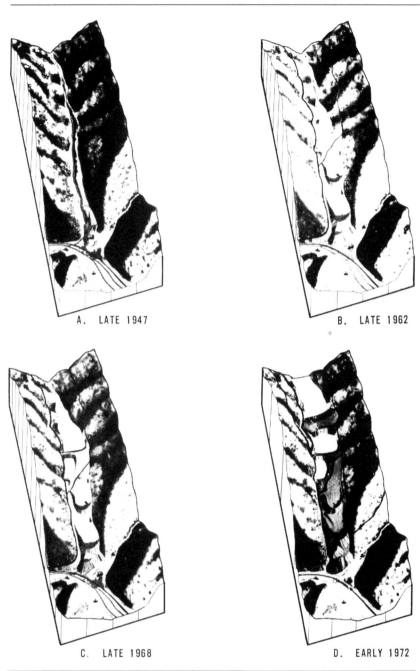

A.   LATE 1947

B.   LATE 1962

C.   LATE 1968

D.   EARLY 1972

Figure 24.1. Artist's conception of the history of the development of Middle Fork.

before it was discharged into the Middle Fork River.

Though there are few precise records or maps to illustrate the growth history of the refuse bank and dams, they were sufficient to reconstruct the overall sequence of events. The history of refuse structure development in the Middle Fork Valley (1947-1972) is reconstructed in Figure 24.1.

Dam No. 1

The dumping of coal refuse in Middle Fork Valley began in 1947 when the Lorado Mining Company completed its preparation plant on Buffalo Creek, about 1,067 meters (3,500 feet) upstream from Middle Fork. The solid refuse from this plant, amounting to 800 to 1,000 tons a day, was trucked to the mouth of Middle Fork Valley and dumped. Bulldozers graded the piles of dumped refuse to provide roadways and level areas on which the trucks placed the next layer of refuse piles. The refuse bank grew both vertically and up-valley almost continuously until construction of Dam No. 3 began. By November 1962, the refuse bank occupied virtually the entire flood plain of Middle Fork Valley. It measured about 488 meters (1,600 feet) long and 76 meters (250 feet) wide and varied between 46 to 58 meters (150 to 190 feet) in height. At this time (and subsequently), Dam No. 1 was actually a dike or an extension of the refuse bank on the right side of the valley.

Dam No. 2

Dam No. 2 was constructed in 1966 to replace Pool No. 1, which had been rendered useless due to extensive silting in the reservoir behind Dam No. 1. Also, the company needed to find an additional space for disposing of coarse refuse. Dam No. 2, located about 183 meters (600 feet) upstream, was constructed by dumping refuse across the width of the valley on the fine-grained deposits formed in the reservoir behind Dam No. 1, and was completed in late 1967.

Dam No. 3

The progressive upstream damming of Middle Fork Valley continued in May 1968 when Dam No. 3 was started about 122 meters (400 feet) upstream from Dam No. 2. Trucks dumped coal refuse from a wide area in the road on the right of the valley, making a platform from which trucks could dump either down the face or into individual piles for later leveling by a bulldozer. The platform became the right abutment contact of Dam No. 3. The embankment was carried across Middle Fork Valley as a moving front about 20 to 23 meters (65 to 75 feet) high at an approximate elevation of 530 to 533 meters (1,740 to 1,750 feet). At the time Dam No. 3 was started, the top of Pool No. 2 sludge is estimated to have been at about an elevation of 511 meters (1,675 feet), and the total thickness of the sludge (from Pools No. 2 and No. 1) at the center of the valley was slightly greater than 18 meters (60 feet). The only compaction received by any of the dams was from truck wheel loads and bulldozer track loads. Since this compaction was uncontrolled, it varied in intensity and effectiveness, producing dams with random amounts of compaction.

February 1969. In February 1969, a large slump occurred on the advancing face of the embankment. An eyewitness account of this failure notes: "that whole

A.   STARTING CONSTRUCTION — LATE 1968   B.   INITIAL CLOSURE — EARLY 1970

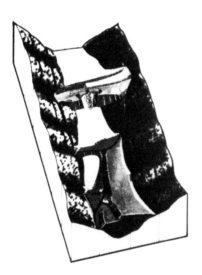
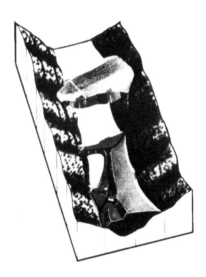

C. FOUNDATION FAILURE — MARCH 1971   D.   PRE-FAILURE — FEBRUARY 1972

Figure 24.2. Artist's conception of the construction sequence of Dam No. 3.

end of the dam, 100 feet of it, disappeared." This failure was not directly related to seepage forces in the embankment, because the embankment did not close off the valley at that time. The phreatic surface throughout the incomplete dam was essentially at the level of the water in Pool No. 2.

The failure in February 1969 was a foundation failure: the sludge foundation simply yielded under the weight of the embankment. Subsurface exploration conducted after the catastrophic 1972 failure found that the sludge (which was underwater) had apparently yielded and flowed laterally, and the embankment sank and also spread laterally. Although the end of the embankment "disappeared," it stopped sinking with its top generally between the water level of Pool No. 2 and the former top of the sludge in Pool No. 2. The slide was "repaired" by simply dumping more refuse on top of the embankment-sludge mixture that resulted from the failure. By early 1970, Dam No. 3 was a true dam, extending completely across Middle Fork Valley.

February 1971. Dam No. 3 reportedly reached its final height by February 1971. After this time, the dam was widened by dumping refuse on the upstream and downstream faces. In February or March of 1971 an ominous shallow failure occurred on the downstream face. This slide was described as being "150 to 200 feet wide across the (downstream) face of the dam and 20 to 30 feet from the face back." This failure was not a second foundation failure; it was a result of high pore pressures (combined with piping within the foundation sludge) in the foundation area near the downstream toe of the dam. The width of the dam (measured from the upstream to downstream toe) was significantly less at this time than prior to the catastrophic 1972 failure. The 1971 failure, coupled with the reported sporadic occurrence during the year of boils of black water in Pool No. 2, downstream left of Dam No. 3, signaled dangerous stability conditions in the foundation of Dam No. 3. Apparently, no one who observed the failure or the boils during 1971 was aware of their significance.

February 1972. By February 1972, Dam No. 3 had reached a substantial size. It measured nearly 183 meters (600 feet) wide at its maximum across Middle Fork Valley. The crest width at the center of the valley measured about 137 meters (450 feet). The estimated volume of the embankment was about 408,300 cubic meters (534,000 cubic yards). The construction history of Dam No. 3 is summarized in Figure 24.2.

Material Properties

Less than a month after the failure of Dam No. 3, a seven-month field investigation of the Middle Fork Valley was begun. This investigation consisted of 60 rotary drill holes, 11 backhoe pits and 2 bulldozer trenches, 14 field permeability tests, 44 field density tests, and 9 penetration and 10 vane shear probes. Based on the field investigation, and the resulting laboratory studies, the following conclusions were developed regarding the embankment and foundation materials.

Embankment materials
Because major portions of the embankments were removed by the failure, the post-failure field investigation was necessarily performed in rather limited areas

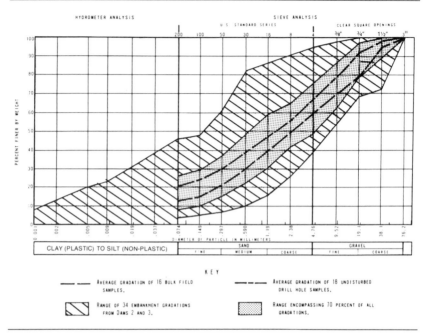

Figure 24.3. Gradation characteristics of the embankment.

defined by remnants of the left and right abutments. Even with these difficulties, however, the correlation of engineering properties between samples obtained by various methods was satisfactory.

Gradation characteristics. The average gradation characteristics of the embankment materials for Dams No. 2 and No. 3 (Figure 24.3) indicate a rather minor variation of the average gradation as determined from 16 bulk field samples and 18 undisturbed 76.2-millimeter- (3-inch-) diameter tube samples. According to the Unified Soil Classification System the majority of the material classifies as a well-graded, gravelly, silty sand.

Specific gravity. In determining the specific gravity of the coarse-grained embankment material, it was necessary for calculation purposes to use a weighted average for the various percentages of materials present because of the difference in specific gravity between the shale and coal fractions of the material. The weighted average of the specific gravity for the embankment material was 1.95 for the typical gradation of 65 percent finer than the No. 4 sieve, 20 percent between the No. 4 and 19.05-millimeter (0.75-inch), and 15 percent between 19.05-millimeter (0.75-inch) and 76.2-millimeter (3-inch).

Dry density. In-place dry densities of the original embankment materials were estimated from both 305-millimeter- (12-inch-) diameter field density tests and 76.2-millimeter- (3-inch-) diameter tube samples. The results of 62 density tests indicate an average in-place dry density of 1.46 tons per cubic meter (91 pounds per cubic foot) for the coarse-grained embankment material.

Permeability characteristics. The permeability characteristics of the embankment materials were determined by evaluating the construction methods, observations at the site, and from field and laboratory test results. The results of the field permeameter and laboratory tests, as well as the coefficient of permeability calculated from the known discharge through Dam No. 3 of 13 liters per second (300,000 U.S. gallons per day), indicate a range of values between 10^{-2} and 10^{-5} centimeters per second with a typical value (horizontal equal vertical) of 2×10^{-4} centimeters per second.

Shear strength. Shear strength parameters for use in stability analyses were determined by testing both undisturbed field samples and laboratory fabricated samples. Also, a special series of undisturbed samples was tested in a portable laboratory in Logan, as well as in the investigators' laboratory, in an attempt to determine if there was any reduction in strength due to conventional shipment and handling of samples. The laboratory fabricated samples were tested and the results compared with those from the undisturbed samples in order to determine what influence, if any, the method of construction may have had on the structure and orientation of individual particles.

Effective stress shear strength parameters were developed, based on consolidated undrained triaxial tests on fabricated and undisturbed samples. These results are summarized in Figure 24.4 in the form of average principal effective stress versus maximum shear stress. The assumed failure criterion used was the point of maximum ratio of major to minor principal effective stress during the course of the triaxial test.

The engineering properties of the embankment materials to which reference is made above do not exhibit exceptionally unusual characteristics when

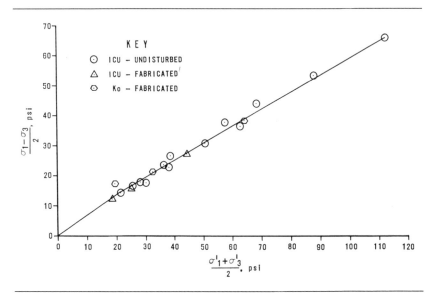

Figure 24.4. Shear strength characteristics of the embankment.

compared to soil-like materials conventionally used for earth dam construction. Furthermore, it was concluded that the engineering properties of the coarse-grained embankment materials were not measurably affected by conventional transportation of field samples, and no significant differences were found between the shear strength parameters for undisturbed and for laboratory fabricated samples at equal densities.

Foundation materials

As mentioned previously, the materials providing foundation support for Dams No. 2 and No. 3 were deposited by discharge of coal waste sludge into Middle Fork from the Buffalo Mining Company's preparation plant. The discharge point into Middle Fork Valley was about 1.6 kilometers (1 mile) upstream of Dam No. 3. The discharge was first ponded behind Dam No. 1 upon a thin layer of alluvium. When the level of sludge approached the crest elevation of Dam No. 1, Dam No. 2 was constructed upstream upon the coal waste sludge retained behind Dam No. 1. This same scheme was employed for Dam No. 3, except that its location was such that coal waste sludge impounded by both Dams No. 1 and No. 2 was contained in its foundation. Based upon historical development of dams and sludge pools in the Middle Fork Valley, both sludge layers of the foundation beneath Dam No. 3 were assumed to have similar behavioral characteristics.

Composite gradation curves for representative foundation sludge materials are presented in Figure 24.5. These materials are described as sandy silt to silty

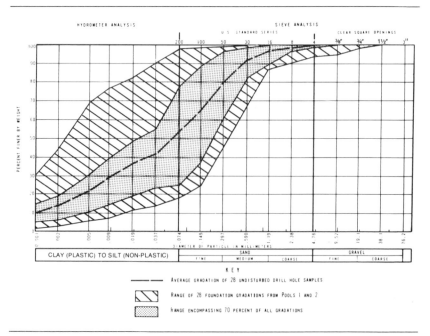

Figure 24.5. Gradation characteristics of the foundation sludge materials.

sand under the Unified Soil Classifications System. The sludge materials were found to have an average specific gravity value of 1.43, with a range of 1.34 to 1.66.

Dry density. The majority of in-place dry densities were determined in the laboratory from undisturbed tube samples. Only a limited number of field density tests were performed because the materials were inaccessible due to the flooding resulting from the dam failure. However, field density test results did confirm the range in dry density of 0.74 to 1.23 tons per cubic meter (46 to 77 pounds per cubic foot) for undisturbed samples. The results of 35 density tests indicate an average in-place dry density of 0.9 tons per cubic meter (56 pounds per cubic foot) for the fine-grained foundation sludge material.

Permeability characteristics. The permeability characteristics of the foundation sludge materials were determined by evaluating the construction methods, observations at the site, and laboratory test results. This evaluation indicated that a significant degree of anisotropy was developed in the foundation sludge materials because of their method of deposition. The sludge materials were found to be highly lenticular, with stratifications varying from a few millimeters to a few centimeters (fractions of an inch to several inches) in thickness.

The fine-grained silts (ML) usually constituted the thinner partings, probably as a result of short periods of relatively quiescent stream flow. The silt partings were determined to have a coefficient of permeability of about 3×10^{-7} centimeters per second, which is about 700 times more impervious than the typical embankment material. The fine- to medium-grained silty sand (SM), which constituted the coarser fraction of the foundation sludge material, had a maximum coefficient of permeability of 3×10^{-4} centimeters per second, which is very nearly the same as that of the typical embankment material.

Clearly from these observations, the sludge foundation was not only more impervious than the embankment, but it consisted of layered materials for which the extreme values of coefficient of permeability varied by a factor of 1,000. In conclusion, a thorough evaluation of the laboratory permeability test results, combined with judgment based on drill hole logs and visual observations of many undisturbed, foundation samples, indicated that the most probable range for the ratio between horizontal and vertical permeability was between 25 to 1 and 100 to 1. These values are referred to later in this discussion as anisotropic foundation permeability ratios.

Shear strength. The shear strength characteristics of the foundation sludge were determined by conventional triaxial testing procedures discussed previously and the results are presented in Figure 24.6.

Development of shearing resistance in the saturated sludge materials was limited by their very low unit weights and the fact that the cohesion component for these nonplastic materials is either very low or nonexistent under effective stress conditions. The frictional resistance of the foundation sludge material is considerably higher than that of an average soil. However, the full benefit of the sludge's ability to resist shear could not be realized because of the material's unusually light unit weight and the saturated conditions acting within the foundation.

The important engineering properties developed for the coarse-grained embankment and fine-grained foundation sludge are summarized in Table 24.1.

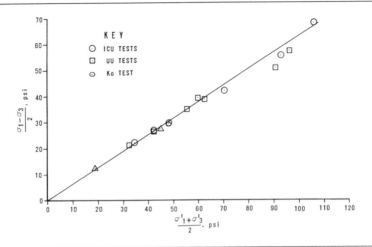

Figure 24.6. Shear strength characteristics of the foundation sludge.

Reservoir Conditions

The condition and level of the reservoirs behind Dams Nos. 1, 2, and 3 on the day of failure were reconstructed by thoroughly reviewing the January and February rainfall records preceding failure, and from eyewitness accounts and visits to the site. These data were utilized, along with data of the surrounding watershed area

Table 24.1. Summary of Engineering Properties

Engineering property	Foundation materials	Embankment materials
Dry density	54 pcf (865 kg/m^3)	90 pcf (1,442 kg/m^3)
Wet density	78 pcf (1,249 kg/m^3)	106 pcf (1,698 kg/m^3)
Equivalent coefficient of permeability ($\sqrt{k_h \times k_v}$)	2 × 10^{-5} cm/sec	2 × 10^{-4} cm/sec
Permeability ratio (horizontal to vertical)	25:1 and 100:1	1:1
Specific gravity	1.43	1.95
Effective stress shear strength parameters	$\phi' = 37°$ $C' = 0$	$\left.\begin{array}{l}\phi' = 41° \\ C' = 0\end{array}\right\}\ 0 < \bar{\sigma} \leqslant 1.12\ \text{kg/cm}^2$ $\left.\begin{array}{l}\phi' = 34° \\ C' = 500\ \text{psf} \\ (0.24\ \text{kg/cm}^2)\end{array}\right\}\ \bar{\sigma} > 1.12\ \text{kg/cm}^2$

and capacities of the reservoirs, to determine the probable pool conditions over a longer period prior to failure.

The normal water surface elevation of Pool No. 3 in the months preceding failure was 529 meters (1,735 feet) or about 5.5 meters (18 feet) below the minimum crest elevation on the upstream face of Dam No. 3. No attempt was made to reconstruct the rise in reservoir level from normal pool to flood stage at an elevation of 534 meters (1,752 feet) because of the complete lack of detailed rainfall records in close proximity to the Middle Fork area. However, an attempt was made to reconstruct the rise in reservoir surface during the 40-hour period preceding failure, using the testimony of Buffalo Creek Mining Company personnel.

Events leading to failure

Based on eyewitness accounts, the water level behind Dam No. 3 was well above the normal pool elevation of 529 meters (1,735 feet) on February 22, 1972. When company personnel became aware of the high water level that morning, they estimated it to be 0.6 to 0.9 meter (2 to 3 feet) below a 0.61-meter- (2-foot-) diameter spillway pipe. The pipe, located near the center of the dam, was reported to be 2.1 to 2.4 meters (7 to 8 feet) below the crest. On Thursday, February 24, water was observed flowing from the 0.61-meter- (2-foot-) diameter pipe spillway, indicating that Pool No. 3 had risen 0.6 to 0.9 meter (2 to 3 feet) since the morning of February 22.

On February 24, one of the mining personnel observed at 4 p.m. that the spillway pipe was underwater and the water was about 1.5 meters (5 feet) below the graded crest. At this time a stick measuring 1.1 meters (3.75 feet) long was placed into the upstream face above the opening of the 0.61-meter- (2-foot-) diameter overflow pipe. Reportedly the top of the stick was about 305 millimeters (1 foot) below the top of the graded crest at an elevation of 534 meters (1,752 feet). Four company personnel continued to read the water level on the crude gauge, and their testimony is the chief record of the rise in Pool No. 3 reservoir level during the 40 hours preceding failure. For 35 hours, the water level rose at a rate of about 25 millimeters (1 inch) per hour from an elevation of 532.8 to 533.7 meters (1,748 to 1,751 feet). The rate then increased significantly, and most likely the water reached its highest level of elevation at 534.3 meters (1,753 feet) just prior to the failure at about 8 a.m. on February 26, 1972 (see Figure 24.7). The significance of this reservoir rise, and the probability of overtopping are discussed below.

Failure Analyses

Although the impoundment was visited repeatedly during the four days preceding failure, there were no eyewitness accounts of the event. It is known, however, that the entire failure occurred in a very short period of time, probably less than 15 minutes. The testimony on the visits contains several conflicting statements about embankment conditions before the failure; however, the bulk of this testimony was critically reviewed to insure that the reconstructed time element of the failure was compatible with witnesses' observations.

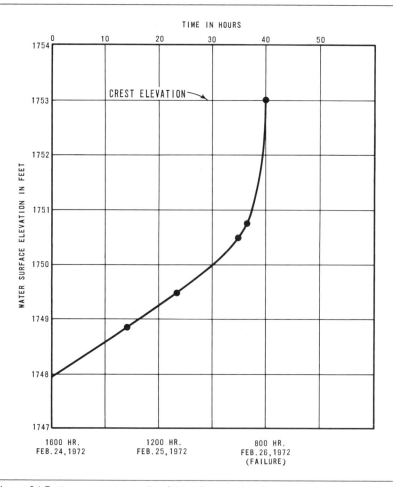

Figure 24.7. Apparent rise in Pool No. 3 water level against time.

To recreate the probable mode or modes of failure, a number of engineering analyses were performed to determine the mode most compatible with the extensive data obtained from the field and laboratory investigations. The complete absence of drawings showing the prefailure embankment configuration was a significant handicap in making the analyses. Although the thickness of the foundation sludge material was reasonably well-defined by the field exploration, it was necessary to reconstruct the embankment configuration of Dam No. 3 from an assessment of the field data and a review of prefailure aerial photographs of the Middle Fork Valley. Longitudinal and transverse sections recording prefailure conditions are presented in Figure 24.8.

As shown on the longitudinal section, the thickness of foundation sludge upon which the central 82 meters (270 feet) of Dam No. 3 was constructed varied from

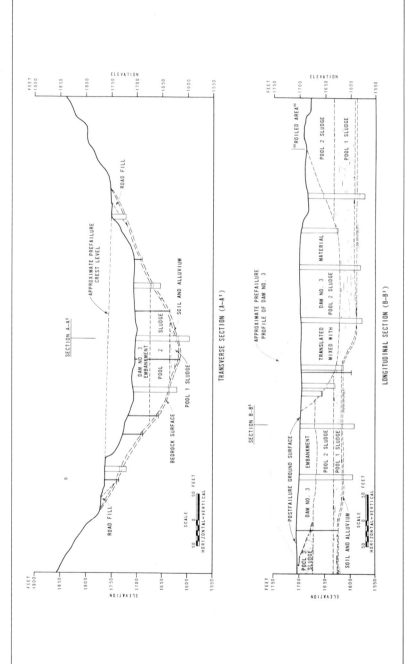

Figure 24.8. Longitudinal section and transverse section across Dam No. 3.

a maximum of 16.8 meters (55 feet) in the center of the valley to 0 meters (0 feet) at approximately 49 and 55 meters (160 and 180 feet) from the left and right crest-abutment contacts respectively. Because of the sloping bedrock surface, however, the maximum thickness of foundation sludge increased beneath the downstream toe area to a maximum of 24 to 26 meters (80 to 85 feet). Two transverse sections were chosen for detailed engineering analyses; only the maximum section representing a 15-meter (50-foot) thickness of foundation sludge beneath the approximate center of the dam is discussed below.

In the performance of the engineering analyses, all reasonably conceivable failure modes were considered. A number of these were rejected on the basis of being so improbable as not to warrant further study. As a result of this elimination process, a total of four possible failure modes were identified. These four possibilities are discussed below and subsequently the most probable mode of failure is identified.

Failure associated with pipe spillway

The postfailure examinations of the site brought to attention two lengths of 0.61-meter (2-foot) diameter uncorrugated steel pipe projecting from the flood debris near the former downstream toe of Dam No. 3. The material in the pipes indicated that they had been carried in the bed load of the flood water after the failure. The two pieces of pipe extracted from the flood debris measured 13.6 and 13.8 meters (44.5 and 45.1 feet) in length. Both ends of both pipes were torch cut and showed no sign of welding. In the attempted reconstruction of Dam No. 3, it appears that the total length of spillway pipe was about 88 meters (290 feet). The pipes pulled from the flood debris account for 27 meters (90 feet), leaving 61 meters (200 feet) most likely buried beneath the flood debris.

From the conditions of the ends of the recovered pipe, it appears that the spillway pipe consisted of several 12- to 15-meter- (40- to 50-foot-) long sections of 0.61-meter- (2-foot-) diameter pipe laid loosely end-to-end and unconnected in a trench across the crest of Dam No. 3 and then buried. If this assumption is correct, water could have readily leaked through the unwelded joints of the pipe and produced local saturation within the dam. The amount of water that the pipe could introduce into the embankment would depend upon the width of the joint openings, the amount of water flowing through the pipe, and the length of time of the flow.

The effect of such a spillway flowing full, or partially full, would be to raise the phreatic surface locally and decrease the stability in the downstream area of the embankment. Additionally, the pipe would have discharged onto the downstream face of the dam, causing erosion and possibly local slumping. It is difficult to assess what contribution, if any, these actions may have had on the ultimate collapse. When compared with other possible modes of failure, it becomes quite obvious that the single spillway pipe was not the principal cause of the ultimate dam failure.

Overtopping of Dams 1, 2, and 3

Dam 1. The fact that all three dams were impounding water at the time of failure required that the possibility of overtopping for each dam be investigated. The probability that Dam No. 1 was overtopped by normal inflow of water into Pool No. 1 is very low. Two 0.61-meter- (2-foot-) diameter spillway pipes were provided for Dam No. 1. The capacity of these pipes, when flowing full, was calculated to be about 1,980 liters per second (70 cubic feet per second, or 5.8 acre-feet per hour). The latter figure indicates that the two discharge pipes could safely accommodate a rise of the water surface elevation of Pool No. 1 (capacity 0.51 hectare [12.6 acres]) at the rate of about 130 to 150 millimeters (5 to 6 inches) per hour. This inflow rate appears highly unlikely, even under heavy rainfall conditions, because the flow into Pool No. 1 consisted only of seepage through Dam No. 2 and any runoff from the small drainage area which immediately bordered the pool.

Dam 2. Dam No. 2 was provided with a 0.76-meter- (30-inch-) diameter spillway pipe with its invert at an elevation of 513.6 meters (1,685 feet), or about 1.5 meters (5 feet) below the crest. The pipe capacity, when flowing full, was calculated to be about 1,405 liters per second (4.1 acre feet per hour). Inflow into Pool No. 2 was limited to a combination of Dam No. 3 seepage, overflow through Dam No. 3 spillway, and runoff from the drainage area along either side of the pool. Under previous winter flow conditions, the seepage through Dam No. 3 could have been as high as 75.4 liters per second (0.22 acre-feet per hour). Even under full reservoir conditions, the seepage probably did not exceed 99 liters per second (0.29 acre-feet per hour). These values are well below the Dam No. 2 spillway pipe capacity, and the probability of Dam No. 2 overtopping as a direct result of storm water inflow is rather low.

Dam 3. Dam No. 3 had a 0.61-meter- (2-foot-) diameter spillway pipe with an invert at about an elevation of 531.8 meters (1,745 feet), about 2.3 meters (7.5 feet) below the low point of the embankment crest near the right abutment. The maximum pipe capacity was calculated to be about 994 liters per second (2.9 acre-feet per hour).

The day-to-day inflow of storm runoff into Pool No. 3 during the months of January and February 1972 was irregular. It is important to point out, however, that the 5.5-meter (18-foot) rise in reservoir level, from normal pool elevation at 528.8 meters (1,735 feet) to flood stage at an elevation of 534.3 meters (1,753 feet), was not due entirely to the storm of February 25-26. The majority of the reservoir rise was due to the natural accumulation of water during the rainy months when the rate of inflow to the reservoir exceeded the seepage rate through Dam No. 3.

Evaluation of rate of rise in water level. To evaluate the rate of rise of reservoir level, it was necessary to consider the approximate 12-hour period preceding failure. According to testimony before the U.S. Congress, an unofficial rain gauge was set up near the Buffalo Mining Company offices near Lorado on February 25, and from 8 p.m. on February 25 through 7 a.m. on February 26, a total rainfall of 44.5 millimeters (1.75 inches) was recorded. Considering that the Middle Fork drainage basin detention time was about 2 to 3 hours and the near-saturated soil conditions resulting from two consecutive months of above average rainfall, it is reasonable to assume that 75 to 80 percent of this rainfall actually reached the reservoir before failure. This amount of rainfall would produce between 8.7×10^7 and 9.4×10^7 liters (71 and 76 acre-feet) of inflow. As shown in Figure 24.7, the

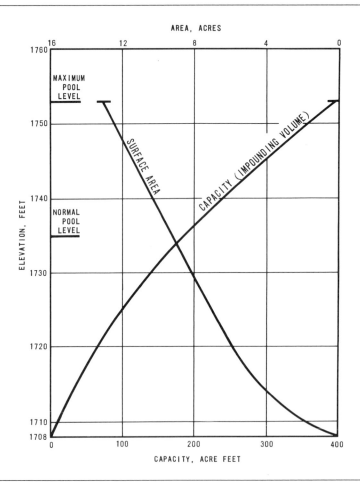

Figure 24.9. Area capacity curve, capacity and area versus elevation for Pool No. 3 are shown in this graph.

water level was at an elevation of about 533.4 meters (1,750 feet) at 8 p.m. on February 25. The available capacity within Pool No. 3 between elevations 533.4 and 534.3 meters (1,750 and 1,753 feet), as shown on the area-capacity curve in Figure 24.9, was approximately 4.3×10^7 (35 acre-feet).

Assuming that the spillway pipe was functioning at full capacity, the total flow through the pipe was calculated to be about 4.3×10^7 liters (35 acre-feet) during the 12-hour period preceding failure. The total seepage through the dam was calculated to be about 4.9×10^6 liters (4 acre-feet) during the same period. A water balance relationship indicates that the sum of inflow, minus seepage, minus pipe discharge, must be equal to or less than the remaining reservoir capacity in order to preclude overtopping. If the inflow amounted to 9.4×10^7 liters (76 acre-feet), then the reservoir capacity would have been exceeded by 2.4×10^6 liters (2 acre-feet) and overtopping of the dam would have begun.

It must be pointed out that the uncertainty of both the rainfall data for the Middle Fork area and the amount of reservoir inflow and outflow up to the time of failure prevents any definitive conclusions regarding overtopping on the basis of reservoir hydrology and hydraulics alone. However, using the values cited above, and the analyses of the other possible modes of failure discussed below, it was concluded that failure of Dam No. 3 did not occur by overtopping. It is highly probable that, because of insufficient spillway capacity, an overtopping failure would have occurred had the dam not failed by other means as described below; however, an overtopping type of failure would have occurred less rapidly than did the actual failure.

Foundation piping

As previously related, boils were noticed in Pool No. 2 downstream of Dam No. 3 after the dam's completion, and apparently persisted in an intermittent fashion up to the time of the failure. These boils were described as the emergence of black water in the relatively clean pool of water downstream of the toe of Dam No. 3. The presence of boils is usually associated with the threat of progressive piping. Seepage forces which develop when water flows through a soil are resisted by the effective grain-to-grain contact stress. If this stress is insufficient to resist the seepage forces, then an imbalance of forces results in the direction of flow. Usually this imbalance of forces is greatest at the exit point of the seepage and results in physical movement of soil particles. This process of gradual internal erosion is known as "piping."

Evaluation of piping potential. In an attempt to evaluate the potential for piping within the foundation, flow nets were drawn for two different cross sections both at normal pool level at an elevation of 528.8 meters (1,735 feet) and at the estimated flood stage prior to failure at an elevation of 534.3 meters (1,753 feet), using foundation anisotropic ratios of 25 to 1 and 100 to 1. Two of the eight flow nets analyzed are presented in Figure 24.10. Both represent a near maximum foundation thickness and a foundation anisotropic ratio of 25 to 1; seepage conditions are at normal pool and at flood stage.

For this anisotropic ratio, the ratio of embankment to foundation permeability (k_E/k_F) was taken to be 10 where

$$k_E = k_v = k_h = 2 \times 10^{-4} \text{ cm/sec}$$

$$\text{and } k_F = \sqrt{k_v k_h} = \sqrt{(4 \times 10^{-6})(1 \times 10^{-4})} = 2 \times 10^{-5} \text{ cm/sec}$$

For the above ratio, one flow line or streamline in the embankment is equal to ten flow lines in the foundation when considering quantity of seepage.

The flow nets were originally constructed on transformed sections by expanding the vertical dimension of the foundation. As is evident from the flow nets presented, the flow lines and equipotential lines shown on the true section of a hydraulically anisotropic material are not mutually orthogonal as they would be for a transformed section.

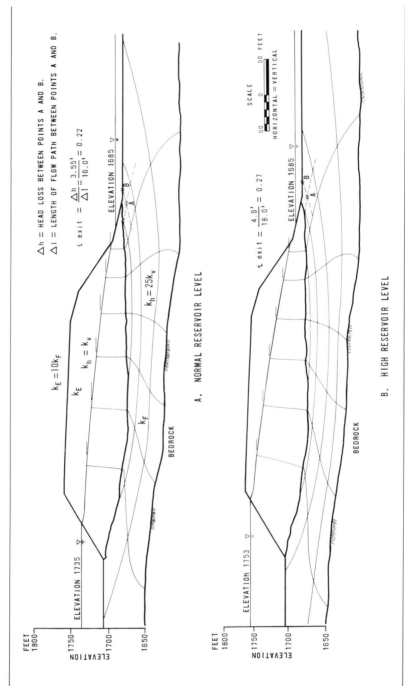

Figure 24.10. Flow net diagrams for normal reservoir level and high reservoir level.

Critical gradient. As shown in Figure 24.10, the exit gradient along the flow path within the foundation sludge beneath Dam No. 3 at the normal Pool No. 3 reservoir elevation of 528.8 meters (1,735 feet), was 0.30. The theoretical critical gradient for the development of piping is defined by the following expression:

$$i_{cr} = \frac{G - 1}{1 + e}$$

where i_{cr} = critical gradient for the development of piping

G = specific gravity
    (average value of foundation sludge = 1.43)

e = void ratio
    (average value of foundation sludge = 0.65)

As shown above, the critical gradient of the foundation sludge is approximately 0.26, as compared with a critical gradient of the embankment material of 0.70. Comparing the theoretical critical gradient occurring within the foundation sludge material with the value obtained from the flow net analysis, the indicated factor of safety against the occurrence of piping is 1.18 for the permeability ratio assumed. A similar analysis for the assumed anisotropic ratio of 100 to 1 resulted in a factor of safety against the occurrence of piping of 0.87.

It is apparent, therefore, that a condition of marginal stability with regard to piping existed within the foundation sludge even at the normal pool elevation of 528.8 meters (1,735 feet). It should be noted, however, that the calculated exit gradients are predicated on the idealized cross sections used for the analysis. It is quite probable that a neat line separating the foundation sludge and coarser embankment materials did not exist in reality, due to minor sloughing of the embankment material near the toe, previous failures of the embankment which displaced the sludge, and differential settlements of the embankment due to variable foundation thicknesses. Any of these conditions would tend to increase the critical gradient because of the presence of more embankment material; however, with such low factors of safety against piping, and a recognition of previous occurrence of piping, there is little doubt that the foundation sludge material existed in a metastable condition.

As the Pool No. 3 reservoir level rose during January and February, conditions favoring the development of piping grew worse as a direct result of the increased difference in elevation between Pool No. 3 and Pool No. 2. The seepage gradient acting within the foundation sludge beneath the toe area of Dam No. 3 for the estimated flood stage at an elevation of 534.3 meters (1,753 feet) is also shown on Figure 24.10. For the two anisotropic foundation permeability ratios considered, namely 100 to 1 and 25 to 1, the exit gradients were computed to be 0.38 and 0.27, resulting in factors of safety with respect to piping of 0.68 and 0.96, respectively. There is little doubt that the entire central 82 meters (270 feet) of foundation sludge material underlying the downstream toe area of Dam No. 3 was actively piping prior to or at the time of failure. However, neither the field evidence nor the engineering analyses substantiate this condition as the principal cause of the catastrophic failure of the dam for the following reasons:

1. If a severe piping condition had developed, large internal erosion channels (pipes) would have been created beneath the embankment. As these pipes began to progress in an upstream direction they would have enlarged and branched in such a manner that large sections of the embankment would have collapsed into the void space created by the piping action. The resulting collapse of the embankment into the foundation would have temporarily halted the piping action until the seepage diverted itself around the blockage and, once again, concentrated in another area to repeat the entire process. The approximate total time of failure of 15 minutes is not consistent with the time required for the full development for such a mode of failure.

2. The field evidence, as determined from the extensive exploration drilling program, clearly indicates that a major portion of the upstream section of the embankment remained intact below the observed postfailure ground surface. As shown by the transverse section A-A' in Figure 24.8, approximately 91 meters (300 feet) of the upstream embankment-sludge foundation contact remained undisturbed, whereas the remaining 67 meters (220 feet) of downstream section to a depth of about 15 meters (50 feet) consisted of mixed embankment and foundation sludge material. Similarly, an interpretation of the field exploration indicated that the contact between the embankment and foundation sludge materials was level and unbroken. The field-determined location and attitude of the undisturbed and mixed materials near the downstream toe of Dam No. 3 indicated that a piping failure in the foundation sludge material could not have produced the distribution of materials observed after the failure.

It cannot be stated with any degree of certainty exactly how extensively the piping condition had developed prior to the time of failure. It is known, however, that the embankment and foundation were extremely vulnerable to more than one potential failure mode because of the high foundation pore pressures associated with the full reservoir condition just prior to failure. Furthermore, regardless of whether or not any piping was in progress just prior to failure, the embankment would have failed in a different mode, as discussed below. This fact, combined with the evaluation of the postfailure physical evidence at the site, leads one to conclude that while piping was undoubtedly in progress just prior to failure this phenomenon was only one of several contributing factors to the principal mode of failure.

Shear failure
Conventional limit equilibrium stability analyses were performed for the downstream face of Dam No. 3 with a pool elevation of 528.8 meters (1,735 feet) to determine the minimum factor of safety for the embankment under normal reservoir loading conditions and also to serve as a check for the assumed engineering parameters and pore pressure conditions. Additionally, because of the horizontal

stratification of the foundation sludge, it was crucial that the stability analyses be performed incorporating pore pressures determined from the flow net diagrams rather than assuming a condition of vertical equipotentials.

More mathematically rigorous stability analyses of the Buffalo Creek failure were performed by Corp, 1974, about one year after the results described herein were published. His approach, which involves finite element simulation, generally confirmed the results of the conventional analyses; however, unlike the conventional analyses, the finite element results appear to define the initial zones of overstressing.

Normal pool level. The results of the limit equilibrium stability analyses for the maximum section of Dam No. 3 for normal Pool No. 3 elevation of 528.8 meters (1,735 feet), are presented in Figure 24.11.

Although results are presented for two anisotropic permeability ratios, the following discussion will concentrate on the results obtained for the ratio of 25 to 1. The range of horizontal to vertical permeability of the foundation sludge is thought to have varied from about 16 to 1 to 100 to 1 with an average value between 25 to 1 and 36 to 1. For this reason, the conclusions regarding the embankment stability were based on the anisotropic foundation permeability ratio of 25 to 1. However, the results obtained for the ratio of 100 to 1 are also presented in order to indicate the sensitivity of the stability results to the assumed permeability ratio.

Three distinct bands or zones or ranges in factors of safety are noted in the results. The upper band No. 1, which includes rather shallow assumed failure arcs confined primarily within the embankment and extending about 6 meters (20 feet) into the foundation, indicates a range of factors of safety of 1.19 to 1.94 against either circular or wedge-type failure. The middle band No. 2, which includes failure surfaces extending 6 to 15 meters (20 to 50 feet) into the foundation, indicates a range of factors of safety of 0.94 to 1.06. Finally, the deep band No. 3, which includes failure surfaces extending 15 to 23 meters (50 to 75 feet) into the foundation, indicates a range of factors of safety of 1.09 to 1.24.

Two additional stability analyses were also performed for Dam No.3. They were as follows:

1. Assuming a nominal 3-meter (10-foot) thickness of foundation sludge material.
2. Assuming the embankment to be underlain only by a relatively thin section of unstripped soil cover overlying the bedrock surface.

The latter analysis actually represents a condition which probably existed for 50 to 60 percent of the crest length of the dam at the left and right abutment contact. The results of these analyses indicate that the factors of safety against either circular or wedge-type failures were 1.11 and 1.29 for cases 1 and 2, respectively.

The stability results for Dam No. 3 at the normal pool elevation of 528.8 meters (1,735 feet) indicate that the downstream face of the embankment was at best only marginally stable. The calculated minimum factor of safety of 0.94 for the condition of maximum thickness of foundation sludge, at first seems

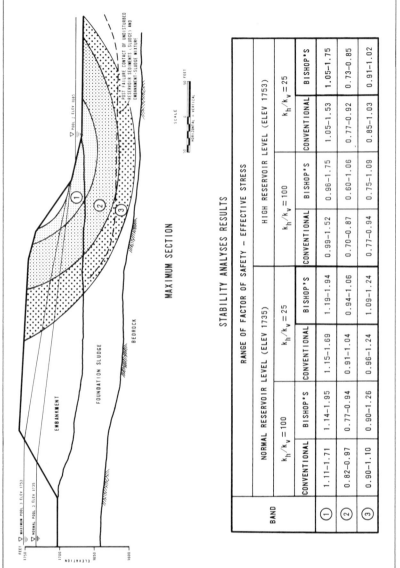

Figure 24.11. Stability analyses results for Dam No. 3.

somewhat paradoxical; however, this factor of safety does appear to be consistent with the assumed parameters and geometric configuration of Dam No. 3 for the following reasons:

1. As indicated by the difference in results for the two anisotropic permeability ratios assumed, the calculated factor of safety is rather sensitive to the pore pressure distribution within the foundation which, in turn, is a function of the assumed permeability ratio. It is possible that local variations within the foundation sludge could result in a permeability ratio less than 25 to 1, thereby reducing the developed pore pressures and increasing the factor of safety.
2. The method of construction by end-dumping and spreading may have resulted in the lower portions of the embankment containing somewhat coarser materials than the upper portions, thereby developing a more pervious zone. The presence of such a pervious zone would tend to act as a horizontal drain near the embankment-foundation contact and reduce the foundation pore pressures, thereby increasing the overall factor of safety in the same manner as discussed in (1) above.
3. As indicated by the results for the two cross sections shown, and the additional cross sections referenced, the factor of safety against the shear failure is increased from 0.94 to 1.29 when the foundation sludge thickness is decreased from approximately 15.2 to 0 meters (50 to 0 feet). Because the foundation thickness was not consistent across the valley, it is possible (although improbable) that the more stable abutment sections of the embankment transmitted some additional shearing resistance by a bridging action (three-dimensional effect) to the central area of lower, more critical stability.

Maximum pool level

The stability analyses for the downstream face of Dam No. 3 with the assumed maximum reservoir condition that existed on February 26, 1972, were performed on the same cross sections described above, except that the reservoir level was assumed to be equal to the field-determined high water elevation of 534.3 meters (1,753 feet). For this analysis, it was assumed that, because of the saturated condition of the foundation due to normal operations, the increase in pore water pressures associated with the higher reservoir level was transmitted instantaneously to the downstream toe area as the reservoir continued to rise.

The stability results for the flood stage at an elevation of 534.3 meters (1,753 feet) are presented in Figure 24.11. In the analysis for normal pool conditions, three distinct bands of factors of safety were noted. These bands, with approximately the same depth of penetration into the foundation sludge as previously discussed, were also noted for the flood-stage analysis. The stability results indicate that the range of calculated factors of safety against shear failure for band No. 1 was 1.05 to 1.75; for band No. 2 was 0.73 to 0.85; and for band No. 3 was 0.91 to 1.02. The two additional analyses for assumed foundation sludge thicknesses of 3.05 and 0 meters (10 and 0 feet) indicated a minimum factor of safety against shear failure of 1.00 and 1.19, respectively.

As a result of these detailed stability analyses, the following conclusions were deduced regarding the overall stability of Dam No. 3:

1. The stability of the dam at normal reservoir loading conditions was at best only marginal.
2. The calculated factors of safety for the highest reservoir loading at an elevation of 534.3 meters (1,753 feet), which probably occurred just prior to failure, indicate a gross instability of that portion of the dam underlain by foundation sludge material. This condition applies to the central 82 meters (270 feet).
3. The location of the failure surface represented by the minimum factor of safety for the high reservoir loading condition coincides reasonably well with the contact, located during the field investigation, between the undisturbed reservoir sludge and the overlying mixture of embankment-sludge material.
4. Because the unit weight of the saturated embankment material was about 35 percent greater than the foundation material, and because a condition of incipient liquefaction existed in the sludge near the downstream toe, a failure of Dam No. 3 could have occurred so rapidly that a large section of the foundation sludge material in Pool No. 2 would have been displaced as a flood wave.
5. The failure mechanism discussed in (4) above would have left an oversteepened downstream face with the phreatic surface emerging near the crest. Stability analyses for such a condition indicate that essentially rectangular blocks up to 6.1 meters (20 feet) in width would be rendered unstable under the imposed loading conditions. The resulting progressive failure of successive wedges would be drastically hastened by the emergence of the phreatic surface high on the exposed face. This progression of failure towards the upstream face of the dam could have occurred rapidly.

Most Probable Mode of Failure

Initial failure

The overwhelming field evidence, in addition to the engineering analyses, indicates that the initial failure occurred in the downstream section of Dam No. 3 and consisted of a massive slide movement involving approximately 99,400 cubic meters (130,000 cubic yards) of embankment material. This slide, depicted by band No. 2 in Figure 24.11, occurred in such a manner that the slide mass physically displaced Pool No. 2 sediments, which were acting as a semiviscous fluid because of the relatively high internal pore water pressures, and translated a large block of these sediments onto the left side of Dam No. 2. Associated with this massive displacement into Pool No. 2, was the initial overtopping of Dams 2 and 1 by the reservoir water displaced from Pool No. 2. This surging of water over the crest of Dam No. 2, which had perhaps only 1.22 meters (4 feet) of freeboard, most likely initiated the breach near the right abutment of Dam No. 2.

Progressive failure

Immediately after the initial failure, Dam No. 3 continued to fail rapidly by progressive action. Because the initial failure undoubtedly created a relatively steep head scarp, as depicted by the typical failure surfaces shown in Figure 24.11, that portion of the embankment not involved in the initial failure was left standing, with the phreatic surface emerging high on the exposed face. The resulting condition of the embankment was very unstable and the remaining portions of the embankment probably commenced to slide into the void created by the initial failure. It is impossible to state exactly how long such a progressive failure mechanism took to develop, but the total time required to complete this mode of failure appears compatible with the approximate 15-minute period within which it is estimated the complete failure occurred.

Final failure

When the failure had progressed upstream until only 30 to 37 meters (100 to 120 feet) of the embankment remained standing as measured from the upstream toe, the analyses indicate that the remaining section of the embankment then failed violently, thereby allowing the first rush of Pool No. 3 reservoir water to start its destructive action. The initial release of water was apparently confined, or nearly so, toward the right side of the valley as it progressed downstream. As water flowed through the breach of Dam No. 3, embankment materials that had slumped as a result of the progressive failure were transported into the Pool No. 2 area. As the heavily laden flood waters hit Dam No. 2, its breach, started by the initial overtopping, was probably widened and deepened. The initial flood wave then continued downstream, overtopping and destroying the small Dam No. 1 until the water reached the narrow portion of the valley formed between the refuse bank and the adjacent road. The initial surge of this flood wave as it hit the burning refuse bank probably caused the explosions reported by numerous observers.

After the flood wave reached the refuse bank, the constrictions in the valley cross section caused a backup of water and the high water lines downstream of Dam No. 3 were formed. As water continued to flow through the initial breach of Dam No. 3, the failure of the remaining portions of the dam progressed toward the right and left abutments.

The field evidence supporting the conclusion that the major flood wave was confined to the right side of the valley includes the fact that the flood deposits of the embankment materials from Dams Nos. 2 and 3 were generally confined only to this and the Pool No. 1 area. Furthermore, the roiled area on the left side of Dam No. 2 contained sludge deposits at elevations ranging from 0.6 to 0.9 meters (2 to 10 feet) higher than the crest of Dam No. 2. Because the original structure of the sludge was preserved in detail in this highly erodible material, it is inconceivable that any major rapid flood flow ever occurred in the roiled area, although it was undeniably inundated by relatively quiescent flood waters associated with the development of the prominent high water line left after flooding.

After developing the mode of failure described above, the remaining Pool No. 3 water continued to flow through the ever-widening breach in Dam No. 3. Relatively minor readjustments of major translated blocks of sediment and embankment materials probably occurred at this time, followed by the final emptying of Pool No. 3.

To aid the reader in following the sequential nature of the most probable failure mode described above, a series of diagrammatic sketches is presented in Figure 24.12 showing the major elements of the collapse of Dams 1, 2, and 3.

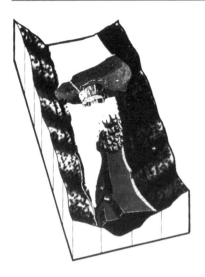

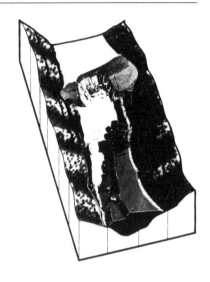

A.　INITIAL MASSIVE SHEAR FAILURE　　　B.　PROGRESSIVE EMBANKMENT FAILURE

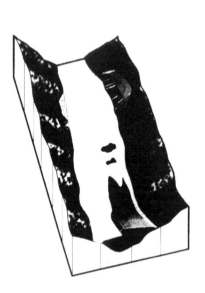

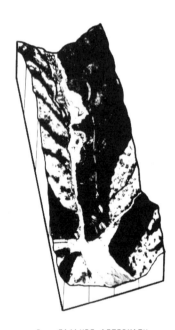

C.　HIGH WATER LINE　　　　　　　　D.　FAILURE AFTERMATH

Figure 24.12. Artist's conception showing probable sequence of failure of Buffalo Creek refuse dam.

Acknowledgments

Though the work summarized herein was performed for the United States Department of the Interior, Bureau of Mines, Washington, D.C., the statements and conclusions presented in this paper are those of the author and in no way should be interpreted as representing official government concurrence or approval.

The author is indebted to his co-workers J. G. Wulff and R. E. Tepel who actively participated and collaborated throughout the field, laboratory, and engineering investigation.

References

Bishop, A. W., 1966, "Geotechnical Investigation into the Causes and Circumstances of the Disaster of 21 October 1966." Her Majesty's Stationery's Office, London.

Corp, E. L., 1974, "A Finite Element Model for Stability Analysis of Mine-Waste Embankments." Ph.D. Dissertation, University of Idaho.

U.S. Congress, Subcommittee on Labor of the Committee on Labor and Public Welfare, 1972b, Hearings on Buffalo Creek Disaster, Appendices A, B, C, and D: U.S. 92d Congress, 2nd Session.

W. A. Wahler & Associates, 1973, "Analysis of Coal Refuse Dam Failure. Middle Fork, Buffalo Creek, Saunders, West Virginia." National Technical Information Service, Washington, D.C.

West Virginia Ad Hoc Commission of Inquiry into the Buffalo Creek Flood, 1972a, The Buffalo Creek flood and disaster, Charleston.

West Virginia Ad Hoc Commission of Inquiry into the Buffalo Creek Flood, 1972b, Hearings Transcript, Charleston.

Design of Coal Slurry Impoundments and the Impact of MESA Legislation

25

William F. Brumund, Principal,
Golder Associates, Inc., Atlanta, Georgia, USA,
and James R. Murray, Senior Engineer, Western Region,
Consolidation Coal Company, Englewood, Colorado, USA

Introduction

This chapter briefly reviews some of the types, uses, design considerations, and construction techniques for surface impoundments used in the surface bituminous coal industry in Illinois. Special emphasis is given to the geotechnical aspects of the design of slurry impoundments. The impact that the Mining Enforcement and Safety Administration (MESA) legislation has had on the design and construction of these impoundments is reviewed.

Types and Uses of Impoundments

Today, in Illinois impoundments are generally constructed and maintained for the following uses:

1. Slurry disposal or storage.
2. Gob disposal.
3. Freshwater supply.
4. Sedimentation control.

Slurry impoundments are constructed to provide a permanent storage location for the slurry waste product from coal preparation plants. The waste slurry is usually pumped by pipeline from the preparation plant to the disposal pond at a solids content of approximately 10 percent by weight. The solids consist of fine gravel, sand, silt, and clay-size material with some fine coal particles. Generally, this waste is in the minus-6.4-millimeter (0.25-inch) size range. Once in the impoundment, the solids are allowed to settle. The water is decanted and then generally returned to the preparation plant for reuse.

Slurry impoundments in Illinois have typically been of the following types:

1. Diked ponds above grade.
2. Incised ponds below grade.
3. Cross-valley impounding fills.

Gob or coarse coal refuse disposal has historically been accomplished in several different ways. The gob is typically a mixture of low-grade coal, rock, and pyrite having a gradation usually in the minus-12.7-millimeter (5-inch) to plus-6.4-millimeter (0.25-inch) size range. Although in the past it has occasionally been pumped with the slurry into slurry ponds, for the most part it is transported to disposal sites using rear dump trucks or by track haulage. Gob has typically been disposed of in Illinois in one of the following ways:

1. Buried below grade in active strip mine pits.
2. Deposited in nonimpounding valley fills using rear dump trucks.
3. Placed above grade in nonimpounding self-supporting waste embankments using trucks or track haulage.

Freshwater impoundments are generally cross-valley fills or incised pond structures, and are used to retain and store water for use as a nonpotable water supply for shop and bathhouse use, or as preparation plant process water. When used for plant process water, these freshwater impoundments may receive the discharge from a slurry pond decant system to form a closed circuit water system for the preparation plant.

Sedimentation ponds are generally cross-valley or incised pond structures. They are used as sedimentation basins for pit discharge or for runoff from an affected mining area. A slight variation in the sedimentation pond may be used in conjunction with an acid treatment plant to neutralize acid runoff before it is discharged into streams and rivers.

Prelegislation Construction Techniques

Part 77, Subchapter O, Chapter 1, Title 30 of the Code of Federal Regulations became law in November 1975, giving MESA the power to approve, disapprove, and inspect all coal refuse impoundments which met certain criteria, and/or were thought to be a possible threat to the safety of the coal miner. Prior to that time most coal refuse disposal areas and impoundments were constructed in the simplest possible way, in the most convenient area, and with the least possible expenditure.

A diked slurry impoundment might be constructed using a small dragline to excavate surficial soils from the inside of a proposed pond, and side casting these soils to form an uncompacted embankment which would be roughly shaped with a bulldozer. This embankment might later be raised to increase the impoundment capacity by dumping gob on the embankment, using side dump cars or narrow gauge track haulage.

A cross-valley impounding fill might have been constructed using bulldozers to push material from each side of the valley to form a crude dam in the center of the valley forming an impoundment.

A self-supporting above grade gob pile might be constructed by progressively dumping coarse refuse on the top of the pile and letting the material roll down the slope to finally stand at its natural angle of repose.

The above examples are not meant to be a criticism of the methods employed. Most of the impoundments and self-supporting waste piles functioned as intended and were constructed at minimum expense.

In all the above examples, however, little attention, if any, was given to the geotechnical engineering fundamentals of embankment design or construction. Seldom was any attention given to systematic compaction of the placed fill, to clearing or grubbing the ground under the fill, to any provision for seepage control, or to the hydrology of watershed and the hydraulic capacity of the decant system. Frequently no assessment was made of the overall stability of the waste dumps or slurry impoundments.

Geotechnical Design Considerations

The MESA criteria for approval of an existing structure or the design of a proposed refuse structure are aimed basically at insuring the safety of the structure under adverse operating conditions. Two areas requiring considerably more time, effort, and expense to satisfy the current legislation are in the hydrologic assessment and decant system design and in the assessment of slope stability.

The requirement for comprehensive stability studies for all impoundments within the purview of the legislation is profound. Not only are more sophisticated analyses required but more elaborate testing and field sampling techniques must be employed. These more sophisticated procedures are well known and are often used in the design of large dams, but sometimes seem inappropriate when applied to the design of a 7.6-meter- (25-foot-) or 9.1-meter- (30-foot-) high slurry impoundment.

Considering that under present legislation an assessment must be made of slope stability under a variety of conditions including steady-state seepage and seismic loading, the realistic assessment of strength parameters and piezometric levels is most important. A brief examination of the determination of only the soil strength parameters will help illustrate the impact of the MESA legislation.

For the variety of conditions that must be investigated an effective stress analysis is most often used in the stability analysis. This permits an assessment of various piezometric conditions in the dam and provides a method for checking the design assumptions in the field by monitoring the water levels.

The assessment of effective stress shear strength parameters is difficult and, in the authors' experience, often done incorrectly. For embankments that will be constructed in lifts and systematically compacted, the use of the triaxial test to determine the strength parameters is highly recommended. In evaluating the design strength for a proposed embankment, laboratory strength tests are usually run on compacted samples. These samples are partially saturated; however, it is probable that as slurry is impounded and seepage develops, the embankment materials, particularly along the upstream face and lower portions of the impoundment, will become saturated. Water percolating through the fill will tend to "flush out" some of the air entrapped during compaction. Also the hydrostatic pressure will tend to dissolve the air remaining in the void spaces of the fill. Since the pore pressures present in the partially saturated and compacted fill are negative they tend to increase the apparent strength of the fill soils. This is unconservative because under steady-state seepage conditions much of the fill will be saturated and the negative pore pressures eliminated.

It is considered, therefore, that the effective stress strength parameters to be used in the design of coal slurry impoundments should be based on results where the sample was initially saturated. In the laboratory this can be done by "flushing" water through the sample and then applying a backpressure to the water in the void space to dissolve the remaining entrapped air in the sample. Backpressures in the range of 0.53 to 0.70 kilograms per square centimeter (70 to 100 pounds per square inch) or higher are often required to achieve saturation of compacted samples. It is recommended that the effective stress strength parameters be determined from a series of isotropically consolidated triaxial specimens that have first been backpressure-saturated and then sheared in the undrained condition. A low compliance, pore pressure measuring system is essential if an accurate assessment is to be made of the results.

For some impoundments, the adoption of a more conservative design might obviate the need for some of the sophisticated testing and analysis techniques, and would reduce the time required to develop an impoundment design.

Post-legislation Construction Technique

A typical impoundment constructed today is described below.

Initially a site would be selected and an extensive drilling and sampling program undertaken. Based on the geotechnical data collected in the field and laboratory, analyses would be made and a final design for the impoundment structure developed. A complete set of plans and specifications for the impoundment would be prepared by a registered professional engineer and submitted to MESA for approval.

After obtaining MESA approval actual construction of the impoundment can begin. Typical foundation treatment includes the stripping of all organic soils and may include provision for limiting or controlling seepage through the foundation. Fill soils are usually placed in 203- to 305-millimeter (8- to 12-inch) lifts using scrapers and systematically compacted to some specified density. Generally some provision is made to control seepage through the embankment, using an internal drainage system which must be installed as the fill is placed. Careful attention is given to properly bedding the decant system's outlet piping.

When the embankment is completed, riprap is often added to the upstream slope and the downstream is normally graded, mulched, and seeded. A MESA inspector must approve the completed impoundment prior to its being put into service. Thereafter, an inspection must be made of the structure every seven days and an inspection report filed in the mine office.

Cost Comparison

An attempt has been made to show the impact that the MESA legislation has had on the cost of the design and construction of the typical slurry impoundment. For

Table 25.1. Impact of MESA Legislation on the Cost for Design and Construction of a "Typical" Slurry Impoundment

Item	Cost pre-MESA legislation	Cost post-MESA legislation
1. Geotechnical investigation of impoundment site; drilling, sampling, testing, and analyses	0	14,000
2. Detailed design drawings, specification, and preparation of MESA submissions	0	7,000
3. Clearing and grubbing	0	24,000
4. Fill placement, 600,000 yd^3	330,000	1,140,000
5. Filter placement, 30,000 yd^3	0	90,000
6. Decant system	7,500	15,000
7. Riprap, 1,800 yd^3	0	5,000
8. Seeding and mulching	500	7,000
9. Engineering supervision	15,000	40,000
Total	**$353,000**	**$1,342,000**

Note: Cost data shown does not depict any specific facility but illustrates in a general way the impact of MESA legislation on the cost of design and construction of slurry impoundment.

comparison purposes the "typical" impoundment was a 16.2-hectare (40-acre) aboveground structure, rectangular in shape, with a centerline dike length of 1,829 meters (6,000 feet). The embankment was 9.1 meters (30 feet) in height, had a crest width of 4.6 meters (15 feet), and had downstream and upstream slope angles of 3 horizontal to 1 vertical and 2 horizontal to 1 vertical, respectively. A rough cost comparison for the design and construction of this impoundment prior to and after the MESA legislation is shown in Table 25.1. The cost information given in Table 25.1 is not intended to depict actual costs on a specific facility but rather to illustrate, in a representative way, the impact of the MESA legislation on the cost of the design and construction of a typical slurry impoundment.

As can be seen from Table 25.1, one of the impacts of the MESA legislation is to increase the design and construction cost for a typical slurry impoundment from $353,000 to $1,342,000. Generally, in addition to requiring substantially more lead time, the cost for the design and construction of slurry impoundments has increased three to four times as a result of the MESA legislation. There is no question, however, that these post-MESA impoundments are better designed, better constructed, and safer structures than were built before the advent of this legislation.

Waste Dump Instability and Its Operational Impact for a Canadian Plains Lignite Mine

26

P. M. Bowman, Geotechnical Engineer,
Luscar Ltd., Edmonton, Alberta, Canada,
and H. G. Gilchrist, Principal Partner,
Golder Associates, Ltd., Calgary, Alberta, Canada

Introduction

The M & S strip mine is located in southeastern Saskatchewan, as shown in Figure 26.1, in an area locally typified by a flat to nominally rolling grassland environment. The principal relief is provided by the valleys of the underfit streams and the Souris River, but large areas of the mine property appear to have little or no drainage pattern.

Precipitation in the area averages about 42 centimeters (16.5 inches) annually, with the main concentration occurring during the months of June to August. Typical daily maximum temperatures average about 27 degrees Centigrade (80 degrees Fahrenheit) in July, and average minimum daily January temperatures average minus 22 degrees Centigrade (minus 8 degrees Fahrenheit) (Energy Mines and Resources, 1974).

The main No. 1 production pit is about 1,200 meters (4,000 feet) in length, and is situated as shown in Figure 26.2. The western end of this pit is adjacent to a 21-meter- (70-foot-) deep reservoir which provides cooling water for the nearby thermoelectric station. A smaller, shallower pit (No. 3) runs generally north-south, and the two pits join near the northeast corner of the property. The highwall angles are generally maintained between 60° and 70° with the higher angle quite common. Spoil piles range in height up to 30.5 meters (100 feet) above their base, at overall angles around 35° to 38°. Pit access is provided by ramps which have been developed through the spoil piles. The ground and surface water is collected and removed from the pits using a sump and pump operation.

During the summer months of 1976, spoil pile instability hampered the extraction of coal from the No. 1 pit. These spoil slides were initially confined to areas adjacent to haul ramps and involved only the spoil pile immediately adjacent to the open pit. As the slides continued they became more spectacular and by the spring of 1977 they had increased in lateral extent. Mining techniques were altered in order to maintain coal production, but some of these had an unfavorable effect on spoil stability.

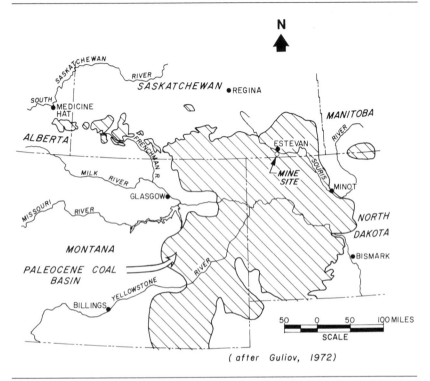

(after Guliov, 1972)

Figure 26.1. The M & S coal mine is located just south of the Souris River in a sequence of poorly indurated Paleocene deposits thinly mantled with glacial drift.

Geology

The lignite coal deposits of southern Saskatchewan are part of the Ravenscrag Formation formed during the Upper Cretaceous and Paleocene epochs (Holter, 1972, Guliov, 1972). In the mine vicinity, this formation contains five lignite seams which are found generally less than 30.5 meters (100 feet) below ground surface. In ascending order these seams are named the Boundary, Estevan, Souris, Roche-Percee, and Short Creek. A total geologic succession of approximately 76 meters (250 feet) is represented by the coal seams and associated partings of interstratified clays, silts, and sands. Only the upper four seams are workable in the Estevan area, and the subject mine recovers only the Estevan Seam.

The Ravenscrag Formation (which is known to extend from the International Boundary north to the 50th Parallel, and from Manitoba to Alberta) is divided into upper and lower facies. The base of the Upper Ravenscrag is assumed to begin with the base of the top coal seam below which point the deposits are of Cretaceous age. The contact of the two formations is considered conformable.

It is generally held that the Ravenscrag Formation in the mine area is represented by an abnormally thick succession of Lower Ravenscrag beds, which

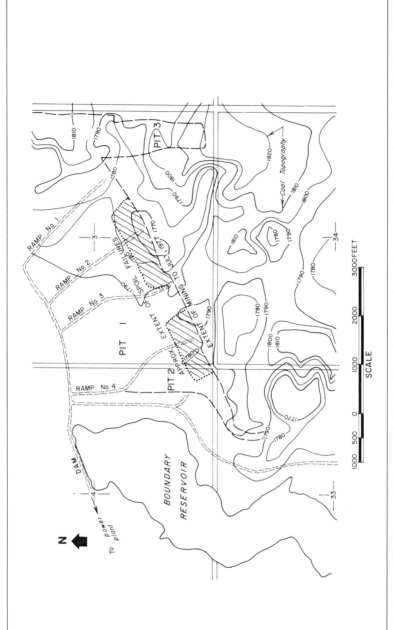

Figure 26.2. Areas of Pit 1 waste failure are highlighted in relation to the coal topography. Boundary reservoir is immediately west of Pit 2.

places the four minable lignite seams in the central part of the Ravenscrag rather than the upper part.

The Upper Ravenscrag consists predominantly of buff colored clays and silt-stones, while beds of sands, sandy clays, and clays containing ironstone bands are not uncommon.

The Lower Ravenscrag and footwall materials are predominantly grayish colored clays, carbonaceous beds, and silts, which are bentonitic in places. Other beds which are characteristic of the Lower Ravenscrag at greater depth include massive sand beds up to 6.1 meters (20 feet) thick, which exhibit cross-bedding and are of medium- to fine-grained particle size.

Drift materials, which overlie the sedimentary strata, consist of glacial till and local sandy outwash or fluvial deposits. The glacial materials are usually less than 6.1 meters (20 feet) thick, and exhibit a high clay content. Small lenses and pockets of silt and sand are commonly found in the till, and clayey lacustrine materials locally overlie the glacial till where positive drainage has not developed.

Groundwater

One of the two main factors influencing the stability of the spoil pile has been the difficulty in controlling groundwater seepage into the open cut. While seepage volumes are not high (generally less than 7.5 to 15 liters per second (120 to 240 U.S. gallons per minute) for every 300 meters (1,000 feet) length of pit, they cannot be neglected since the in situ material underlying the coal is generally highly plastic and quite often bentonitic. This material forms the other key to the stability of the spoil pile. The static groundwater level is located at an elevation of about 563 meters (1,848 feet), being some 1.5 to 4.6 meters (5 to 15 feet) below existing ground level, but about 2.1 meters (7 feet) above the adjacent Boundary reservoir's full water level at an elevation of 561 meters (1,841 feet).

Seepage has been noted to come from the full profile exposed in the highwall, with slightly higher flows from the sand lenses or pockets and the coal seam itself. As the minable seam often contains a parting consisting of approximately 30.5 to 46 centimeters (12 to 18 inches) of highly plastic material, springs have been noted in the pit bottom after removal of the upper seam. Although these springs generally dissipate within a few hours of eruption, they indicate that the coal seam does transmit quantities of groundwater under some head.

Packer tests on the foundation material, prior to construction of the nearby dam, indicated that the mass permeability of the subject coal seam could be as high as 10^{-3} centimeters per second (3.3×10^{-6} feet per second), largely due to cleats and fractures in the coal. It was noted that this value was somewhat influenced by the weathering processes in the valley walls and that it decreased slightly where less weathering was evident.

Seepage has been found to increase as mining approaches the reservoir, although a 152-meter (500-foot) environmental buffer zone is maintained between the mine and the reservoir shoreline.

During the first year of mining, seepage was encountered in Pit 2 which was aligned parallel to the reservoir center line, and adjacent to the buffer. Available pumps could not handle the flow, and the pit was eventually closed in favor of an alternative pit farther east.

Figure 26.3. A photograph of Pit 1 mining activities.

Mining Problems Related to Operating Conditions

Surface conditions

The extremely wet, soft ground which existed over much of the mine property forced the large dragline to be operated on a bench some 6.1 meters (20 feet) below natural ground level as shown in Figure 26.3. The bench was developed by backcutting, with additional leveling being provided by dozers. Prior to backcutting, it was also usual for the wet, highly plastic surface materials to be dozed over the highwall into the pit bottom. This reduced the amount of backcutting necessary, and aided dragline production.

Pit operation

Groundwater seeping into the pits caused considerable operating problems. In the initial years of production, the seepage volumes were small, and were allowed to go relatively unchecked. Occasionally, large portions of the pit would be flooded, or the pit bottom would be too soft to allow coal loading equipment to

Figure 26.4. Specific waste instability.

operate. As the failures became more serious, as illustrated in Figure 26.4, in-pit inventory was reduced to the extent that the coal shovel was operated as close behind the dragline as safety and production would allow. Pit losses were often increased because a larger-than-normal rib of coal was lost under the edge of the advancing spoil.

As the succeeding spoil was placed onto previously failed spoil rows, there were significant consequences. First, additional material was placed onto material controlled by an existing weakened failure surface. Second, spoil room was reduced due to the extra material now in the pit bottom. This condition was compounded when the immediate 6.1 to 9.1 meters (20 to 30 feet) of spoil overlying the top of the coal was chopped down to prevent spoil material from overrunning the coal. The face angle in this region was approximately 45° to 50°, with the overall face angle of the spoil after removal of the coal approaching 42° to 45°. Shallow spoil slides would commence within hours of spoil placement, and failure would inevitably result during, or immediately prior to, removal of the coal which functioned as the toe support.

During the coal loading phase, in-pit sumps and pumps are operated to maintain pit drainage and to avoid development of unnecessarily poor pit bottom conditions. Limited manpower and equipment necessitate that sumps and pumps progress with the coal-loading operation along the pit. This often results in portions of the pit becoming flooded after the coal is removed, and consequently spoil placed on the softened base material subsequently slumps. This type if instability is primarily due to uncontrolled drainage accumulation. The next spoil windrow may experience similar foundation conditions, except that the now-sheared foundation material offers less foundation strength. The result is that the new spoiled windrow also fails, with this failure condition perpetuating in subsequent windrows.

General Solutions Proposed

There appeared to be two main courses of action available for diminishing the detrimental effects of the pit water and the weakened spoil foundation problems. First, immediate action could be implemented which would provide some relief and improvement of the pit working conditions. Second, a geotechnical investigation of the spoil base would be needed to provide parameters upon which the nature of further operational adjustments could be decided.

The immediate remedial action hinged entirely around an improvement in control of water in, and adjacent to, the pit. A surface drainage scheme was implemented which, combined with favorable weather conditions, resulted in a much improved bearing capacity of the upper till layers. This allowed the dragline to be placed at natural ground level, and the time-consuming backcutting operation was avoided. In addition, the dragline boom was lowered for increased reach at the expense of some reduction in bucket capacity.

Next, a scheme was proposed whereby sumps would be excavated in the coal at regular intervals, and particularly at points of low coal topography. Dewatering wells ahead of mining were considered but discarded due to the additional costs of operation. These sumps were complemented by cutting shallow pit bottom ditches during and after removal of the coal in order to maintain the pit bottom in a relatively dry condition. To date, the total success of this operation is suffering from lack of sufficient manpower.

It was also proposed that any wet, highly plastic drift materials should be placed by the dragline against the back of the spoil pile and not dozed into the foundation portion of the pile.

Finally, it was suggested that the toe support supplied by the coal seam should be maintained as long as possible. This could be achieved by initiating coal extraction at the base of the highwall, and removing the coal in long strips parallel to the toe of the waste dump.

These operational improvements are based on the geotechnical conditions which have been observed at the site, and the importance of the observed conditions to the principle of stability is discussed in the following sections.

The Influence of Foundation Conditions on Waste Dump Stability

Geotechnical reasoning

The spoil slides only occur in specific areas in Pit 1, and the general positions of these areas of instability have been outlined in Figure 26.2. In making a comparison between the nature of these spoil dump slides and the nature of the slides which have occurred on the spoil materials of an adjacent pit farther east, it is apparent that the mode of failure is quite different. An example of the slide in the east (No. 3) pit is presented in Figure 26.5, and it appears that the failures are relatively continuous. In Pit 3 the slipping mass typically extends only part way up the spoil bank, and it appears that the failures are shallow and directly related to spoil undercutting for recovery of the coal rib. By contrast, the spoil slippage which has occurred on the waste dumps of the former pit can extend beyond the crest of the last dumping line. The toe of these failures has advanced for distances up to

Figure 26.5. Shallow failures in Pit 3 waste materials appear related to undercutting for rib recovery.

about 23 meters (75 feet), and failed overall waste slope angles of as little as 18°
have been produced. The slide areas have surficial expression, which is quite
different from the shallow failures illustrated in Figure 26.5. Accordingly, it has
been concluded that foundation conditions for waste dump placement at some
locations in the No. 1 pit area are quite different from those of the eastern pit.
Simple index testing, therefore, may serve to highlight the case being advanced
from field observation, by identifying the presence of highly plastic foundation
material whose strength would be highly sensitive to moisture content.

Laboratory test results
Results of limited laboratory testing are presented in Figure 26.6, and it is apparent that the Lower Ravenscrag footwall samples recovered are highly plastic and
of high clay content. This degree of plasticity is suggestive of substantial bentonitic content and indicative of low shearing strength parameters. A value of
residual friction angle φ'_r of less than 7° is suggested by the graph in Figure 26.7,
which was developed from a literature study made concerning stability studies
for another mining project. Further, at low stress levels, the ratio between peak
and residual strength parameters for a highly plastic material is often large
(Sinclair and Brooker, 1967), and consequently shearing displacement during
slope failure would be expected to be greater than for the less plastic, clayey
materials.

Stability analyses
The first part of this chapter has described the pit conditions existing during spoil
handling, which provide opportunity for both footwall softening and weakening
as well as the placement of loose, high moisture content materials in the lower

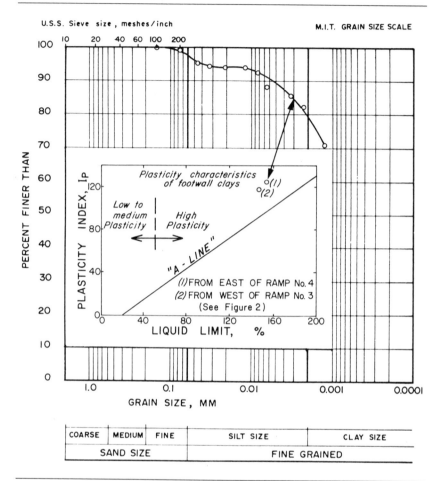

Figure 26.6. Index test data for two footwall samples.

portion of the spoil. Footwall softening and weakening occurs both from normal rebound on stress release and from the effects of local pit traffic. It is, therefore, instructive to examine the sensitivity of waste dump slope angle to foundation strength, where foundation strength is the product of operative friction angle and foundation porewater pressure conditions.

The compound wedge analysis model shown in Figure 26.8 has been used because it highlights the importance of foundation strength to stability (Coates, 1972). It is a useful analytical method, because the purpose of the computations is to demonstrate slope angle sensitivity to the deficient strength of a weakened footwall, rather than to provide absolute values of safety. However, in the analysis, trial and error is necessary to obtain solution of the included interwedge force E, and thus the factor of safety for the wedge pair.

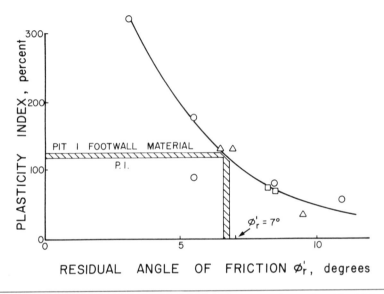

Figure 26.7. A possible relationship between plasticity index and residual friction angle for clay shales (from literature survey).

In the analyses, the concept of effective stress has been used where shear strength $\tau = c' + (p-u) \tan \varphi'$

where: c' = cohesion intercept or stress independent strength
         p = normal stress
         u = porewater pressure
         ϕ' = angle of internal friction

It has been considered appropriate to utilize strength parameters for the body of the waste dump of $c' = 0$ and $\varphi' = 35°$ because of the remolded nature of the material, and the observed 35° to 37° dump slope angles being achieved in areas where no failure is occurring.

Two conditions of foundation pore pressures have been examined; the zero pressure condition, and pore pressure equal to one-half the vertical total stress as shown in Figure 26.8. No allowance is made in the porewater pressure distribution for the effect of shear strength (Bishop and Henkel, 1962). In both cases, the analyses have assumed that there is no porewater pressure in the body of the

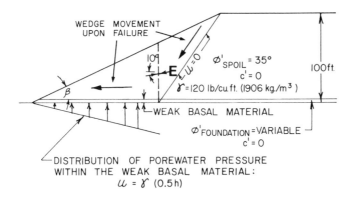

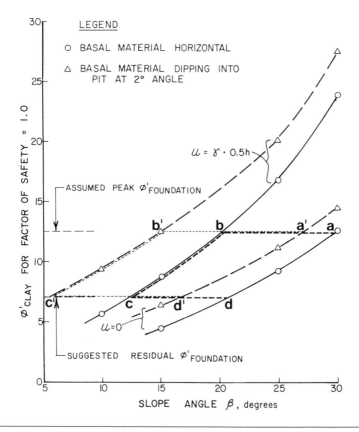

Figure 26.8. The influence of foundation porewater pressure and friction angle on waste dump slope angle at factor of safety = 1.0 using the compound wedge method of analysis and conditions indicated.

waste material, and consequently the comparative analyses are easily made.

A range of interwedge force angles was examined, and it was shown that, for the purpose of this study, use of angles between 5° and 15° did not mask the value of the basic comparisons being sought.

In Figure 26.8 it can be observed that, if the foundation strength is typified by a 12½° peak friction angle assuming no porewater pressure, the statics of the model are in balance for a slope angle β of 30° (graph point a). If foundation zone water pressure is introduced without reducing the friction angle of the foundation material, the slope will fail and the angle β must be reduced to about 20° before static equilibrium is restored (graph point b). If, however, failure simultaneously reduces the foundation material friction angle from the peak value of 12½° to a residual value of 7°, as suggested by Figure 26.7, the waste dump slope will not be stable until the angle β is reduced to about 12° (graph point c). The graph trace abc, therefore, depicts the influence of foundation porewater pressure and reducing foundation friction angle on the waste dump slope angle at $F = 1.0$.

In recovering from the failed condition, it could be assumed for illustrative purposes that the troublesome foundation pore pressures are dissipated before the next windrow of spoil is placed. Therefore, if zero pore pressure is operative with the residual strength condition in the foundation created by the earlier failure, recovery to graph point d (β = 21°) is indicated as shown by Figure 26.8. The net loss of waste slope angle between a and d is due to foundation strength loss, resulting from sliding initially precipitated by foundation porewater pressure. Recovery to point a will require rehandle until the potential zone of foundation failure for the waste placed subsequent to initial failure completely excludes the old shearing zone. During the recovery phase, it is possible that further failures could occur, and for these failures the crest of the failed zone would extend some distance behind the crest of the newly placed spoil row. Therefore, it should be immediately recognized that a program of substantial rehandle is the only method of recovery to point a, unless something can be done to improve foundation strength in advance of spoil placement. If, in the process of recovery from the failed waste dump condition, care is not taken to keep the bottom of the pit dry, a similar cycle will be repeated.

It is possible to extend the simple stability model to examine the influence of footwall topography on comparative waste dump stability. A gently rolling coal seam, or one which has been emplaced on variable preexisting topography, results in the spoil being placed on either an adversely or an advantageously dipping foundation plane during mining.

In Pit No. 1, it appears that the attitude of footwall topography, as shown in Figure 26.2, would influence stability. The simple stability model, therefore, has been modified to examine the comparative impact of a marginally inclined footwall foundation plane on spoil slope angle. Using the same example as previously, and a footwall angle of 2° (angles of about 1° to 3° are possible in the field), it is seen from Figure 26.8 that its effect can be described by the trace a'b'c'. Its influence is to require a 3° flatter slope in the unfailed condition (a' versus a), a 5° flatter slope when pore pressure is present (b' versus b), and a 7° flatter slope for residual strength and pore pressure together (c' versus c).

The comparative losses of spoil room at points d' and d, after failure has occurred and the pore pressures have dissipated, are illustrated by the 13° versus 9° loss of waste dump face angle (inclined versus horizontal foundation attitude). It is clearly evident that, in the case of a rearward dipping footwall, there would be considerable improvement to spoil pile stability for an otherwise similar set of porewater pressure and foundation strength circumstances.

Applicability of analyses

The purpose of the stability analyses has been to demonstrate the dramatic influence of both porewater pressure and reducing shear strength parameters on waste dump angle required for a factor of safety $F = 1.0$. It also demonstrates the factors which must be appreciated when planning recovery from such a failure. It is realized that the demonstration provided by the compound wedge model is influenced by the assumed constant pile height of 30.5 meters (100 feet), while in reality failure of the waste dump creates a rear scarp to the sliding mass, and thus the effective height of the slope is reduced. In consequence, the slope of the face angle-friction angle relationship necessary for stability under conditions of instantaneous increase of porewater pressure, or reduction of foundation friction angle, will be somewhat steeper than suggested by the graphs in Figure 26.8.

It should also be noted that development of foundation zone pore pressure is more complex than the model suggests, and in most cases its magnitude will be strongly influenced by the imposition of shearing stresses. Such stresses produce porewater pressure because of the tendency for shear to increase the density of initially loose materials. If the materials are near saturation, the shear-induced densification will precipitate a dramatic increase in porewater pressure and a rather spectacular rate of failure.

Conclusions

It is apparent that one of the major factors for ensuring stability of the spoil pile is the post-coal-removal condition of the pit bottom. Accumulation of water, and the subsequent placement of soft material from the top of the highwall on to the pit bottom provide a saturated, weak foundation for the spoil subsequently placed by the dragline. In addition to implementing positive pit bottom drainage measures, other steps can be taken to improve waste dump foundation strength.

Spoil dump foundation preparation

It is considered that some advantage could be gained from adopting a program of pre-spoiling foundation preparation. Rather than leaving dry areas of mined-out pit untouched after coal haulage and loading equipment traffic has subsided, the surface could be ripped to disrupt any slickensided surfaces which have been produced by local failure beneath haulage vehicle wheels. Ripping should be parallel to the highwall. The objective is to thoroughly disrupt the slickensided zone and to leave a broken base upon which the spoil will be placed. The advantage of better dump foundation drainage would also be gained. However, it is important that this foundation treatment does not proceed too far in advance of spoil placement, because it does afford the opportunity of substantial foundation softening if the area is allowed access to water.

Removal of coal

Because the failure plane is at the base of the coal, some toe support reaction will be provided by the in situ coal at the base of the spoil. It is beneficial, therefore, to commence coal removal at the highwall and to work toward the spoil toe. Toe excavation of the previous spoil could be delayed until rib excavation is completed, and thus over-steepening of the spoil toe would occur only shortly in advance of total coal recovery.

Selective waste placement

Selective waste placement could direct the placing of relatively competent and coarse-grained materials from mid-highwall height to the base of the advancing waste pile. This procedure would provide major improvement over a foundation presently composed of highly plastic surface material.

Pit floor subexcavation

Careful examination and geotechnical testing of the upper 30 to 60 centimeters (12 to 24 inches) of the footwall materials may reveal that the low shear strength zone is a single thin stratum. Removal of this discrete stratum would greatly reduce the problem of inadequate foundation strength, and it may only be necessary to consider selective removal through use of index testing to identify particularly weak areas. The above-described operational expedients can be utilized on a trial basis pending detailed geotechnical investigations. It is considered that better control of pit bottom drainage is the key to immediate stability improvement because, as shown in Figure 26.8, its effect on slope angle is rather dramatic. Once failure is initiated, an unavoidable reduction in operative friction angle occurs, and this greatly compounds the problem of recovery to stable waste dump conditions.

Acknowledgments

The authors are indebted to their employers for the opportunity of working on this project, and for permission to publish the paper. Mr. Gilchrist is also indebted to his colleagues, N. Skermer for providing study information from other projects and S. McKeown for undertaking much of the computation which preceded preparation of Figure 26.8.

References

Bishop, A. W., and Henkel, D. J., 1962, "The Measurement of Soil Properties in the Triaxial Test." Edward Arnold (Publishers) Ltd., London, Second Edition.

Coates, D. F., editor, 1972, "Tentative Design Guide For Waste Embankments in Canada." Department of Energy, Mines and Resources Technical Bulletin TB 145, Ottawa.

Fremlin, G., editor in chief, 1974, "The National Atlas of Canada." The MacMillan Company of Canada Limited, Toronto, Fourth Edition.

Guliov, P., 1972, "Lignite Coal Resources of Saskatchewan." Proceedings First Geological Conference on Western Canadian Coal, Edmonton, Alberta.

Holter, M. E., 1972, "Coal Seams of the Estevan Area, Southwestern Saskatchewan." Proceedings First Geological Conference on Western Canadian Coal, Edmonton, Alberta.

Sinclair, S. R., and Brooker, E. W., 1967, "The Shear Strength of Edmonton Shale." Proceedings of the Geotechnical Conference, Oslo, 1967, Volume I.

Performance of a Waste Rock Dump on Moderate to Steeply Sloping Foundations

27

D. B. Campbell, Principal, Golder Associates, Ltd.,
Vancouver, British Columbia, Canada,
and W. H. Shaw, Planning Engineer,
Fording Coal Limited, Elkford, British Columbia, Canada

Introduction

Extensive coal reserves occur within the mountain regions of eastern British Columbia and western Alberta. Metallurgical coal is currently being produced at six properties, two in British Columbia, and four in Alberta. Five of these six operations produce a total of approximately 12 million tons annually by open pit mining. By 1985, coal production from reserves in Alberta and British Columbia is expected to increase to between 30 and 40 million tons per year. A significant portion of this expected future coal production will be obtained by open pit mining. Construction of large disposal piles of waste rock, referred to as waste dumps or waste piles, will be an attendant part of the development of future open pit mines. Since many of the mines will be located in mountainous regions, at least some of these waste dumps will be developed on moderate to steeply sloping foundations.

The Clode waste dump at the Fording Coal operations represents a typical waste dump on a moderate to steeply sloping foundation. Development of this waste dump commenced in early 1972, and has continued to the present time. Since the Clode waste dump comprises material having geotechnical properties similar to the waste rock that will be generated at many of the proposed future open pit coal mining operations, the observed performance of the Clode waste dump since inception of its development is believed to be indicative of the performance that may be expected at future waste dumps constructed on moderate to steeply sloping foundations.

Description of Clode Waste Pile

The Clode waste pile at the Fording Coal operations has been developed on the west face of Eagle Mountain. In cross section, the waste dump is triangular in shape, the sides of the triangle being formed by the upper surface of the dump which extends horizontally out from the original ground contours, by the original

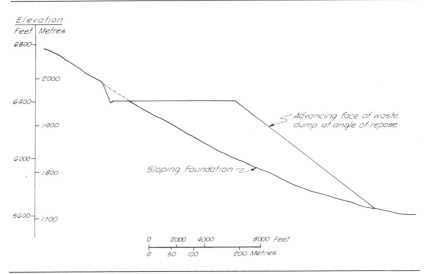

Figure 27.1. Cross section showing Clode waste dump at an intermediate stage of development.

ground profile on the west face of Eagle Mountain, and by the advancing face of the waste dump which is inclined at approximately 37°, the angle of repose for the waste rock. The waste pile presently extends approximately 1,220 meters (4,000 feet) along the face of the mountain, the upper surface of the waste pile extends horizontally an average distance of approximately 244 meters (800 feet) from the original slope, and the waste pile has a maximum vertical height of approximately 244 meters (800 feet). A typical section through the Clode waste dump at an intermediate stage of development is shown in Figure 27.1.

As indicated on the cross section, the original ground profile within the foundation area of the waste dump is concave upward. Segments of the slope at higher elevations are steeper than those at lower levels. Beneath the eastern limit of the waste dump, the original ground profile slopes at approximately 30°, with short localized segments in the 30° to 35° range. Beneath the toe of the waste pile, the slope of the original ground surface is of the order of 10°.

The original ground surface within the foundation area of the Clode waste dump was forested with mixed pine, spruce, and fir. The foundation area was cleared before development of the waste dump commenced.

Within the foundation area of the Clode waste dump, the original ground surface was mantled by a veneer of weathered soil ranging from 0.6 to 1.5 meters (2 to 5 feet) in thickness. At higher levels within the foundation area of the dump, the weathered soil mantle is underlaid directly by shale bedrock. Within the lower regions of the slope, the weathered soil mantle is underlaid by very dense glacial till. Laboratory shear strength tests on representative samples showed that the effective angle of internal friction of the weathered soil mantle ranged between 32° and 35°.

Development of the Clode waste dump commenced by end dumping of waste rock to form a bench extending outward from the face of the mountain slope. The surface of the bench was extended horizontally in a direction perpendicular to the ground contours corresponding to the design surface elevation of the dump. The face of the dump was permitted to assume the angle of repose for the waste rock, and the dump is advanced by the process of gradual accretion of material on its face.

Base drainage
The waste rock consigned to the Clode waste dump consists of fragments of sandstone, siltstone, and shale ranging from dust-size particles, to boulders several cubic meters (or cubic yards) in size. End dumping of this material at the crest of the waste pile results in segregation of particle sizes as the material rolls and slides down the face of the pile. As a result, the largest and most durable rock fragments segregate and accumulate at the toe. Figure 27.2 shows a typical example of segregated coarse rock at the toe of the waste pile. Examination of the face of the waste pile shows a general reduction in the median particle size with increasing elevation between the toe and the crest of the pile. As the face of the pile advances by the process of gradual accretion of material, the accumulation of segregated coarse rock at the toe becomes covered, and forms a coarse, pervious, filtered drainage layer at the base of the waste dump.

The original access road to the open pit was located adjacent to the eastern (uphill) limit of the waste pile. This access road, complete with a side drainage ditch, served as an interceptor for surface runoff water from the mountain slopes

Figure 27.2. Photograph showing segregated coarse rock at toe of waste pile.

above the waste pile. The drainage layer comprising the coarse segregated rock at the base of the waste dump, together with the fact that the pile receives only that precipitation which falls within its perimeter, precludes development of pore water pressures within the base of the dump. Hence, the segregation of particle sizes attendant with end dumping onto the face of the waste pile, which is maintained at the angle of repose, is a factor contributing to waste dump stability.

Crest movements

From the outset of its development, the crest region of the Clode waste pile has been subject to relatively large movements in the downslope direction. These movements are the combined result of compression of the waste rock under self-weight, and of shearing strains at shallow depth beneath the face of the waste pile.

The waste rock consigned to the waste dump is essentially cohesionless, although the presence of fines contributes to minor apparent cohesion near the crest. When cohesionless material is at its angle of repose, the factor of safety on potential failure surfaces located at shallow depth below the face of the slope is close to 1.0. That is, at shallow depth beneath the face of the slope, the maximum available shearing resistance is only marginally greater than the applied shearing stresses. Under these stress conditions, shearing strains occur within the zone located at shallow depth below the frontal slope of the waste pile which result in slight bulging within the lower third region of the face of the pile, and in downslope displacement at the crest. Bulging on the face of the waste pile, combined with localized short segments of steeper-than-average topographic slopes beneath the lower portion of the frontal slope pile have resulted in occasional failures on the face of the dump. Such failures could pose a potential hazard to men and equipment engaged in dumping operations at the crest of the pile.

A shear failure at the crest of the waste pile does not occur without warning. It is preceded by a period during which the displacements at the crest of the pile increase at a progressively faster rate. At the Clode waste dump, records of the rate of crest movement have been maintained from the outset of dump development, to provide advanced warning of impending instability on the face of the waste pile. The devices used to monitor crest movements are illustrated schematically in Figure 27.3. The device consists of a pin driven into the face of the slope at a point slightly below the crest of the waste pile. A light flexible wire attached to the pin is run through pulleys attached to light portable stands. The stand farthest from the crest is located on stable ground and a weight is suspended from the end of the wire. As displacements occur on the face of the pile, the end of the wire attached to the pin is pulled downslope, raising the suspended weight. A record of the rate of movement is obtained by taking periodic measurements of the vertical distance between the suspended weight and a reference point on the base of the stand.

The measuring devices are crude. However, the records show that in many instances, rates of movement in excess of 1.5 meters (5 feet) per day, and total crest movements in excess of 9.1 meters (30 feet) have occurred which did not lead to mass sliding on the face of the pile. Considering the large magnitude, and the rates of the movements that precede failure, the crude monitoring devices have provided data of sufficient accuracy. The operating principles of the movement

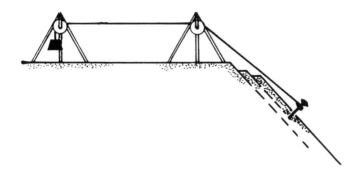

Figure 27.3. Sketch illustrating device used to monitor movements at the crest of Clode waste pile.

devices are simple. They are inexpensive, can be installed quickly, and they can be easily moved to a new location when required. The movement readings can be taken by one person (a survey crew is not required) and the movement progression can easily be watched by the dump foreman during periods when rapid crest movements are occurring at a particular location, and close monitoring of the rate of movement is required.

Waste disposal operations at the Clode dump were temporarily suspended at locations when the measured rate of displacement at the crest of the waste pile was greater than 0.46 meter (1.5 feet) per day. Following a period of temporary suspension, spoiling operations were resumed when the recorded rate of displacement was equal to or less than 0.46 meter (1.5 feet) per day, provided also that the recorded rates of displacement were decreasing progressively with time.

Two representative plots of rate of movement versus time are shown in Figure 27.4. In both cases illustrated, the rate of displacement increased gradually to

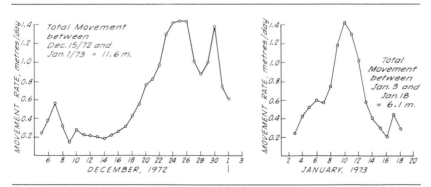

Figure 27.4. Two examples of rate of crest movement versus date of observation. Although the maximum rate of crest movement was nearly 1.5 meters (5 feet) per day, failure on the face of the waste pile did not ensue.

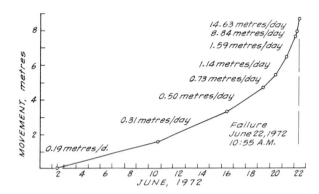

Figure 27.5. Plot of crest displacement versus date of observation prior to failure on face of waste pile.

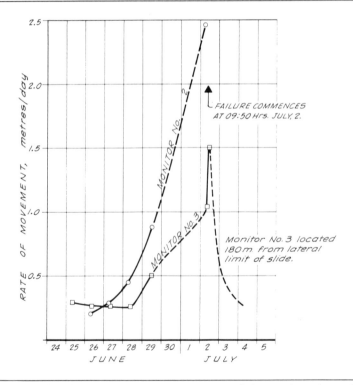

Figure 27.6. Plot of rate of movement versus date of observation prior to failure on face of waste pile.

greater than 1.2 meters (4 feet) per day, and the recorded total displacements were 11.6 and 6.1 meters (38 and 20 feet) respectively. In neither of these cases did the crest movements develop to the point of mass sliding down the face of the waste pile.

Figures 27.5 and 27.6 show examples of recorded crest movements that culminated in mass sliding on the face of the pile. In Figure 27.5, the measured displacements increased progressively over a period of approximately three weeks. Total displacement was approximately 8.5 meters (28 feet), and prior to failure the measured rate of displacement at the crest of the pile was 0.61 meter (2 feet) per hour. In Figure 27.6, the recorded rate of movement increased progressively over a period of six days, during which the recorded displacement was 8.2 meters (27 feet). Monitor No. 3, which was located approximately 122 meters (400 feet) from the lateral extremity of the failure mass, shows that the movements at this location were influenced to a significant degree by the slide activity. Within 24 hours after the slide occurred, the rate of movement at monitor No. 3 had dropped from approximately 1.5 meters (5 feet) per day to approximately 0.3 meter (1 foot) per day.

Figure 27.7 shows a plot of measured rates of displacement at the crest of the waste pile as recorded during 1973. During this interval the toe of the waste pile

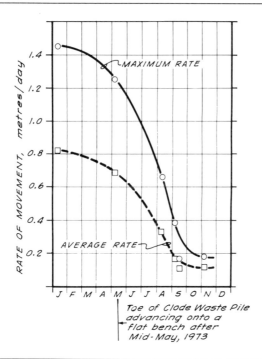

Figure 27.7. Graph showing reduction in rate of displacement at the crest, as the toe of the waste pile advanced onto a flat bench.

advanced progressively onto lower ground and, as a result, the vertical height of the waste pile increased from approximately 162 to 198 meters (530 to 650 feet). During the same interval, the toe region of the waste pile advanced over progressively flatter foundation slopes. The data demonstrate that the inclination of the foundation beneath the toe region of the waste pile had a significant influence on the rates of movement that occurred at the crest of the waste pile.

To provide greater detail of the time rate of displacement at the crest of the waste pile, one of the movement monitoring devices was equipped with a continuous recorder. This continuous recorder was in operation for approximately six weeks during the summer of 1973. The chart, which provided a continuous record of displacement with time, showed that the movements at the crest of the pile progressed at a steady rate; there was no evidence of any "slip-stick" type of movement. In addition, the continuous record on the charts showed that nearby blasting at the Clode pit did not have any discernible effect on the rate of displacement at the crest of the waste pile.

Safety Beyond Toe of Waste Pile

In some instances, when mass sliding develops on the face of the waste pile, the material involved in the slide attains a high degree of mobility, and the slide debris extends surprisingly large distances beyond the toe of the waste pile. Similar occurrences have been observed in nature as a result of rock falls from mountain slopes. In nature, such an occurrence is referred to as a sturzstrom, a term used by Heim* to denote a falling-streaming motion. The large distance between the point of maximum advance, or the distal portion of the slide debris, and the toe of the waste pile, together with the low vertical angle between the center of gravity of the sliding mass prior to failure and the center of gravity of the slide debris indicate that the frictional resistance of the mobile mass is temporarily reduced to a value of considerably less than its static friction. Several investigators have postulated mechanisms to explain this temporary reduction in frictional resistance. These postulations include the presence of water within the slide mass, vapor generated by heat due to friction, and air entrapped within and beneath the mobile slide debris. However, a surface feature which has been interpreted as a sturzstrom was photographed on the surface of the moon near the Apollo 17 landing site. Clearly, on the surface of the moon, neither air nor water can play a part in the sturzstrom mechanism. The low mobile friction angle of the sturzstrom may be the result of reduced effective stress, caused by particle agitation due to kinetic energy transfer within the mobile mass.

At least some of the failures that have developed on the face of rock waste dumps have occurred under dry conditions, and it is doubtful whether air played a significant role in the mobility of the slide debris. A sturzstrom-type slide developed on the face of the Clode waste pile during the early morning of November 8, 1974. The difference in elevation between the crest and the toe of the pile was approximately 213 meters (700 feet), and the lower third of the face

* Albert Heim (1849-1937), Professor of geology, Swiss Federal Institute of Technology. Commissioned to study the 1881 rock fall at Elm which killed 115 people. Over the next half century, Heim continued to study rock falls, and particularly the sturzstrom phenomenon.

of the pile was located above a localized area where the foundation slope was steeper than 30°. The slide debris advanced rapidly over relatively flat ground beyond the toe of the pile, and the distal portion of the slide was located a maximum distance of 564 meters (1,850 feet) beyond the toe of the waste pile. The vertical angle subtended from the crest of the waste pile to the distal portion of the slide was 17°.

In the event that a sturzstrom-like slide develops, the distance that slide debris may travel is an important consideration with respect to location of facilities and operations within areas beyond the toe of a waste dump. Analyses of sturzstroms by Scheidegger, 1973, and by others suggest an empirical correlation between the slide volume, and the apparent mobile friction angle during movement. The data indicate that the mobile friction angle decreases with increasing volume of slide.

Scheidegger produced a log-log plot of apparent mobile friction coefficient versus volume of the slide, and suggested that the resulting correlation can be used to predict the reach and the velocity of an imminent slide provided the volume of the slide can be estimated. The friction coefficient as employed by Scheidegger in his analyses is numerically equal to the tangent of the vertical angle from the distal portion of the slide debris to the crest of the slope prior to the failure. Scheidegger's plot, together with the data points representing four sturzstrom-type slides from the face of rock waste dumps is shown in Figure 27.8. The solid line on Figure 27.8 represents the mean of the data points for rock slides in nature. The two dashed lines represent one standard deviation above and below the mean for natural rock falls. The data points representing the waste dump slides all plot below the dashed line representing one standard deviation below the mean for sturzstroms in nature. Figure 27.8 indicates that failure masses originating from the faces of waste rock dumps have mobile friction angles that are

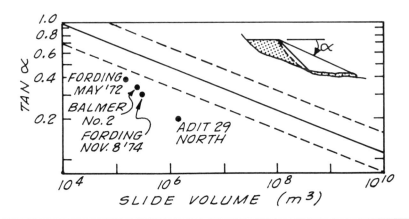

Figure 27.8. Scheidegger's correlation between landslide volume and the vertical angle from the breakaway point to the distal portion of the slide debris. The data points for four sturzstrom-like slides from the faces of waste rock dumps indicate that the run-out angles are flatter than for natural rock falls.

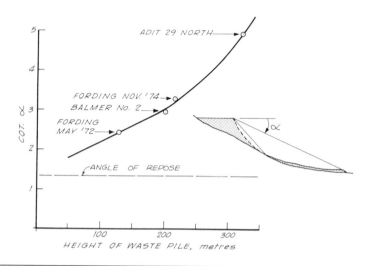

Figure 27.9. Correlation between height of waste rock dump and cotangent of run-out angle.

lower than the mobile friction angles for rock falls in nature. The flatter run-out angles for failures from the face of waste rock dumps may be due to the fact that the dump material is already broken so that less energy is expended in breaking up the mobile mass as compared to rock falls from natural slopes. The data suggest that Scheidegger's correlation does not apply to failures from the faces of waste rock dumps.

Data from four sturzstrom-type slides originating from the faces of waste rock dumps in eastern British Columbia indicate an empirical correlation between the vertical height of a waste rock dump and the fahrboschung, or vertical angle below the horizontal from the crest of the dump, before failure to the distal portion of the slide debris. The data are plotted in Figure 27.9 in the form of the vertical height of the waste dump versus the cotangent of the fahrboschung. The shallowest angle so far observed is approximately 11°. In this case, while the waste rock itself was sufficiently dry to preclude reduction of shear strength as a result of pore water pressures, the slide debris advanced onto a wet soil mantle within the region beyond the toe of the slide. Rapid loading of this saturated soil mantle may have contributed to the distance that the slide debris traveled beyond the toe of the waste pile before coming to rest.

The face of the Clode waste pile below the crest remains approximately at the angle of repose. However, the face of the completed waste pile will be graded to approximately 2 horizontal to 1 vertical. During the final stages of development, successively lower benches (wraparound dumps) will be constructed on the face of the waste pile. As part of the final reclamation work, the stepped face profile formed by the wraparound benches will be graded to produce an approximately uniform slope on the face of the waste pile.

Summary

The observed performance of the Clode waste dump indicates the following:

1. When a waste rock dump containing siltstone, sandstone, and shale is developed on a sloping foundation, and the frontal slope during development is maintained at the angle of repose for the material, occasional mass instability may develop on the face of the waste pile.
2. A failure on the face of the waste pile is preceded by an interval of several days when the rate of displacement at the crest increases progressively.
3. Displacements at the crest of the waste pile are influenced by the inclination of the foundation slopes beneath the toe region of the pile, by the rate of loading at the crest, and by the height of the waste pile.
4. It is advisable to maintain a convex scalloped configuration along the crest of the waste pile. Such a scalloped configuration tends to limit the lateral dimensions of the zone of mass movement in the event that instability develops on the face of the waste pile.
5. The operating crest of the waste dump should be as long as practicable, to avoid concentrated dumping within short segments of the crest, and to provide optional dumping areas in the event that the rate of crest movements indicates that dumping operations should be temporarily suspended within a particular segment of the crest.
6. Occasionally, a failure on the face of a waste dump may develop into a sturzstrom-like slide which may travel surprisingly large distances beyond the toe of the waste pile. An estimate of the potential travel distance can be made with reference to Figure 27.9

Reference

Scheidegger, A. E., 1973, "On the Prediction of Reach and Velocity of Catastrophic Landslides." Rock Mechanics, Volume 5.

Mine Waste Disposal with Economic Stability

28

E. S. Smith, Chief Engineer, Geotechnical Division,
B. H. Dworsky, Geotechnical Engineer, Geotechnical Division,
International Engineering Company, Inc.,
San Francisco, California, USA,
and R. A. Tinstman, Director, Mining Engineering,
Industrial and Mining Engineering Division,
Morrison-Knudsen Company, Inc., Boise, Idaho, USA

Introduction

In 1966, nearly one-half of a cone-shaped coal waste pile slid out and down from the hillside position it occupied. The flow slide attained a velocity of about 32 kilometers (20 miles) per hour and traveled a considerable distance before the waste material stabilized. This failure was only one in a succession of movements experienced by that pile and its adjacent spoil piles. Like the preceding events, it would normally have attracted only local attention. It received widespread attention, however, because this ultimate slide had the momentum to overrun part of the small village at the base of the hill (Davies et al., 1969).

In 1950, an eminent soils engineer described the various types of failure most frequently observed in earth embankments (Casagrande, 1950). Certain of these types, either alone or in combination, were even then at work in the vicinity of the 1966 flow slide. They were to continue for 16 additional years. The basic knowledge existed for recognizing and averting these mine waste stability problems; those for whom it would have been most useful, however, were not aware of their need.

Stability in waste disposal requires familiarity with:

1. The possible forms of failure.
2. Design procedures which will minimize, in advance of disposal, the risk of failure.
3. The characteristic "distress signals" which may foreshadow a potential failure in the waste pile.
4. The remedial measures which will restore the initial stable condition.

The geotechnical engineer possesses, by training and experience, a disciplined awareness of potential waste disposal problems, often unimagined and unanticipated by those preoccupied with actual mine production. Mine operators today, therefore, have good reason to seek assistance from the geotechnical specialist.

Economy Without Stability

The Aberfan disaster

The colliery at Aberfan, South Wales, employed antiquated waste disposal procedures. Since about 1916, the hillside above the village had served as a spoil storage area. Over the years, a series of cone-shaped spoil piles or "tips" had risen. In 1958, Tip 7 was begun. A conspicuous bulge at the base of Tip 5, a prominent scar on Tip 4 from a massive slip-induced slide in 1944, and irregular features on the slopes of Tip 7 all attested to the inherent instability of these piles.

This hillside spoil pile complex presented a seldom-equaled exhibition of man-made landslide phenomena. Only one display was absent: a slope failure of catastrophic proportions. In October 1966, this final exhibit was added.

Site selection for Tip 7 was determined entirely by economic considerations: the presence of an available area in a convenient location. Geologic investigations would have disclosed that the surface soil cover was underlaid by a fissured, jointed sandstone. Soils investigations would have found that the surface mantle on the lower slopes of the hillside site was a relatively impermeable boulder clay. Visual surface examination would have noted that natural drainage channels were incised down the hillside through the site. A geotechnical review would have concluded that groundwater and surface drainage conditions rendered this site unsuitable for a spoil pile. None of these studies were conducted.

Economy was achieved for a brief eight years. Man's fiscal books were balanced, but nature's books were not. In October 1966, nature rebalanced its books.

Results of investigation

Most significant of the findings, lessons, and recommendations of the governmental authority (Davies et al., 1967) appointed to investigate the disaster were:

1. Waste disposal was pursued without any guidance, either from government, industry, or the engineering profession.
2. All spoil piles should be regarded as potentially dangerous.
3. Spoil piles should be treated as civil engineering structures.
4. Mine personnel engaged in the daily management and control of waste disposal should be trained for their responsibilities.
5. Mine managers and engineers should be made aware of the fundamentals of soil mechanics and groundwater conditions.

These statements are reemphasized here because they have not always been observed since their initial presentation.

Stability by Economic Design

Compared with the other applied sciences, geotechnical engineering represents an inexact technology. The materials with which it deals vary widely in character. The external conditions acting on these materials also cover a wide range. In the narrow view, the soil or rock mechanics specialist appears handicapped by his inability to clearly define these initial conditions. Usually, uncertainty induces a conservative analysis and design. The competent, innovative geotechnical engineer, however, is predisposed to the broader view: economic benefits may be realized by implementing an initially flexible design, subject to continuing evaluation and revision.

Coincidental with and implicit in this initial flexible design is the planning and implementation of on site monitoring of the designated concept or construction procedure. The monitoring program itself need be no more elaborate than circumstances require: periodic visual inspection may be as satisfactory as intensive instrumentation (Smith et al., 1977). The initial design may possibly entail a "calculated risk." The consequences of failure, however, will be clearly defined and limited in extent. Remedial measures, subsurface drainage or slope-flattening, for example, will be identified in advance and will be available for instant use (Casagrande, 1965).

Prerequisite safety considerations, however, place a limitation on use of this "observational" approach. Where natural, uncontrollable events, such as earthquake or flood, may be reasonably anticipated, design must necessarily follow a more conservative approach. Where developed and inhabited areas lie in proximity to the project site, the design method must leave minimal allowance for error.

Certain phases of mining operations are especially well-suited to the observational method of design. Excavation and waste disposal is effected by a wide variety of equipment, functioning as single units or in combination, with varying capabilities. Trained mine personnel are usually on site for visual observation. Reliable instrumentation exists for monitoring soil and rock structures, either in their natural or disturbed states. Mines are often located in remote areas away from the general population.

Waste disposal at Aberfan did not conform to a preconceived plan or design. No effort was expended to observe the waste structures, and no remedial measures were employed or held in reserve to improve their stability. Tip 4 experienced a massive slide in 1944 and was abandoned. Tip 7 experienced a large slump in 1963. The resulting bowl-shaped depression was filled in and waste disposal was resumed. Subsequent surface movements were ignored. The mine managers assumed that lives and homes were not endangered and continued their established disposal procedure. The observational method of design was not workable at Aberfan because a willingness to "look and learn" did not exist. Mine managers pursued their form of calculated risk in the absence of geotechnical guidance.

The following section will examine the economic effects of variable waste material properties and alternative waste disposal procedures. These features of waste handling offer a potential application for the observational method of design. Although the simplified example presented pertains to overburden excavation at a strip mine, the geotechnical implications are equally relevant elsewhere in waste handling.

Productivity and Waste Handling

Excavation and waste disposal are necessarily subordinate to the total mining program. Waste handling can be viewed properly only within the broader frame of overall activities. Handicaps imposed on other production activities may counterbalance the economic benefits otherwise obtained from a specific excavation

Figure 28.1. Overburden excavation by dragline in a strip mine located in the eastern part of Montana.

or materials handling procedure. Regardless of method, however, geotechnical studies retain their usefulness for identification of spoil material and waste structure properties and characteristics, either in their naturally placed condition or in an altered, reworked condition.

The mining engineer determines the type of equipment and mining procedure most suitable for a specific mining project. Each type of excavation unit develops a highwall, waste dump, or spoil pile with unique characteristics. The highwall excavated by dragline (Figure 28.1) presents different potential stability problems from that excavated by truck and shovel (Figure 28.2). Spoil placed by dragline has material properties different from spoil placed by shovel-loaded end dump truck. A compactive effort applied to the surface level of the dragline spoil pile will have little effect on the lower levels of the material. Proper equipment, however, can obtain relatively efficient compaction of a waste dump created by end dump truck.

The major structural features of geotechnical interest in a coal surface mine are the highwall, waste dump, and spoil pile. Safety, production reliability, and economics are the primary considerations in a slope stability evaluation for these structures. Although the relation between slope stability and safety is important, the initial concern of mine planning and design usually focuses on the production reliability and economic aspects.

For any new surface mine, an evaluation is made to assess the angle at which the highwall, waste dump, or spoil pile will be stable. It can range from a quick judgment factor based on experience to extensive drilling, testing, and assessment. This evaluation is quite often more critical than is usually realized.

Figure 28.2. Overburden excavation by truck and shovel in an open pit mine in southwestern Wyoming.

Case 1

A strip mine which utilizes a large dragline presents a good example. Figure 28.3 shows a strip mine with 30.5 meters (100 feet) of overburden, 3 meters (10 feet) of coal, and a 46-meter- (150-foot-) wide cut. Consider Case 1, with the spoil pile created from the dragline cut having a stable spoil slope of 1.25 to 1, or 38.7°. Assuming the overburden will have a 30 percent net swell, a dragline with a 91-meter (299-foot) operating radius is required to strip the overburden and place the spoil in the configuration shown with no rehandle. An 84-cubic-meter (110-cubic-yard) dragline with a 102-meter (335-foot) boom will accomplish this objective. Using

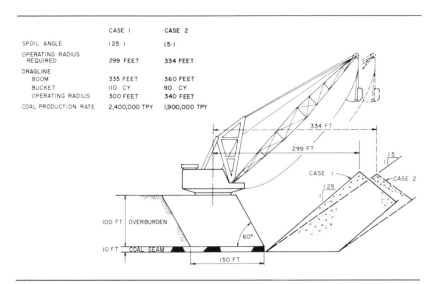

	CASE 1	CASE 2
SPOIL ANGLE	1.25:1	1.5:1
OPERATING RADIUS REQUIRED	299 FEET	334 FEET
DRAGLINE:		
BOOM	335 FEET	360 FEET
BUCKET	110 CY	90 CY
OPERATING RADIUS	300 FEET	340 FEET
COAL PRODUCTION RATE	2,400,000 TPY	1,900,000 TPY

Figure 28.3. Effect of stability of spoil pile slope on productivity.

typical production capabilities for this unit and a 90 percent coal recovery factor, 2.2 million metric tons (2.4 million tons) of coal per year could be mined.

Case 2

Now consider the same mining operation in Case 2, with identical parameters except that the spoil is stable at a 1.5 to 1 slope, or 33.7°. For this situation, a dragline with a 102-meter (334-foot) operating radius is required to strip the overburden and place the spoil in the configuration shown with no rehandle. By utilizing the same basic dragline unit as in Case 1, but with a boom length extended to 110 meters (360 feet) and a lower boom angle, an operating radius of 104 meters (340 feet) can be achieved. By extending the operating radius to 104 meters, the maximum recommended bucket size will be reduced to 69 cubic meters (90 cubic yards). Because of the smaller bucket size and different swing cycle time for Case 2, the rate of overburden removal will be decreased, and resulting coal production will be reduced to 1.7 million metric tons (1.9 million tons) per year. Since the dragline units of Case 1 and Case 2 are essentially the same, the operating cost per ton for stripping will be approximately 25 percent higher for Case 2, due to the difference in quantity of coal uncovered.

Although, in Case 2, there are numerous other approaches and techniques which could be implemented to reduce the impact of the flatter slope, this simplified example explicitly points out these conclusions:

1. If the slope evaluation predicts optimistically steep slopes for the highwall and spoil and these slopes can be maintained, obvious economic benefits can be realized. However, if after operations commence, the actual slopes are flatter than predicted, production demands will not be met and the cost per ton will increase.
2. If the slope evaluation predicts a conservatively flat slope for the highwall and spoil, a unit will be selected which will meet the production requirements. However, if the predictions are too conservative, then the unit cannot readily be revised and the operations will be limited by the unit selected.

These general conclusions also apply to an open pit mine, except that a typical truck and loading unit operation is more flexible because additional units can be added more readily. Although operating costs will still be affected, this flexibility makes it possible to achieve original production goals.

The above examples indicate that slope evaluations are critical to the overall production and cost goals established for the mine operation. The soil or rock mechanics specialist must be aware of the economic impact of his conclusions and should not form an overly conservative analysis. Similarly, the mine operator must recognize that he should not use highly optimistic slope values which could result in higher cost, lower production rates, and possible slope stability problems.

Summary

Waste disposal in coal mining occupies an essential but nonproductive phase of overall mine operations. Until recent years, economic considerations have governed siting and disposal placement procedures. Failure of Tip 7 at Aberfan,

South Wales, in 1966, has focused both professional and public attention on the safety aspects of waste disposal. Use of geotechnical engineering procedures provides mine operators with their required measure of safety at minimum increase in overall operating costs. The observational method of design applied to mine waste disposal permits consideration of less costly design alternatives. The observational approach is well-suited to waste disposal and is already widely practiced in the mining industry.

Waste disposal with economic stability challenges the skills of both the mining and the geotechnical branches of the engineering profession. With each specialty exercising its own competence, with each employing its own tools, but with both efforts directed toward a common goal, mine management can responsibly relieve the hazards associated with mine waste handling. A coordinated effort can insure safety as well as economy.

References

Casagrande, A., 1950, "Notes on the Design of Earth Dams." Journal of the Boston Society of Civil Engineers, Boston, Mass., Vol. 37, No. 4. •

Casagrande, A., 1965, "Role of the 'Calculated Risk' in Earthwork and Foundation Engineering." Journal of the Soil Mechanics and Foundations Division, American Society of Civil Engineers, New York, N.Y., 91, SM4.

Davies, H. E., Harding, H. J. B., and Lawrence, V., 1967, Report of the Tribunal Appointed to Inquire into the Disaster at Aberfan on October 21st, 1966, Her Majesty's Stationery Office, London, U.K.

Davies, H. E., Harding, H. J. B., and Lawrence, V., 1969, A Selection of Technical Reports Submitted to the Aberfan Tribunal, Her Majesty's Stationery Office, London, U.K.

Smith, E. S., Poindexter, D. R., and Bleikamp, R. H., 1977, "Observational Approach to Tailings Dam Enlargement." Geotechnical Practice for Disposal of Solid Waste Materials, American Society of Civil Engineers, New York, N.Y.

Stability Considerations During Abandonment of Sludge Ponds

29

Dr. James M. Roberts, Director of Business Development,
Ackenheil and Associates Geo Systems, Inc.,
Pittsburgh, Pennsylvania, USA

Introduction

In September 1975, new Mine Safety and Health Administration (MESA) regulations established additional requirements for the design, construction, and operation of coal waste dams. These requirements, in some cases, required the construction of a large emergency spillway. For some rugged sites, this created a potentially great expense. In the case study presented in this chapter such site conditions existed. Coal company management decided to abandon the existing slurry pond and adopt a "closed system" preparation process. This particular plant had a coal refuse production of about 1,147 cubic meters (1,500 cubic yards) per day and a sludge or slurry production of 25 tons per hour of solids. The anticipated equipment deliveries for the preparation plant modification indicated that the existing slurry pond was required for another 12 months.

Purpose

The purpose of the study was twofold. First, to evaluate the existing pond capacity and recommend necessary construction in order for the pond to be in service for 12 months. Second, to select and design a site for continued refuse disposal (consisting of coarse refuse and filter cake material) for a period of 25 years.

Existing Pond Capacity

Using existing mapping, a pond capacity-elevation curve was developed. Based on the results of routing the design storm, allowing for reasonable freeboard, and considering the rate of slurry production, the existing pond was found to have an available storage capacity sufficient for only four months' future use. Accordingly, further study was necessary to determine what modifications were necessary in order to allow continued use for 12 months.

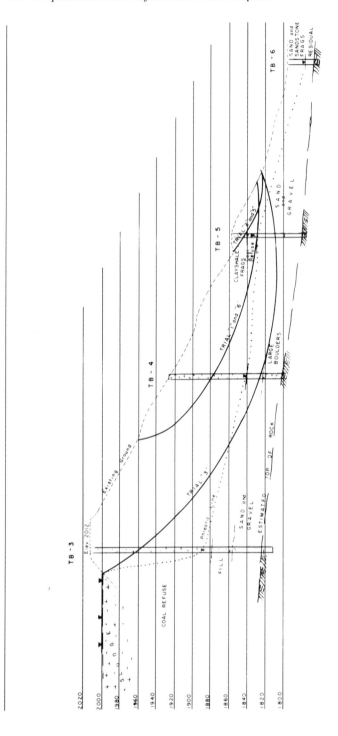

Figure 29.1. Geologic cross section of the materials encountered at the pond site.   The five different trial circles for the stability analysis of the existing embankment are also shown.

Table 29.1. Initial Strength Parameters

Type	c (psf)	ϕ	Wet density (pcf)
Coal sludge	0	32°	83
Coal refuse	300	36°	97
Sand and gravel	400	39°	120

Initial Investigation

Drilling phase
Six test borings were drilled along the axis of the original stream channel. A geo-logic cross section of the materials encountered is shown in Figure 29.1. The materials encountered were generally coal refuse, and compact sand and gravel. Clay shale fragments were encountered in Test Boring 5. Coal sludge was sampled at the surface of the pond using Shelby tubes.

Testing phase
Due to the granular nature of most of the materials encountered, only direct shear tests were performed on the samples obtained. Since only a small amount of clay shale fragments was obtained in Test Boring 5, insufficient material was available for laboratory testing. Based on the results of multiple direct shear tests, the strength parameters presented in Table 29.1 were used in subsequent stability analyses. Strength values were assumed for the clay shale fragments.

Stability analysis (existing embankment)
Stability analyses using the conservative Swedish Circle method were performed. Five different trial circles, as shown in Figure 29.1, were considered. With Trials 5 and 6, reduced shear strength parameters were used to assess the potential effect of actual conditions, and was less than that measured by laboratory test results. The results of the stability analyses are shown in Table 29.2, and indicate that the existing embankment had a factor of safety in the range of 1.6. This

Table 29.2. Results of Stability Analyses on Existing Embankment

Trial circle	Factor of safety		Remarks
	Without earthquake	With earthquake	
1	1.9	1.6	
2	2.5	2.1	
3	2.4	2.0	
5	1.6	1.4	Used $\phi = 20°$, $c = 300$ psf
6	1.7	1.4	Used $\phi = 36°$, $c = 0$ psf

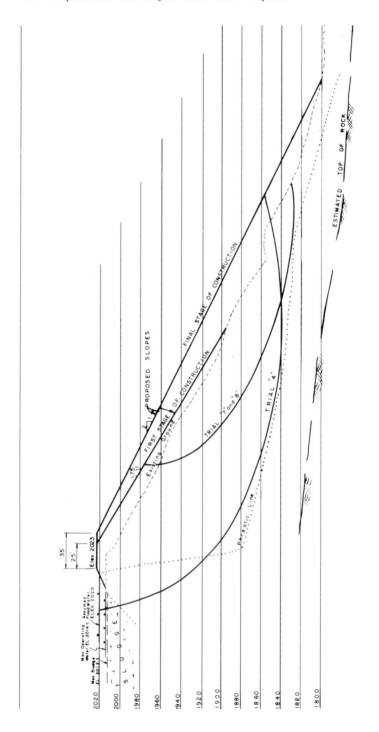

Figure 29.2. Proposed embankment modifications. The trial circles for the stability analysis of the modified embankment are also shown.

factor of safety was achieved primarily because of the relatively high strength characteristics of the refuse and underlying sand and gravel and the relatively low position of the phreatic line.

Stability analysis (proposed modification)
Due to the need of maintaining a disposal site for slurry for the next 12 months, several alternative plans were developed. The most economical plan was modification of the existing embankment. The crest elevation of the modified embankment was determined by flood routing. Two different downstream embankment slopes were considered (2 horizontal to 1 vertical and 1.75 horizontal to 1 vertical). The construction time estimates for these two embankment slopes were approximately five and three months respectively. The proposed embankment modifications considered are shown in Figure 29.2. The trial circles used for the stability analyses are also shown, and the preliminary results are given in Table 29.3. The results indicated that either a 2 to 1 or 1.75 to 1 embankment slope could be used. The coal operator selected the 1.75 to 1 slope in order to complete construction in three months. Subsequently, the slope was flattened to 2 to 1. Additional stability analyses were performed prior to the development of plans and specifications. Compaction testing and shear strength tests indicated that the field density specification for compaction of the coal refuse could be 85 percent of the Standard Proctor maximum density. A field representative from Ackenheil and Associates was on site during embankment construction to measure compaction results.

Pond Abandonment

Background
The second phase of the study included selection of a site for long-term disposal of coal refuse. Based on the extremely adverse terrain and property restrictions around the plant, the existing hollow needed to be used for future refuse disposal. An aerial tram system had been selected by management for long-range disposal at this site.

The first step was to review the economics of disposal of material with the aerial tram system. It became apparent that the lower the refuse pile elevation, the lower the disposal cost. Accordingly, the possibility of disposing of refuse immediately over the existing slurry pond was considered. This location provided

Table 29.3. Results of Stability Analyses on Modified Embankment Slopes

Trial circle	Factor of safety		Remarks
	Without earthquake	With earthquake	
4	2.1	1.7	2:1 slope
7	1.9	1.6	1.75:1 slope
8	1.7	1.4	1.75:1 slope, c = 0 psf

favorable economics for coal refuse disposal, but several questions needed to be answered. How would the existing coal slurry behave during loading with great heights of coal refuse? What would be the strength parameters of the future coarse refuse and filter cake mixture? What refuse pile configuration could be used and still meet the regulatory requirement of maintaining a factor of safety of 1.5? To answer these questions, additional laboratory testing and additional stability analyses were performed.

Additional laboratory testing

Future coal refuse. Based on the preparation plant flow sheet, the new coal refuse would have, on a weight basis, 10 to 15 percent filter cake material. Laboratory tests were performed on mixtures of the existing coarse refuse plus 10 and 15 percent filter cake size material. A reliable estimate of the moisture content of this future refuse material could not be made. It was felt, however, that based on experience, the moisture content would be in excess of the optimum moisture content of the new mixture.

Accordingly, samples were prepared for direct shear testing at 4 and 9 percent over the optimum moisture content. At 9 percent over the optimum moisture content the refuse mixture was almost fluid in nature. The samples were placed in the direct shear box and compressed under a static loading of only 1,465 kilograms per square meter (300 pounds per square foot) prior to quick shearing under a range of normal loads. The Standard Proctor test results and direct shear results are shown in Table 29.4.

Coal sludge testing. Due to the increased importance of the coal sludge strength parameters, triaxial shear tests were performed to supplement the slow direct shear tests performed in the initial phase of the investigation. The triaxial shear test results indicated a friction angle at about 32° and little or no cohesion (same as slow direct shear). Permeability tests were also performed which indicated the coefficient of permeability of the coal sludge was about 2.5×10^{-6} centimeters per second (0.08×10^{-6} feet per second).

Table 29.4. Compaction and Shear Strength Parameters of Future Refuse

	Standard Proctor		Shear strength properties		Test conditions	
Mixture	Maximum dry density (pcf)	Optimum moisture content (%)	ϕ (degrees)	c (psf)	Water content (%)	Dry density (pcf)
Coal refuse with 10% filter cake	98	9	39	20	13	68
			37	100	18	83
Coal refuse with 15% filter cake	96	13	39	160	17	75
			39	300	22	81

Consolidation tests were performed on the coal sludge in order to estimate the compression characteristics of the sludge. The results of this testing were used in calculating the excess pore pressures generated when the proposed refuse pile would be placed on the covered sludge pond. The initial void ratio of the coal sludge was calculated to be about 0.89. The compression index was found to be about 0.2 and the coefficient of consolidation was estimated by the Taylor method to be about 0.06 square centimeters per second.

Stability analysis (pond abandonment phase)

Background. The characteristics of the coal sludge were such that when the proposed fill would be placed on it, an excess pore water pressure would instantaneously develop in the sludge. With time, these pore pressures would dissipate as the sludge consolidated and the weight of the overlying coal refuse would in turn be transferred to the solid particles within the sludge. This process would occur concurrently with refuse placement over the sludge pond. Stability analysis of this area required that the excess pore pressure along various failure circles be estimated so that the effect of pore pressure due to the added coal refuse could be analyzed.

Excess pore pressure. The Massachusetts Institute of Technology Computer Package (ICES-SEPOL) was used to calculate the excess pore pressure at various locations and depths. Using the maximum possible refuse loading rate of 3.05 meters (10 feet) per month, pore pressures were evaluated after 4.5, 9.5, 14.5 and 19.5 months of construction.

Factors of safety. Using the Swedish Circle method, factors of safety were calculated for the four construction periods cited above. The position of the trial circle corresponding to the lowest factor of safety computed for each loading increment is shown in Figure 29.3. The soil strength parameters used in the analyses are listed in Table 29.5. It should be noted that values lower than the test results were used for the new coal refuse. The results of the stability analysis are summarized in Table 29.6. The lowest factor of safety without an earthquake loading was 1.6, and with an earthquake loading of 0.7 g, it was 1.3. As noted in Figure 29.3, the final proposed slope was flatter than the slope analyzed. Additionally, the sludge was loaded at a rate slower than the analyzed 3.05 meters (10 feet) per month.

Final design and specifications

A final design and typical specifications were prepared and submitted for MESA and state approval. After securing approval, and after completion of the closing of the plant circuit, the slurry pond ceased to be used. Backfilling of the pond itself did not begin until about three months later.

Field observations

The technical specifications for fill placement allowed for variable lift thickness. It was the intent to work with as thin a lift of material as could be placed over the coal sludge without subgrade or foundation failure. At the location of the discharge pipe (coming from the plant) and along the embankment, a relatively thin

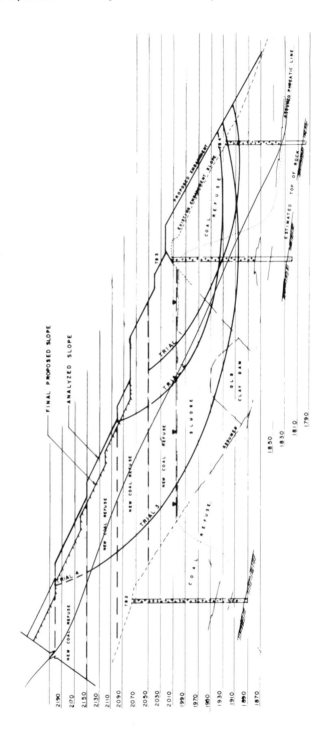

Figure 29.3. Trial circles for the stability analysis at the pond abandonment phase. The final proposed slope is also shown.

Table 29.5. Soil Parameters

Type	c (psf)	φ	Wet density (pcf)
Coal sludge	0	32°	83
Coal refuse	300	36°	97
New coal refuse	0	36°	90
Existing clay	200	20°	100

(1.2-meter [4-foot]) layer of coal refuse could be placed. However, after placement started to extend 15.2 to 22.9 meters (50 to 75 feet) from the former shoreline, small localized bearing-capacity-type failures occurred at the interface between the coal slurry and the coal refuse. These failures caused equipment operators to have concern about their safety during the backfilling process.

Explanation of the failure mechanism by Ackenheil's field representative during this period only somewhat relieved the equipment operator's concern. With time, the thickness of the backfill layer continued to increase as material was placed. With an increasing thickness of initial backfill layer, the area extent and depth of the bearing capacity failures increased. While it was apparent that the greater the lift thickness the greater the depth of failure, equipment operators continued to increase the lift thickness through the final covering of the impoundment. Pore pressures were measured periodically in the coal sludge during and subsequent to the backfilling. These measurements generally conformed to that predicted using theoretical techniques.

Table 29.6. Results of Stability Analysis on Loaded Pond Area

Trial circle	Factor of safety		Remarks
	Without earthquake	With earthquake	
1	2.0	1.6	After 4.5 months of fill placement
2	1.7	1.3	After 9.5 months of fill placement
3	1.7	1.3	After 14.5 months of fill placement
4	1.6	1.3	After 19.5 months of fill placement

Abstract:
International Overview of Coal Refuse Disposal Problems and Trends

Michael J. Taylor, Project Manager,
D'Appolonia Consulting Engineers, Inc., Denver, Colorado, USA

Coal refuse disposal is governed by a number of technical and nontechnical factors including:

1. Operational considerations.
2. Regulatory requirements.
3. Engineering practices.
4. Environmental considerations.

The disposal method adopted for a site may be governed by one dominant factor or a combination of several factors. An international overview of the elements which make up these factors, how the factors affect actual practice, and why certain elements dominate at certain sites was presented.

Operational considerations contribute to the economics of a mine operation. However, the least effort is often expended in planning and optimizing for purchase and operation of expensive refuse transport, equipment, and for disposal embankment location. Besides, preparation plant operations determine the physical characteristics of refuse, and significantly affect the technical factors.

Regulatory agencies have become more active in stipulating disposal practice because of disasters caused by embankment failures. Regulations, however, often specify in a conservative, broad-scope manner operational, engineering, and environmental practices not economical or technically feasible at all sites.

Engineering and environmental considerations may dominate regulatory requirements and operational considerations. Yet, general knowledge of coal refuse behavior remains limited. The potential for unacceptable leachate and burning varies depending on the characteristics of the mined coal. Standard control practices are difficult to establish for industry-wide implementation. The generalized approach of dams engineering and environmental planning provides a basis for establishing coal refuse disposal practices, but the technical problems remain significantly different.

Each site is a challenge to the operator and the professional designer, but the problems are being solved to provide economically and environmentally sound disposal areas.

Note: The publishers regret that the full text and illustrations for Mr. Taylor's presentation at the symposium were not available at the time of printing.

Reclamation

State of the Art Surface Reclamation, Implications of Legislation, and Potential Stabilization Programs

30

I. H. Reiss, President,
Meadowlark Farms, Inc., Sullivan, Indiana, USA

Introduction

At the present time, there is a controversy in the United States which centers around reclamation and the ability of mining companies to restore surface-mined coal lands to their former productive capacity. There are people in the United States who would not only like to severely regulate surface mining, but would actually outlaw this method of mining coal. Consequently, an educational job has to be performed.

Almost any business today exists in a social and political context in which it must be responsive to external issues. In fact, that which is social today becomes political tomorrow and economic the day after. Products and processes are being continually challenged by lawyers, politicians, or environmentalists. Usually it is a combination of all three.

The public relations function, the educational function, or greater social awareness are factors to which reclamation specialists must address themselves with increased vigor.

A number of questions may be asked. How is stability achieved in a controversial business? Why is it controversial in the first place? Who makes it controversial? Do laws make for stability or instability?

Is part of the problem the public's lack of knowledge about the industry? Or is the industry guilty of not doing something?

Stability is contingent on an educational awareness of the need to produce coal by surface mining and also how the industry reclaims disturbed lands.

Even though coal is now in an enviable position in the total energy picture, the social demands to reclaim mined lands to a higher degree of productivity than that prior to mining cannot be ignored. The United States has recently suffered a strike-induced coal supply crisis which should have made people more aware of their dependence on coal. However, these same people are not prepared to compromise on any reclamation requirements.

The new mining and reclamation laws provide for citizen participation at the local level. Unless measures are taken to keep citizens educated and informed about business, then expensive periods of shutdown will be encountered. There are many people with an anti-business bias. They don't trust big business. Many don't like surface mining either, and now some laws are giving them the strength to do something about it. What can be done?

Handling Mining and Reclamation

Mining and reclamation have to be handled so that surface mining is only an interim land use; productive capacity must be restored to mined-out areas so that land can be a renewable natural resource.

It is not an impossible task if common sense and judgment prevail. The extreme heterogeneous character of surface-mined lands must be recognized, and it must be admitted that not all disturbed lands should be restored for row crop production. Practical and realistic standards must be adhered to which recognize the technological and economic limitations of the finite resources with which the industry has to work.

What is very worrying, however, is not so much the statutes but their interpretation, and the rules and regulations that are written by individuals who may not be operations-oriented. The same rules have to be interpreted again and again to apply to many different circumstances. Many judgments, therefore, have to be made by fallible people; therein lies the uncertainty. In the meantime, coal must be competitive or yield to more reliable sources of energy.

The reclamation history of Meadowlark Farms, Inc., a subsidiary of Amax Inc. and an operating unit of Amax Coal Co., may be reviewed to observe how the state of the art has evolved. It began with trial and error methods and with little academic reference, and has grown to be a fairly sophisticated science that is being increasingly supported by basic research.

Meadowlark Farms, Inc.

Meadowlark Farms, Inc. was created over 30 years ago as a separate entity to productively farm lands acquired for coal interests, and to reclaim the lands disturbed by surface mining. This was an unprecedented move in the United States at that time.

A land use map of the USA shows that mining involves 0.25 percent of the total land area. At the present time, AMAX operates mines in the states of Illinois, Indiana, Kentucky, and Wyoming.

Meadowlark Farms and Amax Coal have a sophisticated surface overburden evaluation program. Through core analysis the following overburden conditions are determined:

1. Total sulfur profile.
2. Acidity/alkalinity (pH).
3. Potash content.
4. Presence of toxic elements.

Once the strata have been identified, draglines offer greater flexibility in strategically placing the materials. However, topsoil and subsurface soil may be placed on top of the spoil piles with an electric shovel having a variable-pitch dipper.

Reclamation

Reclamation involves dealing with different strata varying from extremely rocky material to glacial till with excellent chemical and physical properties. In the early days of reclamation, Meadowlark Farms was mainly concerned with reforestation—12 million trees were planted—or seeding without grading. Very little grading was carried out because the costs involved had to be offset against very thin profit margins in a highly competitive coal business. Most importantly at that time, the free market did not include the impact on the environment in the cost of the product. Grading of reclaimed land is now required by law.

A reclamation project completed 15 years ago by Meadowlark Farms was very successful; the land shows no evidence of the overburden having been removed to a depth of 29 meters (95 feet) and the coal mined. Good reclamation goes unnoticed on mined land.

Meadowlark Farms uses a disc instead of a moldboard plow to break up the reclaimed ground because of the rock problem in seedbed preparation. When the mined-out land has been satisfactorily reclaimed it is rented to tenant farmers. Meadowlark believes in the concept of the family farm, and there are about 200 tenant farmers involved in the total utilization program before and after mining.

Crops farmed

The reclamation program includes the growing of wheat and alfalfa. Wheat is grown both as a nurse crop and as a grain crop. Alfalfa in many instances is grown without having to add fertilizers. Prior to mining, much of the land would not support the growth of alfalfa without large applications of lime and fertilizer. Grasses

After strategic placement of overburden in the mining process, the next task is to grade to a farmable topography with bulldozers and scrapers.

Wheat growing on graded mined land within 12 miles of mining. Active mining in background. Prior to mining this was a submarginal agricultural area.

and legumes like alfalfa not only condition the soil but they also yield a good cash crop (hay and pasture).

Some of the early reclamation projects were very successful. In 1951, a field was graded after mining had finished, and in 1952, 31.5 cubic meters (894 bushels) of wheat were harvested from 12.1 hectares (30 acres). In some more recent instances, 1.6 cubic meters (45 bushels) per 0.41 hectare (1 acre) have been harvested as a grain crop. Up to 2.1 cubic meters (60 bushels) of corn and 1.06 cubic meters (30 bushels) per 0.41 hectare (1 acre) were also grown immediately after mining. In 1977, 3.9 cubic meters (110 bushels) of corn per 0.41 hectare (1 acre) were grown on reclaimed land after several years in legumes.

The company is trying to reduce the time interval that land is out of agricultural production. Wheat was growing on reclaimed land at one mine less than 12 months after the mine was opened. In addition, compatible grading with the existing landscape is achieved, so there are no abrupt changes between the mined and the undisturbed land.

Over 405 hectares (1,000 acres) of water (lakes) have been created by the mining process, and in Illinois, Meadowlark Farms is cooperating with the state to provide food and refuge for birds and small animals.

Reclaimed mined land showing diversity of uses—lush pasture, beautiful water impoundments, and reforestation.

Combines harvesting wheat on graded mined land.

One of AMAX's newest mines is the Belle Ayr surface mine in Wyoming. The land is primarily short grass prairie grazing land. The coal seam at Belle Ayr is between 21.3 and 24.4 meters (70 and 80 feet) thick. Because of the fragile environment the topsoil is removed and then replaced after mining. Seeding, although seasonal, is carried out as soon as possible to stabilize the soil to prevent wind and water erosion.

Mining and farming can coexist; the philosophy at Meadowlark is that corn, coal, and cattle are compatible. The new federal regulations require 100 percent topsoiling, and on prime agricultural lands a 1.2-meter (4-foot) soil profile must be reconstructed. Can this be justified on an economic basis, especially in the light of past reclamation experience and accomplishments?

Conclusions

The future stability of surface mining is contingent upon knowing how to deal effectively with people and their feelings about land and the human environment. All of this can be put under the heading of social awareness.

Restoring land to its former productive capacity is an aspect of the problem that has currently received legislative action in the United States. Land is used for functions other than food production, and so society's demands in the future may emphasize or de-emphasize some of the current reclamation requirements. If the coal industry is to survive and enjoy stability, then it must deal with society's demands. This does not necessarily mean that it should yield to what the general public thinks it should do in all cases.

If its point of view is to prevail, the industry must be ready to perform an educational function not only with the usual vocal minority of antagonists but also with the lay public. People may be misinformed, or they may lack information about the roles that coal and coal mining are playing in their lives, and how efficiently and effectively reclamation is being carried out.

Reclamation legislation may be a basis for total industry compliance, minimum standards, and uniform procedures, but it should also provide for the use of new technology and adjust to changing economic and social circumstances.

Coal Mine Reclamation Problems and Practices in the Southwestern United States

31

Gerald D. Harwood, Research Associate, Tika R. Verma, Professor, and John L. Thames, Professor, Watershed Hydrology, School of Renewable Natural Resources, University of Arizona, Tucson, Arizona, USA

The Importance of Reclamation

In the western United States, low sulfur coal deposits underlie large areas of land. Substantial amounts of this resource occur under land of delicate ecology and great scenic beauty. The damage that would result from strip mining in some areas would be virtually irreparable. But given the facts of an ever growing world population, the pending depletion of petroleum reserves, and increasing world political pressure for the equitable distribution of the earth's resources and wealth, it appears ultimately to be a question of when and how, rather than whether, many of these reserves will be mined. The development and implementation of an effective reclamation technology and improved mining methods are essential if many of the western coal reserves are to be developed without significant damage to the land.

The potential for the development of politically and socially explosive situations is great as questions arise concerning the precise timing and methodology of the exploitation and distribution of the earth's remaining nonrenewable natural resources. Should all resources be developed and consumed as rapidly as demand dictates and technology permits? Should decisions whether or not to reclaim mined lands be made strictly upon a cost-benefit basis? How much weight should short-term needs of and benefits to a single generation be given in decisions which will have relatively permanent consequences?

Problems Related to Reclamation in the Southwest

Reclamation of coal lands in arid and semiarid regions of the world such as the western United States promises to be more difficult than in areas which enjoy abundant precipitation. Box et al., 1973, have suggested that "restoration" of these lands to their premining conditions will rarely be possible. More likely options will be "reclamation" of the land, producing life habitats at least analogous to those that existed prior to strip mining, or "rehabilitation" of the land to some other useful and ecologically stable form.

Some people argue that the present urgency of our energy needs and economic factors should outweigh all other considerations with respect to what lands shall be mined and the stringency of reclamation and environmental quality regulations. However, at this point in history we still have an incomplete understanding of the functional importance of undisturbed (or restored) natural lands or of man's possible need thereof (Harwood, 1978). Ultimately, exactly how, when, and to what degree these resources will be developed and the degree to which the lands will be restored, reclaimed, or rehabilitated are questions which will be decided upon the basis of priorities set by society (Box et al., 1973). The determination of these priorities in the United States may be a difficult process due to the fact that American society is complex and encompasses many value systems. For example, as Box et al. point out, the land values of Anglo Saxons are significantly different from those of many American Indian tribes whose land overlies some of the larger western coal deposits.

Factors affecting reclamation

A number of ecological, political, and social factors contribute to the problems of reclaiming strip-mined lands in the western United States. Not the least of these is our incomplete understanding of the ecology of arid and semiarid lands. Commonly, low precipitation in these areas results in sparse vegetative cover and high soil erosion. Precipitation in the intermountain region varies from less than 101.6 millimeters (4 inches) in some of the desert valley floors to more than 50.8 centimeters (20 inches) in the higher mountains. Much of the precipitation occurs as a result of intense storms of short duration and, in cold areas, a good deal of it falls as snow during the period of dormancy for many plants. Many desert plant and animal species have developed a variety of adaptive mechanisms and/or structures to permit survival under conditions of restricted water availability. One common trait found in desert vegetation is slow growth, and plant communities are typically of low density resulting in sparse ground cover. Accordingly, arid lands are normally subject to high erosion under the forces that can develop in the brief but intense storms common in these regions, and those which are disturbed are characteristically slow to revegetate and are subject to severe erosion and sedimentation.

Revegetation efforts

Efforts to revegetate strip-mined land have been disappointing on the Black Mesa in northeastern Arizona (Verma and Thames, 1975). Even the native species which were present prior to the mining operation are hard pressed to survive. Part of the problem derives from the fact that the vegetation which preceded the mining operation was the product of centuries of successional site preparation. Also, for the past 100 years, heavy grazing pressure had resulted in a set of edaphic and climatic factors to which plants in the community had become adapted. The initial establishment of native plant communities probably occurred over long periods of time and possibly under climatic conditions that no longer exist on the Mesa. Unless methods can be found for bringing about suitable conditions for the reestablishment of native species or the establishment of

Figure 31.1. Unassisted revegetation process on Black Mesa coal spoils, two years after recontouring and seeding with alfalfa.

introduced species, the short- and mid-term outlook for successful reclamation of these strip-mined semiarid lands is poor.

Experience on the Black Mesa has shown that strip-mined lands in the Southwest do not necessarily revert to the aesthetically pleasing and stable plant communities that preceded the mining operations. Rather, the recontoured lands tend to become overgrown with weed species such as Russian Thistle, which is inferior as a revegetation species and is palatable as forage only in certain stages; alternatively, the land can remain largely barren and subject to wind and water erosion (Figure 31.1). Such conditions are not conducive to rapid reestablishment of native vegetation (even if seeded or planted) which, depending upon the location, varies from desert shrubs to range grasses to juniper and/or pine stands.

Restricted availability of water and associated reclamation problems often limit the options for post-mining land use. Reclaimed mining areas may be poorly suited for many crops, and of limited value as a source of forage for grazing livestock. In many cases, strip mining of an area will necessitate economic and social readjustment to post-mining land use limitations. Land that once supported grazing may not be able to do so after it has been strip mined, at least not for an extended period of time. In many areas, low water use options such as recreation are being explored for post-mining land use. Box et al., 1973, also note that surface mining can disrupt groundwater flow patterns and has the potential for interrupting traditional water supplies.

Hodder, 1978, reminds us that an important aspect of dealing with reclamation or rehabilitation of western surface mines is that the potential for reclamation is "critically site specific." The reclamation potential depends upon such variables as: existing ecological and physical conditions, the purpose of reclamation (dependent, in part, upon the projected land use), the mining techniques and equipment used, and the skills applied in technological efforts. For example, adjacent mines using differential surface mining techniques may present quite distinct situations from the viewpoint of reclamation.

Some Possible Solutions

In the western states, most of the reclamation-related problems discussed above are closely related to the magnitude of available water resources. If the water problem could be overcome, many new reclamation options would open up for treatment and post-mining land use. For example, the handicap introduced by a less than optimal environment for the growth of the new seedlings could be significantly reduced if sufficient moisture could be made available to the plants, at least on a temporary basis, until they can establish themselves. As previously noted, many problems, particularly those involving water availability, have social, political, and economic facets, as well as environmental. There are many possible approaches to solving reclamation problems associated with arid lands, and only a few are discussed here.

Verma and Thames, 1975, have suggested that, based upon preliminary data from the Black Mesa, the potential of the recontoured spoil material for reclamation would be good if sufficient water were available. However, water would be required in an abundance that is not present or available on the surface. Groundwater is present under the Mesa, as it is in many parts of the arid and semiarid West, but such water is expensive to obtain and questions of water rights are complex. Short of using groundwater for the initial reestablishment of vegetative cover through supplemental irrigation, surface manipulation techniques and water conservation measures can enhance the effectiveness of available moisture. Additionally, water harvesting and runoff agriculture may provide means of increased development of on-site precipitation. The potential of this application of water harvesting is presently being investigated on the Black Mesa.

Surface manipulation
Surface manipulation can be used for erosion control and on-site water conservation. Pre-mining planning of reclamation can help to reduce costs of the surface manipulation phase by minimization of spoil material movement. Erosion control can be brought about by specialized methods of grading and shaping of the recontoured spoils and topsoil. These processes can go far to reduce erosion while simultaneously improving the potential for vegetative growth. Verma and Thames, 1978, have addressed the topic of grading and shaping for erosion control and vegetative establishment in arid regions. They indicate that the application of refined grading and shaping techniques to recontoured spoil materials, and the construction of various structures such as small basins, contoured terraces, furrows, and trenches, if carefully designed, can greatly enhance on-site conservation of precipitation and erosion control, particularly if used in conjunction with additional measures such as topsoil dressing and mulching.

Plant species selection

The selection of plant species for revegetation of a strip-mined area will require consideration of a number of factors. Ideally, regional criteria will be developed for reclamation project species collection, but it is likely that the final selection of potentially successful species will often be site specific. Generally, species that were native to the pre-mining site will have the best potential for successful revegetation. But this may not always be the case, particularly where the site supported a post-climax plant community. Factors which must be taken into account will include: plant water requirements; planned land use; whether or not supplemental irrigation is feasible until the plants have become permanently established; local climatic conditions; water rights and various social, legal, political, and economic considerations. Most important will be the selection of plant species that will have the long-term potential to be able to survive local drought conditions without intervention.

In determining the best plant species to use for revegetation, differentiation should also be made between the amounts of water required for minimal establishment, long-term subsistence (allowing for periods of drought), and maximum growth or production. A program of supplemental irrigation should be considered until the plants become sufficiently established to survive on their own. If supplemental irrigation is contemplated, it will be necessary to determine and demonstrate optimum irrigation methods, based on water availability and cost-effectiveness. It is likely that water is the major growth limiting factor, although it is possible that edaphic and microhabitat conditions may also inhibit the development of certain species.

Supplemental irrigation

Data presented by Verma and Thames, 1975, and additional unpublished data from the Black Mesa indicate that the potential of recontoured spoils for the support of vegetative growth is good if supplemental water can be supplied by some means to support the plants until they can become permanently established. (See Table 31.1.) Once the plants have reached a sufficient size and density providing ground cover and an organic layer, it may be possible to withdraw the irrigation with only partial detriment to the survival of the new plant community. The ultimate success of such a program will in large measure depend upon the species which compose the community. The data indicate that, although the spoils themselves lack structure and are subject to crusting, they offer more available water to plants than does the local topsoil and have higher amounts of total (not necessarily available) nitrogen, phosphorus, and potassium. A soil pH of 6.9 in both natural and recontoured test watersheds was found. However, the soil water levels in the recontoured spoils were near the permanent wilting point.

In the arid and semiarid Southwest, surface water is rare. As previously discussed, groundwater is expensive and might not be considered to be a cost-effective option as a water source for purposes of reclamation. Water harvesting and runoff agricultural techniques offer a potentially less expensive option. Although water harvesting is by no means a new technology, it is not well known at this period in history. Studies are currently underway on the Black Mesa and

Table 31.1. Water Quality Analyses, Natural and Surface-Mined Watersheds, Black Mesa, Arizona

Item	Runoff water		Soil extract		Wells	
	Natural	Mined	Natural	Mined	Tucson*	Yuma
	(ppm)					
Soluble salts	177	310	112	293	254	1,448
Calcium	24	41	17.0	38.5	36.1	116
Magnesium	3.6	13.0	2.7	16.0	15.4	29
Sodium	21	22	3	26	44	328
Chloride	3.0	3.0	6.5	9.5	19.2	246.0
Sulphate	65	100	26	152	153	460
Carbonate	0	0	0	0	0	7.2
Bicarbonate	80	129	56	50	155	259
Fluoride	0.2	0.3	–	–	0.6	0.9
Nitrate	2.4	1.6	0.3	0.4	–	2.4
pH	7.6	7.4	7.7	7.5	7.9	7.6

*Average of 36 wells in Tucson

elsewhere to "rediscover" the old technology and to test its potential in the context of strip mine reclamation. A brief discussion of its history follows.

Water harvesting and runoff agriculture. Water harvesting offers a mechanically and economically feasible means of providing water for supplemental irrigation on strip-mined land in a semiarid region such as the Black Mesa. In areas where moderate precipitation occurs, water harvesting offers a potential water supply for additional uses beyond the simple reclamation of mined land.

Water harvesting and runoff agriculture are very old technologies, dating back thousands of years. The process is simple in concept, involving the collection, conveyance, and storage of water derived from surface runoff or precipitation. The watershed may either be left in its natural condition or treated by various methods to make it less permeable to water. Systems designed for this purpose were used by the peoples of the Negev Desert perhaps 4,000 years ago (Evenari, Shanan, Tadmar, and Aharoni, 1961). Hillsides were cleared of vegetation and smoothed in order to increase runoff; the water was then channeled in contour ditches to agricultural fields below. Similar methods were used for irrigation in the Southwest, near what is now the Mesa Verde area of Colorado, some 400 to 700 years ago (O'Bryan, Cooley, and Winter, 1969).

Chiarella and Beck, 1975, have reviewed the use of various water harvesting methods on Indian reservations in the western United States. A number of experiments have been attempted over the years, and a few have worked well. A catchment at Shongopovi on the Hopi reservation has provided supplemental domestic water since the early 1930s; this catchment was constructed with very little disruption of the natural topography. On the Navajo reservation, butyl rubber catchments and storage bags are successfully used by stockmen. McBride and Shiflet, 1975, have described the construction, placement, and use of larger catchments in the Safford district of southern Arizona. Myers, 1961, began experimentation with the waterproofing of the soil itself to serve as a catchment structure. His group tested materials such as asphalt and various kinds of films that impart hydrophobic properties to the soil when bonded to it.

Aldon and Springfield, 1975a, investigated the suitability of water harvesting techniques for growing shrubs on coal strip mine spoils in New Mexico. They found that paraffin-treated ground increased runoff and increased growth of two-month-old saltbush and Siberian peashrub transplants; the soil moisture content of treated plots was increased by about 20 percent. There have been a number of studies on the use and suitability of water harvesting as an agricultural technique. Fogel, 1975, has discussed various aspects of runoff agriculture, including techniques for computing runoff and water spreading systems. Morin and Matlock, 1975, have computer-modeled a system of water harvesting and strip farming, based upon data collected in the Tucson area. Their results suggest that, while conventional irrigation techniques produce greater crop yields, significant short season grain sorghum crops could be produced from harvested water in four out of five years.

Conclusions

It seems certain that the pressure to develop coal resources will increase in the future, perhaps rapidly as petroleum reserves are depleted. Unless new mining technologies are rapidly developed, the effect of increased strip mining will be deleterious to large areas of land in the western United States and other arid sections of the world where a scarcity of water renders reclamation efforts only partially effective. New reclamation methods are needed and many of the social, political, and economic problems that affect reclamation need to be resolved. Protection of the land for future generations dictates that we find solutions to both technological and nontechnological problems encountered in surface mine reclamation. Possibly, the "rediscovery" of older practices, such as water harvesting, and the application of wisdom and foresight in defining our priorities will provide hope for a bright future.

References

Aldon, E. F., and Springfield, H. W., 1975, "Problems and Techniques in Revegetating Coal Mine Spoils in New Mexico." Practices and Problems of Land Reclamation in Western North America, M. K. Wali (Editor), The University of North Dakota Press, Grand Forks, N.D.

Aldon, E. F., and Springfield, H. W., 1975a, "Using Paraffin and Polyethylene to Harvest Water for Growing Shrubs." 1975. Proceedings of Water Harvesting Symposium, Phoenix, March 1974.

Box, Thadis W., et al., 1973, "Rehabilitation Potential of Western Coal Lands." Environmental Studies Board, National Academy of Sciences, National Academy of Engineering, Cambridge, Mass., Ballinger Publishing Company.

Chiarella, J. V. and Beck, W. H., 1975, "Water Harvesting Catchments on Indian Lands in the Southwest." Proceedings of Water Harvesting Symposium, Phoenix, March 1974.

Evenari, M. L., Shanan, L., Tadmar, N., Aharoni, Y. , 1961, "Ancient Agriculture and the Negev." Science 133.

Fogel, M. M., 1975, "Runoff Agriculture: Efficient Use of Rainfall." Watershed management in arid zones: a prototype short course, Saltillo, Mexico, March 1975, School of Renewable Natural Resources, University of Arizona, Tucson, Arizona.

Harwood, G., 1978, "Reclamation—Our Legacy to the Future." Proceedings of the Congress on Energy and the Ecosystem, Grand Forks, N.D. (June, 1978) M. K. Wali and J. L. Thames, Eds. Univ. of North Dakota Press. (in press).

Hodder, R. L., 1978, "Potentials and Predictions Concerning Reclamation of Semiarid Mined Lands." The Reclamation of Disturbed Arid Lands. R. A. Wright (Editor), Albuquerque: Univ. of New Mexico Press.

McBride, M. W., and Shiflet, L. W., 1975, "Water Harvesting Catchments in the Safford District, Southeastern Arizona." Proceedings of Water Harvesting Symposium, Phoenix, March, 1974, ARS W-22 February, 1975.

Morin, G. C. A., and Matlock, G., 1975, "Desert Strip Farming—Computer Simulation of an Ancient Water Harvesting Technique." Proceedings of the Water Harvesting Symposium. March, 1974, ARS W-22 February, 1975.

Myers, L. E., 1961, "Waterproofing Soil to Collect Precipitation." Soil and Water Conservation Journal 16.

O'Bryan, D., Cooley, M. E., and Winter, T. C., 1969, "Water, Population Pressure, and Ancient Indian Migrations." Bulletin 10, International Hydrologic Decade. June, 1969.

Verma, T. R., and Thames, J. L., 1975, "Rehabilitation of Land Disturbed by Surface Mining Coal in Arizona." Soil and Water Conservation Journal 30(3).

Verma, T. R., and Thames, J. L., 1978, "Grading and Shaping for Erosion Control and Vegetative Establishment in Dry Regions." Reclamation of Drastically Disturbed Lands, Schaller, F. W., and Sutton, P., eds. Madison, Wisconsin: American Society of Agronomy.

West Virginia's Controlled Placement Method of Surface Mining: Maximum Recovery—Minimum Disturbance

32

Benjamin C. Greene, President,
and William B. Raney, Vice President,
West Virginia Surface Mining and Reclamation Association,
Charleston, West Virginia, USA

Introduction

Little could one realize, some five years ago, that West Virginia's "controlled placement" methods of surface mining would become personifications of Fredric Taylor's "one best way" time and motion theory. Today, however, it has become just that. What began as an alternative to "shoot and shove" mining has become a widely accepted practice characterized by productional efficiency and environmental effectiveness. As the title of this chapter indicates, the concept assures the maximization of mineral recovery while minimizing land disturbance, both of which are important considerations to the industry as well as to regulatory agencies. Thus the bold and revolutionary step taken by West Virginia in 1973 has become a template for surface mining programs throughout the United States. As a matter of fact, it served as the foundation for the recently enacted Surface Mining Control and Reclamation Act of 1977, which was passed by the 95th Congress of the United States, and signed by President Carter on August 3, 1977.

Although a number of organizations and individuals will take credit for this revolutionary methodology, it was a West Virginia product cooperatively conceived and initiated by the state's Division of Reclamation and the surface mining industry. It was a result of practical policy analysis combined with pertinent adaptations of surface mining methods, material handling technology, sound engineering principles, and economics.

Historical Background

Prior to a discussion of the actual application of the methodology, the historical background of its origin and acceptance in West Virginia will be established. The 1971 amendments to the West Virginia Surface Mining and Reclamation Act (originally enacted in 1967) established one of the most comprehensive laws in the nation, and its implementation was most effective in eliminating numerous detrimental off-site effects of surface mining activity. Despite these beneficial aspects, however, it was incomplete in addressing all the problems inherent to steep slope land disturbances. As a result, in 1973, two years following implementation of the 1971 changes, a thorough reassessment of West Virginia's program was conducted which indicated several significant shortcomings. Most critical of these was the need for additional control of downslope overburden placement on extremely steep outer slopes. Slopes in excess of 50 percent (26° 34′) are considered extremely steep.

Control of steep outer slopes
Even though notable progress, through research and experimentation, had been made in the stabilization of steep outer slopes, the uncontrolled spillage remained an obvious problem. It was concluded that the bench width restriction, based on percentage steepness of the original slope, was no longer a dependable guideline for assuring long-term stability and effective reclamation. Consequently, with the problem identified, attention was directed to alternative solutions. The encompassing guidelines and strong enforcement program of 1971 had offered new approaches to the ultimate stabilization of problematic outer slopes. However, they had minimal impact on the long-term condition of such areas.

The most effective way in which to control such slopes was determined to be the prohibition of their initial creation. This "prevention rather than cure" philosophy was projected in a 1973 administrative guideline which dictated the use of controlled placement mining methods on all areas where the original steepness of slopes exceeded 50 percent. In essence, it prohibited the uncontrolled spillage of overburden on slopes of such defined steepness.

Guideline principles
Because of the revolutionary nature of such a directive in a state renowned for its mountainous terrain as well as its coal production, valid bases for guidance were necessary. To this time, the control of overburden, as practiced in modern area mining operations, had never been applied to contour sites. Therefore, to enhance implementation, the industry adapted material-handling practices employed on highway construction projects and the long-standing, proven engineering principle that a 2 to 1 slope is controllable and stable to their needs.

Controlled Placement

As a result, contour mining operations adopted controlled placement methodologies by employing one or a combination of the following variations:

1. Valley fill.
2. Mountaintop removal with valley fill.
3. Lateral movement mining, which is the epitome of controlled placement mining.

Since the initiation and refinement of these operating methods, the thoughts that a modified approach to contour mining would bring about higher quality reclamation, while eliminating detrimental offsite effects, have been substantiated.

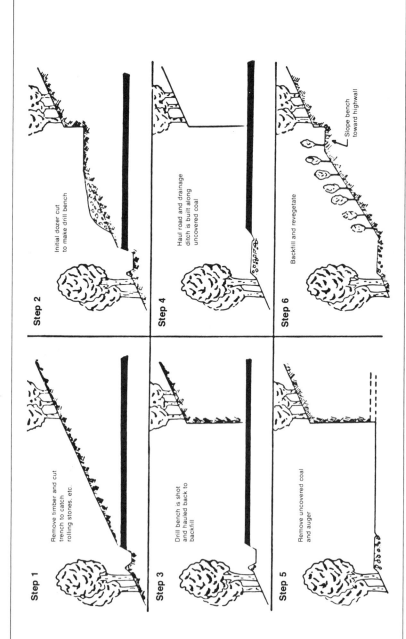

Figure 32.1. Steps in West Virginia's "controlled placement" method of surface mining.

Also, it has accredited the long-time philosophy that the protection of the environment and the continuation of surface mining are goals, which are directly related and supplementary, not conflicting.

In addition to maximizing offsite protection, controlled placement mining develops several other beneficial effects to the industry as well as the environment. By its very nature, this methodology provides the ability to mine valuable reserves which were heretofore unavailable due to slope restrictions. By not creating outside fill slopes, productional opportunities were significantly increased, particularly in the metallurgical coalfields of southern West Virginia.

Coincident with this increased availability of reserves, the industry has realized explicit improvement in long-term operational efficiency. As previously mentioned, the handling of material only once, as compared to twice or three times on conventional operations, is much more efficient as well as effective in achieving "concurrent" reclamation.

Preplanning

Another benefit is the increased attention to and scope of the quality of preplanning. In order to be totally effective, a controlled placement operation must be well-planned in detail prior to initiation (Figures 32.1 and 32.2). It also dictates that these planning functions be translated to productional vernacular and passed on to the operational personnel. The advantage to this concept should need no further explanation as it relates to ultimate efficiency and quality of a surface mining operation.

Many times a controlled placement operation will result in level, or near level, terrain, which is very important to the rugged areas of West Virginia. The need for such land with increased utility is critical in these areas. This has brought about more comprehensive preplanning, as well as a transition in the approach to material handling and final disposition. The direct benefits of an operation which "knows where it is going and what it is to be doing" should be evident to everyone. Surface mining, consequently, has become a true interim land use which is viewed by landowners and operators as a method of capitalization for future development. This is contrary to the old attitude that mining was the ultimate income-producing effort for a portion of real estate; in many cases, the final act before returning to a property tax sale.

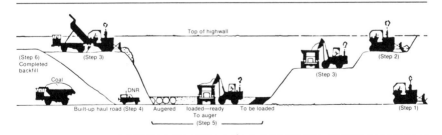

Figure 32.2. Steps in the "controlled placement" of soil in West Virginia's surface mining operations.

Benefits and advantages

Cumulatively, the benefits and advantages of controlled placement are significant to both the interests of industry and regulation, with additional attributes becoming visible each succeeding day of use. Having approximately five years' experience with the concepts, West Virginia's industry feels strongly that the principles are general enough and adaptable enough so as not to have its employment limited by manmade boundaries. The controlled placement attitude should be encouraged and utilized in every area where surface mining exists. It is the primary reason that the people of West Virginia supported and endorsed the Surface Mining Control and Reclamation Act of 1977, as it was passed by the United States Congress.

This was the first time that West Virginia had advocated federal legislation, but, with its experience, it was felt that it would be a progressive step for all mining areas, and would demonstrate much-needed initiative for universal adaptation. It is time to equalize, as much as possible, the operating standards, enforcement procedures, and environmental expectations among the several states of the USA, as well as many of the countries and continents of the world. Environmental concerns should not be jeopardized by the weaknesses and lack of initiative of a government or one of its agencies. A state or nation should not nurture competitive inequality at the marketplace, which, of course, it does by disregarding controlled placement mining. Hopefully, the federal program, which has such a principle for its operational foundation, will alleviate this critical situation.

Following is a discussion of the specific variations of controlled placement which have been most widely accepted and successful in West Virginia. Each possesses unique characteristics, but there are certain operational components common to all. For example, each one requires blasting to be extensively controlled, the use of different equipment, extreme care in the removal and placement of all overburden materials, appropriate topographical and hydrological conditions, comprehensive preplanning, and knowledgeable operational personnel. Of course, the entire spectrum of application is dictated by evaluation of the economic dimensions of coal to overburden ratios and market trends.

Valley Fill

The valley fill variation of controlled placement can be employed individually, but is most often used in conjunction with one of the other methods. Basic modifications of the method have been used for several years in the Appalachian region of the United States. Only recently, however, was it technically refined to the point of being acceptable as an alternative method of operation. In order to comply with the strict environmental standards of today, as well as our requirement for high-quality reclamation, a valley fill operation must meet detailed specifications. The guidelines were developed in accordance with proven engineering principles of fill slope and drainage control.

As the name implies, valley fill (also known as head-of-hollow fill) entails the methodical filling of the head of a designated hollow with overburden generated by a contour operation immediately adjacent to the valley or hollow.

Site selection and drainage control

In addition to the coal to overburden ratio, the selection of an appropriate hollow is governed by several factors. Ideally, a valley should be narrow, V-shaped and have steep sides. The designated valley *must* be of such a size that its head will be completely filled by overburden generated from the adjacent operation.

Once the site is selected and evaluated, provisions must be made for comprehensive drainage control from the area. The drainage control facility must be of adequate size to accommodate and treat drainage from the initial clearing operation as well as the area disturbed in construction of the fill's first lift. Also, all drainage control facilities shall be installed and certified prior to any disturbance in the valley.

Filling procedure

Following the installation of the drainage control system, the designated portion of the valley must be progressively cleared of all organic material, as construction advances. Overburden from the adjacent contour mining is then deposited in the valley in layers. Layers should not exceed 1.2 meters (4 feet) in thickness with each subsequent layer being compacted with conventional hauling and mining equipment. The ultimate height of the fill is determined by the elevation of the adjacent desired topography. To achieve such a result requires precise engineering particularly as it relates to the starting points and elevations.

Throughout the filling process, a rock core is constructed through the entire center length and height of the fill area. In order to minimize potential blockage, it is imperative that the largest rock available be used for the construction of this drainway core. The constructed drainway must be at least 4.6 meters (15 feet) wide at all points throughout the fill (Figures 32.3 and 32.4). As construction of this rock "chimney" progresses, drainage control for the site is continually improved.

Once the desired height of the fill is reached and the rock core drainway has been adequately constructed, the final bench of the valley fill is sloped toward the head of the original hollow. The maximum pitch of this sloped area should be 3 percent.

Sump construction. At this point on the bench of the fill, a sump is usually constructed to capture all runoff from the disturbed area prior to its release into the rock core drainway. The capacity of this sump should not exceed 283.2 cubic meters (10,000 cubic feet). The sump, rock core, and the water control facility below, provide total control of drainage from the operation. They also provide a perennially dependable means of conveying surface water through the area after mining is completed.

In addition to these drainage control measures, the final exposed slope of the fill must be maintained at a 2 to 1 ratio. To further retard and control runoff, benches are constructed at vertical intervals of 15.2 meters (50 feet) on the face of the fill. These benches, which must be a minimum of 6.1 meters (20 feet) wide, act as diversion ditches spilling into the rock core chimney. These benches should provide a 3 to 5 percent slope toward the face of the fill and the rock core. Also, to increase stabilization, the face area is progressively revegetated as each lift is completed.

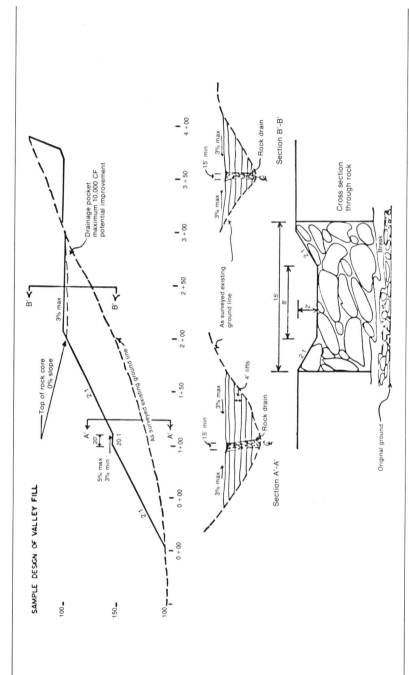

Figure 32.3. Sample design of valley fill.

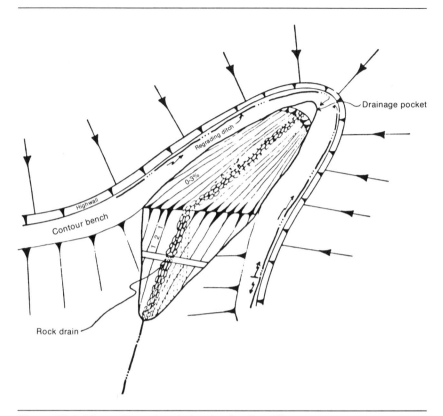

Figure 32.4. Three-dimensional sketch of valley fill.

Advantages of valley fill

Instead of uncontrolled spillage into or around natural drainways, the valley fill provides a plateau-like area, with total slope and drainage control. An evident and significant advantage of a valley fill is that steep-sided hollows are transformed into relatively level plots of ground offering developmental potential. This is very important in West Virginia's mountainous terrain, as it would be in any steep-sloped region.

As well as providing increased utility, this operation offers greater accessibility than did the previous methods of contour mining.

Environmentally, the valley fill is desirable because it reduces the threat of extensive erosion and the subsequent blockages of natural drainways. By providing overburden storage space, valley fills tremendously reduce disturbed area.

The valley fill also offers productional advantages and provides the opportunity to mine valuable reserves which would be otherwise unrecoverable. West Virginia and federal laws prohibit any disturbance within 30.5 meters (100 feet) of either side of a natural drainway unless specific approval has been granted by the regulatory authority. By providing means for conveyance of water, using the valley fill, the operator, with prior approval, can mine through the hollow without threat to the natural drainway.

It also increases efficiency in achieving a desirable end product. This is because overburden materials are handled only once, as opposed to twice in the reclamation of conventional contour.

Mountaintop Removal with Valley Fill

In the mid-1960s, coal recovery in mountainous terrain benefited from the introduction of the mountaintop removal (also known as flat-top mining or skylighting) method of surface mining.

At its onset, however, mountaintop removal could not be considered a viable variation of controlled placement mining. Incorporation of the valley fill concept, five years ago, transformed it to an environmentally sound surface mining practice, while maintaining its production benefits.

Original mountaintop removal

Originally, mountaintop removal was conducted as a series of conventional contour mines. These sequential strips occurred one behind the other, until the ridge line was reached and the entire mountaintop was eliminated. It added new dimensions to mineral recovery and provided a more desirable reclamation effort, but it also magnified the inherent problems of overburden stabilization on steep outer slopes.

Since the operation made no provisions for the control of overburden from the first cut, it was spilled onto the steep outer slopes surrounding the mountaintop. This practice was particularly prevalent in the case of multiple seam operations.

Valley fill variation

In order to correct this undesirable situation, the valley fill variation was recommended as an operating supplement to mountaintop removal. Valley fill construction sites provided an acceptable storage area for the overburden generated from the initial cut, while permitting the general concept of mountaintop removal to continue as planned.

By providing large areas of relatively flat land, with the controlled placement of overburden on steep outer slopes, this dual method of operation has significantly enhanced reclamation efforts in West Virginia. Aside from the evident benefits of higher quality reclamation, such an operation offers increased potential for future land use, more dependable drainage control, the minimization of detrimental offsite effects, and, in addition, results in increased operational efficiency.

Following mountaintop removal with valley fill, a larger percentage of the disturbed area is accessible to conventional grading and seeding equipment. Such accessibility permits more even distribution of upper horizon materials which are required to be stockpiled and returned to the regraded surface, a finer degree of grading, seedbed preparation, and more effective application of revegetation materials.

Economic advantages

In addition to the environmental and reclamation benefits, this variation of controlled placement offers economic advantages to the industry. Mountaintop removal permits the recovery of 100 percent of the reserves. This is in contrast to the conventional methods of mining which may offer limited recovery. Also, if the operation is adequately planned and takes full advantage of the valley fill supplement, backfilling and regrading should be automatic. That is, the overburden from each subsequent cut should fill the previously mined pit. Consequently, material is moved only once and reclamation is "concurrent." Experience has proven that a rapid reclamation sequence enhances the ultimate product. This significantly reduces the odds that an area will have to be perennially retreated in order to meet the stringent standards of acceptable revegetation.

Advantages

As mentioned, the valley fill is employed in conjunction with other variations of controlled placement, providing the needed area for the storage of materials from initial cuts. The use of valley fill in conjunction with these other methods is more fully discussed in the following section.

Regardless of how it is employed, this method of operation is a most desirable alternative to the conventional types of contour mining in the steep hollow regions of mountainous terrain. Reclamation efforts are significantly enhanced and the long-term stability is measurably increased.

Ironically, the federal government recommends fill techniques which utilize rock flume channels around the fill at the point where the front slope ties into the original ground. Correspondingly all surface water is diverted to the sides. This recommended technique has never been effectively employed in any steep slope region of the United States. West Virginia has practiced valley fill construction for approximately five years. There are dozens of examples throughout the state and none have failed in any manner. Surely, a proven methodology is more dependable and practical than one which has not been extensively employed or has been employed with less than desirable results.

Fills are built in West Virginia with longevity in mind. When a fill is constructed on slopes of 65 percent or greater steepness, it is intended that it will stay there with no future maintenance. The record supports that approach.

Restrictions by natural conditions

Although widely accepted, the employment of mountaintop removal with valley fill is restricted by natural conditions. The majority of these constraints are similar to the characteristics which limit the application of valley fill. Feasibility is further determined by the quality and quantity of mineral available, its proximity to the mountaintop, as well as the nature and amount of overburden.

The results of present operations in West Virginia, as well as those which have been completed, provide graphic illustrations of the advantages of controlled placement relative to the attainment of high quality reclamation. With such successful experiences under today's technology, this mining methodology would more accurately be identified as mountaintop reconstruction. This identification would certainly be more indicative of the means and ends of the process.

Federal restrictions

Mountaintop operations raise another interesting, but puzzling point with the ensuing federal program. Despite all the attributes and progress made with the refinement of the procedure, the federal program discourages its use and application. One of the most significant restrictions is the requirement to retain a 4.6-meter (15-foot) barrier of crop coal around the perimeter of the mountain. This may be justified in theory. It is assumed to be beneficial to the long-term protection of water quality. It would seem to make better sense to remove all the potentially pyritic mineral rather than to leave an outside rim for all drainage to flow through.

In addition to being contrary to good hydrologic practice, this requirement significantly inhibits the procedural efficiency of the entire operation. West Virginia has never experienced post-operation water problems from a mountaintop project. However, the federal program apparently anticipates such problems. This idea is most disturbing to those who have perfected controlled placement mining.

Subsurface dendritic system of drainage

This requirement also negates the use of the most recent innovation relative to mountaintop operations in West Virginia. In an effort to increase on-site drainage control, a number of West Virginia companies initiated the use of a subsurface dendritic system. The system is basically a network of large rock cores constructed continuously from the first cut and following the mining sequence. The water which is collected in the pit, or from saturation of the restored surface, is directed through these drainways to the outside perimeter. There it may be collected in a "toe-berm" diversion and routed to a final sediment control facility.

This concept eliminates the potential saturation of the final graded surface area. By providing an uninhibited subsurface channel for water collection and flow, the utility of the resultant level plateau is significantly increased. Also, the operational control of water quality, particularly from a suspended solids standpoint, is obviously improved as the system is progressively installed. It only stands to reason that, if all water is routed through a massive rock core prior to being discharged to a constructed ditch and finally to a pond, the effluent is going to be superior to that drainage originating on and conveyed over the surface of the regraded area.

Consequently, concern with the new regulations is accentuated as further steps are taken in West Virginia to improve on the existing system, but, at this time, they are disregarded by federal guidelines.

Lateral Movement ("Outbound Haulage")

Because of the precision movement required for its conduct and ultimate success in steep slope mining, lateral movement mining epitomizes the concept of controlled placement. This haulback methodology is the most recently developed variation of our controlled placement program.

Lateral movement mining is referred to as West Virginia's "outbound haulage" because it is an adaptation of current methodology to contour mine extremely steep slopes. While incorporating some basic characteristics of the modified box cut, it is felt that it is unique and justifies consideration as a totally separate method of mining. Since its introduction, this concept has revolutionized West Virginia's surface mining industry, particularly in the southern coalfields.

One of the most environmentally significant differences of haulback is the control of overburden generated from the initial cut. Even though the box cut minimizes uncontrolled downslope spillage, it does not prohibit it.

Modified box cut method
Since the controlled placement program was designed to completely eliminate outslope spillage, it was necessary to modify the basic box cut principle to meet specific needs. Consequently, any lateral movement proposal provided for the adequate handling and storage of this initial overburden material.

In all cases to date, this initial material has been deposited either on an old bench area situated adjacent to the active operation, or in a valley fill. Either situation provides close control of overburden placement. When an old bench area is used, the generated overburden is hauled back in trucks and placed in such a manner that the reclamation of the previously mined site is greatly improved. The nature and degree of such improvements are governed by the original condition of the area. If such an area is not immediately available, a valley fill is utilized. In this event, it must meet all preoperation requirements and be in accordance with all specifications as previously discussed.

One beneficial effect relative to old bench storage is the rehabilitation of previously mined areas to current environmental standards. In the past three years, the industry has eliminated approximately 241.4 kilometers (150 miles) of old highwalls in West Virginia. This is recorded and completed, not proposed. Therefore, mining companies in West Virginia have reason to be proud of their mining methods.

A lateral movement operation actually begins with the construction of a haulageway on or just below the estimated cropline of the coal seam. Since this entire operation is based on the lateral haulage of overburden, the roadway is a critical component. Aside from its evident value, it also acts as a berm to prevent the uncontrolled movement of overburden and water.

Removal of overburden
In order to ensure that no uncontrolled outslope is created, the operator must predetermine the width of cut to be taken on the particular area. Once this has been determined, the removal of overburden is initiated by the construction of a drill bench at a point which would eventually be the location of the highwall or the inside limit of the contour bench. All material excavated in the construction of this drill bench is hauled back and placed in the predesignated storage area. All upper horizon materials are required to be segregated and protected for return to the regraded surface.

The removal of the remainder of the overburden entails the employment of stringently controlled blasting techniques. In lateral movement, maneuverability in the handling of overburden is the key to success. Once fractured by blasting, the overburden is initially removed by an interior cut, as opposed to the conventional exterior cut on the cropline. Consequently, a trough is created along the contour of the natural topography. Conventional equipment removes this material, places it in trucks, and prepares it for outbound haulage.

As the excavation of this interior trough continues, the remaining outer portion of overburden is rolled inward toward the initial cut. This process continues until all the coal in the cut is prepared for extraction. Since there is no movement of overburden material toward the downslope side of the operation, the possibility of uncontrolled spillage is eliminated. As the coal is removed from each cut, the entire procedure is sequentially repeated, along the contour of the mountain, to the end of the permit area.

The mechanics of West Virginia's lateral movement are in direct contrast to those of "shoot and shove" surface mining. As opposed to the modified box cut, the initial cut of a haulback operation is taken in a direction parallel to the cropline of the coal, not perpendicular.

Unique characteristic of lateral movement

The unique characteristic of lateral movement is the specific handling of all overburden materials. Unlike the conventional box cut, in which overburden materials are moved to the immediately adjacent pit, the haulback method, utilizing trucks, permits the controlled transfer of removed overburden a greater distance from the active mining face. This provides greater flexibility in maintaining the efficiency of the operation since supplemental activities, such as augering, can be conducted without inhibiting the progress of production and reclamation.

Such intimate attention to overburden movement and placement provides for maximum backfilling control and significantly enhances reclamation. As the material is hauled back, it is deposited according to its potential effect on regrading and revegetation efforts. That is, large rocks and pyritic material are placed in the bottom of the pit area closest to the highwall. By doing this, undesirable material is adequately buried and has minimal effects on stabilization. In the same manner, upper horizon and favorable substrata materials are placed so as to maximize beneficial characteristics.

Elimination of the highwall

Recently, the highwall has been of major concern to opponents of surface mining. The lateral movement method of operation provides for the elimination of this highwall. Such elimination in less steep terrain is not revolutionary today, but in areas where slopes exceed 65 percent, it is an unprecedented accomplishment which greatly contributes to the aesthetic value of the state. Once the highwall is eliminated, the upper horizon materials are returned to the regraded area, and it is immediately revegetated. To add to pregermination stability, the regraded slope is "tracked-in" and mulch is applied. The continuing sequence of overburden removal, backfilling, topsoiling, and revegetation provides for reclamation which is as current as possible with today's technology. As a case in point, a recent contour operation in West Virginia was totally closed within two days following removal of the last coal.

Environmental advantage

Another environmental advantage of this method of operation is that the original vegetation is not disturbed on either side (above the highwall or below the crop-line) of the active mine. This is one of the most evident differences of the new concept since conventional contour mining required the clearing of all land surrounding a proposed operation. This maintained tree line acts as an effective buffer to present and future subsidence and/or erosion problems. In turn, this increases the life and utility of the required drainage system's impoundments. Also, this undisturbed vegetation significantly minimizes the aesthetic impact of surface mining in steep slope regions.

Concurrent Reclamation

By eliminating the disturbance of nonproducing acreage, the requirements for bonding and revegetation are minimized. Because less unstabilized land will be exposed, the chance of detrimental offsite effects is reduced. This also lowers the ultimate costs of revegetation. Generally this method of mining has reduced the size of a permit in steep sloped areas by approximately fifty percent.

Since its introduction, the lateral movement variation has provided increased opportunities to West Virginia's surface mining industry. Because it was designed for mining on slopes of excessive steepness, where fill benches were prohibited by law, its application has permitted recovery of valuable reserves which were not previously available for surface extraction. It provides for "concurrent" reclamation, which enables the release of an individual permit area more rapidly than was possible in the past. Concurrent operations also project the most desirable reclamation products.

In addition, lateral movement increases the efficiency of contour mining in several ways. First, only producing acreage need be disturbed. Thus, the amount of mineral removed is much greater relative to total disturbance, bonding, etc. Second, the material is removed and immediately placed in the backfill, therefore, it is necessary to handle it only once for the achievement of adequate reclamation.

Cost savings

The industry has also realized cost savings in drainage system installation. Since disturbance is minimized, smaller and fewer impoundments are required. Due to concurrent reclamation, the ponds which are installed require fewer cleanings and less maintenance.

Although the initial costs of lateral movement are greater, inherent benefits and advantages make its application economically feasible.

Application of method governed by several factors

The application of this variation of controlled placement is governed by several factors.

Coal/overburden ratio. The coal to overburden ratio is important, particularly since the resulting bench area must handle increased amounts of equipment and activity. The width of the cut must provide a bench area which is large enough to permit the conduct of haulback procedures. Regardless of bench width, however, the effective conduct of such an operation requires extensive planning and close coordination among personnel and equipment.

Topographical characteristics. Topographical characteristics of the immediately surrounding area are also determining factors. Since the initial activity must have adequate storage area for overburden materials, it is necessary to have an old surface mine in the vicinity, or a valley which is capable of being filled according to the specifications set forth for valley fill.

Since the original application of lateral movement, reclamation has enjoyed progressive success never before thought possible. Once mining is completed, the land is returned to a protective cover immediately. It is no longer necessary to wait long periods of time for regrading and revegetation to be completed. Unlike the mining of mountainous terrain in the past, the finished product is totally accessible to all types of activity and equipment. Since highwalls are usually eliminated, the areas above the walls are no longer isolated from the bench or slopes below.

After-Mining Land Use

In West Virginia, areas which have been mined in this manner are being used for a variety of recreational purposes, such as snowmobiling, hiking, camping, cycling and trailbiking. In addition, these areas provide access for future deep mining activity in remote mountain areas and are useful in forest fire control. Notable increases in wildlife activity around these reclaimed areas have been recorded. Consequently, the benefits of lateral movement will enhance the state's development and maintenance of an extensive wildlife population.

The environmental, aesthetic, and overall reclamation advantages provided by this method of controlled placement make it most desirable for steep slope surface mining. At present, in West Virginia it is in use on practically all areas in the state's southern coalfields. The majority of these have extremely steep slopes in excess of 65 percent. For instance, the first operation to employ this methodology was situated in Logan County, West Virginia. The steepness of the original slopes averaged approximately 66 percent, with some areas having slopes as steep as 80 percent. There was no downslope spillage, and reclamation was usually no more than two weeks behind the removal of coal. With such results, which have been duplicated on other lateral movement operations, mining companies in West Virginia feel that this concept is the key to the future success of surface mining regardless of its location.

National Leaders

For several years, West Virginia and its industry have been recognized as national leaders in surface mined land reclamation. Also during this period, the amount of acreage reclaimed has significantly exceeded the amount which has been disturbed by continuing and new operations. Both are trends of which West Virginia is proud and intends to continue, particularly in light of the ensuing federal standards.

West Virginia's introduction and implementation of this program is particularly significant since the state possesses more high-quality, strippable bituminous coal reserves than any other Appalachian state. Since this is based on 1968 estimates, it is felt that the operational advantages inherent to controlled placement will cause these estimates to be significantly increased.

In retrospect, it seems that the introduction of such a revolutionary program of methodology revision was well timed. With surface mining accounting for 58 percent of the nation's entire coal production and all indicators predicting a continuous increase, it is felt that this program adequately prepares West Virginia and the coal mining industry for any sudden upsurge in surface mining activity without sacrificing environmental protection. Also, West Virginia is better prepared for the federal legislation since its efforts have been utilized as examples for the program.

Conclusions

In summary, controlled placement has offered viable solutions to past and present reclamation problems in West Virginia, while providing increased opportunities for the protection and development of the state's natural resources. What began as a simple, but controversial amendment, which eliminated outside fill slopes, has become a complicated amalgamation of innovative approaches with numerous benefits, both direct and indirect. Cumulatively, its implementation has brought about a change of attitude in the surface mining industry which is emblematic of its present level of sophistication and responsibility.

By maintaining its basic premise, West Virginia's controlled placement maximizes productional recovery while minimizing surface disturbance in providing the nation with much needed energy today, in a manner which preserves and protects the environment for tomorrow.

Multiple Seam Operations at the Navajo Mine, New Mexico

33

W. W. Karna, Chief Mine Engineer,
Utah International Inc., Fruitland, New Mexico, USA

Introduction

Today we are in an energy transition period. The United States can no longer depend on the cheap, convenient, and once plentiful fuels that were used and wasted with such abandon during the last 30 years.

Coal deposits make up 88 percent of the United States energy resources, and 78 percent of its recoverable reserves. Approximately 51 percent of these reserves, or 234 billion tons, are located in the western United States. Other than the Great Plains region, the Southwest is probably the richest region in energy resources. The Navajo mine is owned and operated by Utah International, an affiliate of General Electric. It is located in northwestern New Mexico, near the Four Corners area of the United States in which the four states of Arizona, Utah, Colorado and New Mexico come together at one common point.

The mine was opened in 1963, and now comprises some 12,717 hectares (31,400 acres) of land leased from the Navajo Indians with estimated reserves of 1.1 billion tons. All the coal produced from the mine, which amounts to about 7 million tons annually, is delivered to the Four Corners generating station owned by a group of Southwestern utilities, and operated by Arizona Public Service Company. The plant is tied into an electric power grid that serves the cities of Los Angeles, Phoenix, Tucson, Albuquerque and El Paso providing electricity for use in industries and homes.

Method of Mining

Figure 33.1 depicts the sequencing methodology employed at the Navajo mine. Noteworthy are steps 3 and 4 which indicate the two positions of the stripping equipment. At the mine three draglines are operating, each having a capacity of about 38 cubic meters (50 cubic yards).

When stripping takes place over the upper seams or when overburden is shallow the draglines are placed conventionally as indicated by step 3. Because of the

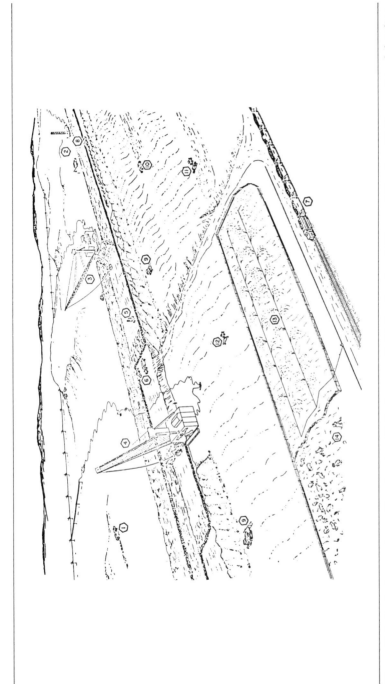

Figure 33.1. Typical operation at Navajo mine: (1) Saving topsoil; (2) Overburden drilling and blasting; (3) Conventional dragline stripping; (4) Spoil side dragline stripping; (5) Coal drilling and blasting; (6) Coal loading and hauling; (7) Rail haulage; (8) Ash disposal; (9) Regrading; (10) Topsoil placement; (11) Discing; (12) Seeding; (13) Irrigation; (14) Experimental controlled grazing.

multiple seam nature of the mine it is also desirable to employ a spoil side position as indicated in step 4. This technique takes advantage of the geometric placement of the machine to aid in overburden removal. The Navajo mine presently operates in areas having three seams, one below the other. The seam thicknesses range from 0.91 to 6.1 meters (3 to 20 feet). The thickness of partings or waste between seams varies from 3.35 to 15.2 meters (11 to 50 feet). The Navajo mine will ultimately have up to eight minable coal seams all of which dip to the east at a grade of about 3 percent.

Environmental concerns
The first physical step which takes place in any new mining area is of an environmental concern, that is the saving of topsoil. Prior to entry into any new mining area an environmental engineer maps out areas in which topsoil must be saved. The engineering department then schedules the necessary manpower and equipment to accomplish this by using three selfloading 25.2-cubic-meter (33-cubic-yard) scrapers.

Removal of overburden
After the topsoil is removed, the overburden or waste material above the coal seam is drilled and blasted. This material is usually a shale or sandstone, and it is too hard to dig without first being shattered. Blastholes nearly 254 millimeters (10 inches) in diameter are drilled and loaded with a mixture of ammonium nitrate, a fertilizer product, and fuel oil. The mixture, called ANFO, is the blasting agent. Each cubic meter/yard of overburden requires nearly 181 grams (0.5 pounds) of ANFO to break up the material for removal.

   The next phase is stripping, the removal of overburden to uncover coal. This is definitely the most important part of the mining operation. Each strip is approximately 30.5 meters (100 feet) wide and varies in length from 1.6 to 4.8 kilometers (1 to 3 miles). The overburden removal is accomplished by the three large walking draglines. Each dragline weighs nearly 2,041 tons and has a boom length of about 91.4 meters (300 feet), the length of a football field. Each bucketful of material that the dragline takes would fill the living room of an average home; it weighs nearly 68 tons. Annually, the three draglines combined move a total of 26,761,000 cubic meters (35 million cubic yards) of material. Each dragline is able to walk from one pit to another, at the rate of about 4.27 meters (14 feet) per minute. The new dragline now being built costs in excess of $10 million and takes 50 men one year to assemble. Stripping, as stated earlier, is the most important part of the operation and all other mining and reclamation activities are planned around the operation of the giant machines. It should be emphasized that draglines are used only for removing overburden and are not used for mining coal; they simply uncover the coal.

Coal handling
After the coal is uncovered, it must be drillled and blasted because it is too hard to dig without being shattered. Each ton of coal requires approximately 91 grams (0.2 pounds) of ANFO to properly break it up. After the coal is blasted, it is loaded into trucks by the use of front-end loaders. Each loader holds approxi-

mately 18 tons of coal. The loader dumps the coal into 120-ton-capacity bottom dump coal haulers. At this point the coal travels either directly to the crushing plant by truck, or it may be hauled to an intermediate railroad stockpile. If the coal is hauled to a stockpile, it will be rehandled later using the same kind of front-end loader. The loader dumps the coal into an 11-car train. Each rail car has a capacity of 125 tons. Approximately 75 percent of the coal produced at the Navajo mine is delivered by train, and the rest is hauled directly from the pits to the plant by truck.

Coal Preparation

At the crushing plant, coal is received at rates up to 40,000 tons per day and is crushed down from a run-of-mine size of 0.61 or 0.91 meter (2 or 3 feet) to 19 millimeters (0.75 inch). Coal is blended, using traveling stocking conveyors, from grades varying from 6,500 to 9,000 Btu per pound to a uniform blend grade of 8,750 Btu per pound, the grade which is necessary for the efficient operation of the power plant. The coal is then reclaimed by the use of traveling bucket wheel reclaimers.

Power Generation

The power plant has a combined generation output of 2,085 megawatts, which is sufficient to supply a city of approximately 1.5 million people, including industrial users. With all units at full operation the power plant will consume 28,000 tons of coal per day.

Waste ash generated by the power plant is hauled away and dumped into the pits from which the coal was originally mined. After ash disposal is completed, the overburden from the next strip is piled on top of the ash so that in most cases it has 9.1 meters (30 feet) of cover and in all cases it has at least 1.2 meters (4 feet) of cover.

Reclamation

The overburden removal process creates spoil piles. The first step in the reclamation program is the grading of the spoil piles to a gently rolling shape that resembles the surrounding terrain.

The primary or rough grading is done with large crawler tractors. The final grading is done by a specially designed grader with a 7.3-meter (24-foot) wide blade. The final grading enables use of farm machinery, necessary to accomplish revegetation.

After grading, new soils created by mining are mapped and analyzed to determine topsoiling and seeding requirements. Topsoiling is done on those areas marked and staked by an environmental engineer, who also checks to see that a uniform 152-millimeter (6-inch) layer is put down. In some cases, rather than being stockpiled, the topsoil is removed from an area before stripping and placed directly on the mined soil. This is one of the most costly operations involved in revegetation work.

Discing is carried out to mix the mined soil and topsoil interface and also to break up overly compacted areas. The equipment used to accomplish this is a special heavy-duty hydraulically controlled disc pulled by a four-wheel drive tractor. After grading, topsoiling, and discing, the soils are ready for seeding. Seeding is accomplished with two seed drills connected in series. Two seed drills are used because of the large range in size and shape of the 10 different seeds involved. The seeding rates differ with each area and all but two are native species which must be hand collected at a rate of about six tons per year. After an area has been seeded, the final step is to mulch the soil with straw or bottom ash to conserve moisture and protect the seeds during the germination phase.

Revegetation

A major effort of the revegetation program has been to supplement natural precipitation by irrigation. The heart of the irrigation operation is a large filtration and pumping station. The main distribution line is a 40.6-centimeter (16-inch) steel delivery line which is fitted with measuring devices that record water consumption. The line runs nearly 32.2 kilometers (20 miles) to provide water to areas being revegetated.

The irrigation program for a given revegetated area consists of frequent irrigation during germination to help the young plants get started, followed by infrequent irrigation to provide adequate moisture through the late summer months. A spring irrigation is applied the following year to complete the revegetation program. Total net water application amounts to about 305 millimeters (12 inches).

Utah International believes mining is a temporary land use and complies with or exceeds all legal requirements. The ultimate reclamation commitment is to return the mined areas to a condition "equal to or better than that before mining."

Reclaiming Mining Lands in Alaska

34

C. N. Conwell, Mining Engineer,
Division of Geological and Geophysical Surveys,
Alaska Department of Natural Resources, College, Alaska, USA,
and Stanley Weston, President,
Weston Agricultural Consultants Limited,
Vancouver, British Columbia, Canada

Introduction

The term "tender tundra" is a myth—a well-published myth. Any area that supports herbivores must have a strong plant base to supply food. The caribou roam in large numbers north of the Arctic Circle. The domesticated caribou, or reindeer, was introduced into Alaska in 1891, and the herd increased to a peak number of about 1 million by 1934. Palmer, 1934, estimated that a small area of western Alaska (near the Seward Peninsula) could easily support 4 million animals without damage to the "tender" tundra.

In 1977, Kaiser Engineers reported on the tender tundra with regard to the northern Alaska Coalfield, but reluctantly admitted that natural reclamation at the Meade River coal mine (near Barrow) was very complete.

Alaska land can be reclaimed after strip mining. The after-mining land use can vary from wildlife habitat to grazing and to agricultural crops.

Plant production studies have been conducted on northern latitudes, including Barrow, the northernmost tip of Alaska. Annual production of sedges, grasses, and other herbs has generally ranged from 91.8 to 165.2 kilograms per hectare (500 to 900 pounds per acre) of dry matter. By applying a management rule of thumb of "take half and leave half," the herbaceous tundra range is estimated to be capable of supplying 55.1 to 91.8 kilograms per hectare (300 to 500 pounds per acre) of leafy forage. By comparison, the alpine meadows of Wyoming or Colorado can supply only one-third to one-half as much (Mitchell, 1974).

Many areas in the state, particularly the interior, have a high agricultural potential. Excellent gardens are grown above the Arctic Circle. Dinkel and Ginzton, 1976, documented the vegetable yield in Alaska, and found that compared to the "Lower-48" average, the "Nation's Icebox" yields two to three times as much per acre.

Alaskan cabbages are famous, of course. Some weighing 22.7 to 31.8 kilograms (50 to 70 pounds) are routinely shown as fair exhibits. The high productivity applies equally to broccoli, lettuce, green peas, etc. Grains are produced commercially near Fairbanks. Wooding, 1977, documented grain production throughout Alaska, including locations with the Arctic Circle, and proves the viability of grain production.

With such a proved basis for agriculture throughout the state, it should be no surprise that land in Alaska cannot only be easily reclaimed but subsequently farmed.

Reclamation

Until 1970, no effort was directed toward reclaiming mined lands in Alaska. The first scientific attempt at reclamation was not at a mine, but on a coal lease accessible only by helicopter; the lease, in the Beluga Coalfield, near Anchorage, was held by Placer AMEX, Inc. AMEX selected Stanley Weston, specialist agronomist, as its agricultural consultant.

The first reclamation project had, unfortunately, a rather sorry result. The seed selected by Mr. Weston was flown to the prepared plots by helicopter and deposited. However, when he arrived at the site to plant the grass, only the sacks were left. Bears had eaten the seed.

Extensive program of reseeding

Undaunted, Stanley Weston returned to Alaska in 1972, this time to Healy, 16.1 kilometers (10 miles) from the entrance to Mt. McKinley National Park. The Usibelli coal mine there started an extensive program of reseeding its mined land.

Test plots were prepared on mine refuse dumps. The nutrients available in the mine spoil dumps were analyzed, and a proper fertilizer was applied by spraying from an airplane. Some plots of grass were planted without fertilizer, and others had varying amounts of fertilizer applied. Certain fertilized plots had an excellent growth of grass by fall. These plots were on the north side of mine dumps, where there would be no snow cover and they would be subjected to the severest of winter conditions.

During the winter, however, wild Dall sheep ate the grass down to the roots, but in spring the plots turned green again.

With the information gained from the test plots, 364.2 hectares (900 acres) of mined land were seeded the following spring. By fall there was a luxurious growth of grass. As the snows came, even more sheep came to winter in the mine area.

This brought on a new problem—hunters looking for an easy kill. At first, mine safety regulations were used to exclude hunters. This caused trouble from a local sportsman's association. Then the mine operator, Joe Usibelli, in cooperation with the Alaska Fish and Game Department, had the mine and surrounding area declared a State game preserve. Today, at least 150 Dall sheep winter in the mine area on Healy Creek and have their lambs near the loaders and huge trucks.

In the winter of 1974-75, the wolf population increased, causing a consequent decrease in the sheep population. Today, more animals are needed to eat the grass and fewer wolves to eat the sheep.

Commercial crops

In Alaska, reclaimed mined land may be restored for sport game management. However, the experimental work at the Usibelli mine has been directed toward new commercial crops, both forage and grains. In 1976 and 1977, high-quality hay was harvested for sale on the local market. Experimental work over the past three years has proved that two "short-season" crops, one an edible oil seed and another a high-protein seed suitable for human consumption, can be produced commercially. Until this work was done at the Usibelli mine, it was believed that crops could only be propagated through the use of native vegetative means and that leguminous plants would not nodulate under Alaskan conditions. It was considered impossible to establish perennial growth of grasses and legumes unless native species were used in vegetative form through use of rooted stolons, or "runners." It has been conclusively proved at the Usibelli mine that nodulating can occur with legumes.

Also, it was thought that individual plant types should be used rather than a design mixture, in which a desired plant succession is nurtured for several years before the permanent plant population is established. This takes place in about five years in this area. "Nurse" crops and "companion" crops are important in their function of providing conditions to carry the permanently desired plant population through the establishment years. However, the Usibelli project proved the ability of crop seed to develop commercial crops in the planting seed mixture. Exciting possibilities exist for the production of margarine and an edible flour.

Thus, it has been demonstrated that grains, hay, and perennial crops can be grown successfully in Alaska and that leguminous plants will thrive and produce nodules.

The Surface Mining Control Act of 1977

The Surface Mining Control Act of 1977 requires the reclaiming of mined land in the United States. Its regulations permit reclamation without the saving and storing of topsoil. The interim regulations provide:

Part 715.16 — General Performance Standards — Section 715.15(a) (4) Selected over-burden materials may be used instead of, or as a supplement to, topsoil where the resulting soil medium is equal to or more suitable for vegetation, and if all the following requirements are met:

(i) The permittee demonstrates that the selected overburden materials or an overburden-topsoil mixture is more suitable for restoring land capability and productivity by the results of chemical and physical analyses. These analyses shall include determinations of pH, percent organic material, nitrogen, phosphorus, potassium, texture class, and water-holding capacity, and such other analyses as required by the regulatory authority. The regulatory authority also may require that results of field-site trials or greenhouse tests be used to demonstrate the feasibility of using such over-burden materials.

(ii) The chemical and physical analyses and the results of field-site trials and greenhouse tests are accompanied by a certification from a qualified soil scientist or agronomist.

(iii) The alternative material is removed, segregated, and replaced in conformance with this section.

Overburden characteristics

The experimental work at the Usibelli mine has been directed toward revegetation without saving and storing topsoil, principally because there is little to save. The sedimentary rocks above the coal seams are clay-siltstone, sandstone and small pebble conglomerates. In mining, the rock disintegrates during the process of loading, hauling, and dumping. The sandstone will breakdown and mix with the clay-siltstone to a base with sufficient clay to hold moisture and sand for drainage. The spoil can be easily reworked during the grading and continuing cycle of reclamation to provide an acceptably smooth surface for revegetation.

On-going soil analyses will provide a basis for selecting the required nutrients, and field and greenhouse trials will provide information on acceptable plants. The six years of trials and experiments at the Usibelli mine are continuing to provide a good data base required for reclaiming the land.

There are certain characteristics peculiar to the Usibelli mine and the north country in general that help in mine reclamation:

1. There is no winter "break" in dormancy, whereby the seed germinates and the plant is killed by the subsequent frost.
2. Winter snows provide spring moisture.
3. Long summer days (nearly total daylight) provide a successful growing season.
4. There is adequate moisture during the growing season.
5. The fall freeze is sudden, but the grain may still be harvested before the snowfall. At the Usibelli mine, hay cut in the fall, too late for baling, was dry, sweet, and ready for baling the next spring.

Conclusion

Experimental work at the Usibelli mine has proved:

1. The overburden above the coal can be reworked during stripping sufficiently to provide a soil with enough silt and clay to hold moisture and enough sand for drainage.
2. Proper nutrients can be added to provide excellent plant growth.
3. The coal lands can be reclaimed for mining without saving topsoil.
4. The fourth conclusion, and possibly the most important, is that properly planned reclamation enables the land to yield more and better crops after mining than before.

References

Dinkel, D. H., and Ginzton, L. M., 1976, "Vegetable Variety Trials: Agroborealis." Volume 8, Number 1.

Kaiser Engineers, 1977, "Technical and Economic Feasibility, Surface Mining Coal Deposits, North Slope of Alaska." Kaiser Engineers, Inc., Oakland, California.

Mitchell, W. W., 1974, "Rangelands of Alaska—Alaska's Agricultural Potential." University of Alaska Cooperative Extension Service.

Palmer, L. J., 1934, Raising Reindeer in Alaska: U.S. Dept. of Agriculture, Miscellaneous Publication.

Wooding, F. J., 1977, "The Grain-growing Process in Boreal Environments—Expanding Agriculture and the Management of Interior Alaska Resources." A seminar series. School of Agriculture and Land Resources Management, University of Alaska.

Constructing Spoil Storage Fills in the Steep Terrain Appalachian Coalfields

35

John D. Robins, Partner, Robins and Associates,
Mechanicsburg, Pennsylvania, USA

Introduction

Surface mining and reclamation technology in the United States has improved dramatically over the past ten years in response to public demands for environmental protection and mined land stability. As the era of "shoot and shove" surface mining was replaced by a conscientious effort to restore the utility and aesthetic integrity of disturbed land, mine operators were forced to modify schemes and eliminate downslope spoil casting, particularly in the steeper terrain common to much of central Appalachia. One such mining technique to evolve during this period was the practice of head-of-hollow fill spoil disposal (Figure 35.1).

Head-of-hollow fill, a very popular spoil disposal technique in Appalachia, refers to the practice of placing excess spoil material from contour mine benches and mountaintop removal mine sites in adjacent hollows, grading to a predesigned landform, and revegetating in accordance with applicable regulations. Although the objective of all head-of-hollow fill reclamation is basically the same, i.e., environmental stability, construction practices have varied greatly throughout the states employing this disposal technique.

The following briefly discusses various construction practices and related potential stability problems within the states of West Virginia and Kentucky. These two states are cited specifically since, between them, they can lay claim to nearly all of the head-of-hollow fills completed to date. Future fill construction practices, as outlined under the United States' newly enacted federal criteria, are also delineated since these requirements are generally considered more stringent than existing state regulations.

At the time of the Second International Symposium on Stability in Coal Mining, Mr. Robins was Associate, Skelly and Loy, Engineers-Consultants, Harrisburg, Pennsylvania. He prepared this paper while employed by Skelly and Loy.

Figure 35.1. Completed hollow fill at a mountaintop removal mining operation.

West Virginia

West Virginia is the undisputed leader in the construction of stable and environmentally sound head-of-hollow fills. This success stems largely from the diligent effort devoted to site preparation, drainage control, and spoil placement. Before construction begins, all vegetation is removed from the fill site and is either burned or windrowed near the toe of the proposed fill, to act as a filter and trap eroded soils. Next, a haul road is constructed within the disposal area to the projected toe of fill where spoil placement operations begin. A rock core is then started at the toe, and is progressively constructed through the fill mass from the original valley floor up to the top of the fill bench. Rock cores are constructed concurrently with fill placement, maintaining a minimum width of 4.9 meters (16 feet) as both the core and fill are brought up together. This artificial drainage system permits surface runoff and natural water percolation to exit without saturating the fill, thereby greatly reducing erosion and landslide potential (Figure 35.2).

Reclamation laws (West Virginia)

In accordance with West Virginia reclamation laws, all spoil placed in the fill is transported to the active lift, where it is placed and compacted in 1.2-meter- (4-foot-) thick layers. The rock core must be maintained at a minimum of 1.2 meters (4 feet) above spoil deposition during all phases of construction. A terraced appearance is created on the fill outslope by recessing each successive 15.2-meter (50-foot) lift, thus incorporating an external drainage scheme into the fill design. The resultant bench (or terrace) is sloped into the core, reducing the time that storm water runoff is in contact with the fill material. Fill face stability is accomplished by regrading the outslope to a maximum of 26°.

Upon completion, the top of the fill is graded to drain back to the head of the hollow. A drainage pocket maintained at this point intercepts surface water run-off and directs it to the rock core (Figure 35.3).

These construction techniques represent the best reclamation practice currently in use in the construction of head-of-hollow fills. However, designed to assure stability under worst conditions, these requirements allow no flexibility for alternative reclamation techniques under more favorable conditions. Consequently, operators are compelled to comply with costly requirements which may be overly restrictive in many instances.

Factors influencing cost of fills

The factors having the most significant influence on the cost of West Virginia fills are fairly obvious: hauling all material to the fill site, maintaining a chimney-type rock core drainage system to the top of the fill, and compacting the material in restricted lift sizes. However, depending upon specific physical factors such as topographic, geologic, and hydrologic conditions, it is possible that none of these costly requirements are actually necessary to ensure stability. Conversely, some areas may require additional precautions to maintain a structurally sound fill mass. For instance, current West Virginia requirements make no provisions for drainage from springs or natural seeps within the fill, which could eventually result in slope failure if the fill mass is composed of poorly draining soils. While stable head-of-hollow fill construction is indeed a complex problem, it is important to realize that there are a number of alternative construction techniques that

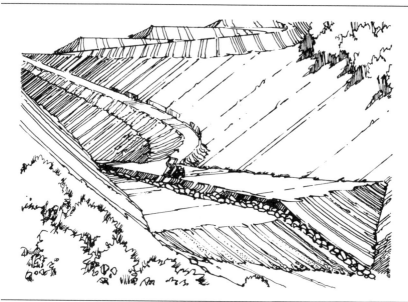

Figure 35.2 Initial stages of a West Virginia head-of-hollow fill showing cleared fill site, first lift construction, and rock core drainage system.

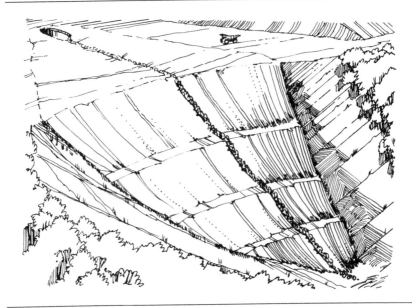

Figure 35.3. Completed West Virginia head-of-hollow fill showing face terraces, rock core drainage system and drainage pocket.

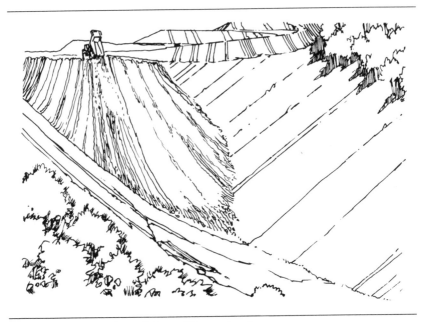

Figure 35.4. Kentucky end-dump fill construction.

can be successfully employed depending upon these site considerations. The construction practices commonly employed in Kentucky fill sites and the recently promulgated criteria of the Federal Office of Surface Mining represent two such alternatives.

Kentucky

Whereas West Virginia provisions require a tremendous effort to assure internal drainage, Kentucky criteria ultimately leave internal drainage pretty much to chance. Spoil materials are dumped over the outslope at the head of the hollow rather than deposited in uniform lifts from the toe, on the assumption that there will be a natural segregation of material with the heavier material rolling to the bottom and creating a natural french drain (Figure 35.4). Under some circumstances this philosophy holds true, particularly in the typically steep terrain of much of eastern Kentucky. Unfortunately, there are two factors that interfere with the successful application of this technique:

1. Friable or easily weatherable rock units (such as shales) roll just as well as durable sandstone rock units and this technique does not allow for their segregation. Consequently, the underdrainage system can be prone to deterioration and failure.
2. Kentucky criteria stipulate that the outslope be pushed down every 48 hours to a maximum slope of 20°. This inadvertently prevents the natural formation of an underdrainage system since, on this reduced grade, rocks tend to remain where dumped and natural segregation and subsequent underdrain establishment does not occur as fill construction progresses. During pushdown, fine spoil which has accumulated near the dump point is pushed to the toe of the fill, and compacted in the process. This, in turn, results in two undesirable features from a stability standpoint:
   a. Intermittently spaced zones of coarse rock and fine-grained spoil in place of a well-formed underdrainage system.
   b. Compacted planes parallel to the final outslope which tend to channel subsurface water and create slip planes for the relatively unconsolidated material above these areas (Figure 35.5).

If spoil is *not* pushed down, compaction is limited to the surface of the fill only, and deep-rooted slip planes and tension cracks can develop. The potential for failure as a result of this practice depends, again, on specific physical factors, but it should be emphasized that the potential for failure is relatively great.

External drainage systems

Kentucky's provisions for *external drainage* systems differ considerably from those in West Virginia. Surface drainage is directed away from the fill site into diversion ditches constructed in the undisturbed slopes parallel to the final outslope. This construction practice is preferable, since much of the water, if directed into or over unconsolidated fill material, would be quickly absorbed, thereby reducing stability (Figure 35.6).

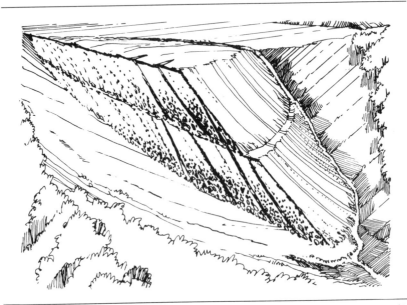

Figure 35.5. Cross sectional view of Kentucky head-of-hollow fill showing slip planes and interrupted underdrain.

Figure 35.6. Completed Kentucky head-of-hollow fill showing crowned surface and side drainage channels.

Federal Surface Mining and Reclamation Act

The Federal Surface Mining and Reclamation Control Act of 1977 recognized some of the problems associated with head-of-hollow fill construction as currently practiced, and attempted to improve on these techniques in its subsequent rules and regulations. The construction methods stipulated in these regulations are modeled, to a large extent, on West Virginia's hollow fill criteria, but are more considerate of the factors affecting stability such as foundation materials, topography, hydrology, and fill size. The principal aspects of the new Office of Surface Mining's requirements, which take precedence over existing state criteria, are as follows:

1. Keyway cuts to stable bedrock or rock toe buttresses must be constructed where the slope in the disposal area exceeds 20°.
2. Underdrains must be constructed along any natural drainway and to any area of potential drainage or seepage. These underdrains are not carried to the surface as in West Virginia fills. The regulations also specify minimum and maximum size limitations for rock used in the underdrain.
3. Underdrain dimensions are dependent on the size of the proposed fill and the type of material used in fill construction. Drains range between 1.2 and 4.9 meters (4 and 16 feet) high and 2.4 and 4.9 meters (8 and 16 feet) wide.
4. Spoil is to be placed in 1.2-meter (4-foot) lifts (or smaller, as specified by the regulatory authority) and compacted.
5. Terraces 6.1 meters (20 feet) wide are to be constructed at 15.2-meter (50-foot) vertical intervals on the outslope.
6. The top of the fill and each terrace will drain surface water to stabilized channels off the side of the fill, thereby minimizing surface water contact with fill materials.
7. All surface drainage from above the fill is to be diverted away from the fill (Figure 35.7).

Summary

The criteria employed by these regulatory agencies, as outlined in the preceding discussion, illustrate the various approaches which are being used to establish stable spoil disposal fills. Unfortunately, there are a number of problems with each of these three programs, ranging from too little to too much control depending upon site specific factors. Under the proper conditions, any of these practices could produce an environmentally and structurally stable fill; however, the potential also exists for failure or unnecessarily high construction costs with each regulation. The point here is that, while these three agencies have established guidelines for hollow fill construction, they are just that—guidelines. There is no one "best way" to construct head-of-hollow fills: each and every mine site is physically different and mining and reclamation strategies must be flexible enough to be responsive to these factors.

, Spoil composition and physical properties are extremely site specific factors; some spoils may require small-size lifts to assure stability while others are inherently stable in much larger lifts. The slope and soil properties at some sites

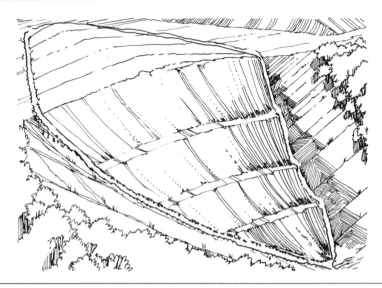

Figure 35.7. Completed hollow fill configuration as per federal construction regulations.

require stabilizing features like rock toe buttresses, or keyway cuts; other sites do not. The effort necessary to assure a specific degree of compaction and/or stability cannot be generalized by a single set of criteria, nor should the criteria be so complex, cumbersome, or expensive that they discourage or prohibit their implementation.

Conclusions

The construction techniques outlined in the preceding discussion present a synopsis of the current state-of-the-art of head-of-hollow fill disposal practices employed in central Appalachia. When considering these guidelines, it is important to remember that, unlike the highway industry, the mine operator is *not* in the business of embankment construction and involved earthwork is a distinct economic liability. Therefore, it is important that the construction techniques employed assure adequate stability for the projected final land use without being unnecessarily restrictive or placing an undue economic hardship on the operator. This will require considerably more effort devoted to pre-mine soils engineering such as foundation seepage and overburden analyses, and more detailed assessments of the physical character of the proposed fill site.

Certainly the design and construction of stable spoil disposal fills is a complex undertaking which requires careful evaluation by qualified soils engineers. However, through application of accepted engineering practices on a site by site basis, spoil disposal areas *can* be designed to ensure environmental protection, while also allowing mining activities to proceed more efficiently and economically.

Reclamation Planning and Mining Permit Applications in Several Rocky Mountain States

36

R. E. Luebs, Chief, Agronomy Division,
and D. M. Stout, Range Scientist,
Woodward-Clyde Consultants, Denver, Colorado, USA

Introduction

Coal production in the western United States has made a markedly increased contribution to the nation's total output in the 1970s. From less than 5 percent of the total production in 1970 to nearly 20 percent in 1977, coal mined west of the Mississippi River has become a significant energy source in the United States. Growing national energy needs, low sulfur content, improved surface mining technology, and the relatively shallow depth of coal deposits have been the paramount factors in the rapid expansion of surface coal mining in the West. A large proportion of western coal is surface mined in the states of the Northern Great Plains and in states generally designated as being the Rocky Mountain region.

Coal Production

This chapter primarily considers surface coal mining and the regulations relating to it in three contiguous Rocky Mountain states; Wyoming, Colorado, and New Mexico. During the 1972-76 period, surface mined coal has contributed over 90 percent of the total tonnage in Wyoming and New Mexico (Table 36.1), and over 40 percent in Colorado. Underground mining contributed a fairly consistent tonnage during this period except for the apparent cessation of underground mining in Wyoming in 1976. In the five-year period, 1972-76, surface mined production nearly tripled in Wyoming from 9 to 26 million tons. Surface mined output has more than doubled in Colorado during the same period while that in New Mexico has remained fairly constant. A rapid expansion occurred in surface coal mining before 1972 in New Mexico: a total output of only 0.9 million tons was reported in 1960 (Tabet, 1977). During the 1972-76 period, tonnages produced by underground mining averaged 9.4 and 7 times greater in Colorado than in Wyoming and New Mexico respectively. Underground production has slowly but steadily increased in Colorado but was surpassed by surface mined output for the first time in 1974. At the end of 1976, surface mined coal production in the states of Wyoming, Colorado, and New Mexico comprised 91 percent of the total output.

Table 36.1. Coal Production in Wyoming, Colorado and New Mexico, 1972-1976 (thousand tons)

Year	Wyoming			Colorado			New Mexico		
	Surface	Underground	Total	Surface	Underground	Total	Surface	Underground	Total
1972[1]	10,487	442	10,929	2,452	3,070	5,522	10,172	0	10,172
1973[1]	14,461	425	14,886	2,834	3,361	6,195	8,336	733	9,069
1974[1]	20,176	526	20,702	3,636	3,260	6,896	8,864	529	9,393
1975[2]	23,369	436	23,805	4,896	3,468	8,364	8,235	500[3]	8,735
1976[4]	29,200	0	29,200	5,650	3,600	9,250	8,800	600	9,400

1. Minerals Yearbook, Metals, Minerals and Fuels, U.S. Department of the Interior.
2. U.S. Bureau of Mines.
3. Keystone Coal Industry Manual.
4. Preliminary estimates, U.S. Department of Energy.

Land Area Disturbed by Mining

The most obvious difference between surface and underground coal mining as observed by the public is the vastly larger amount of land surface affected by surface mining. Estimates of surface minable coal reserves are usually given in tonnages. Less attention has been given to the areas of land to be disturbed in recovering these reserves. Smith et al., 1972, has made acreage estimates in Wyoming based on the following criteria:

1. A minimum coal seam thickness of 1.5 meters (5 feet).
2. An overburden to coal ratio of less than 7.6 cubic meters (10 cubic yards) per ton of coal.
3. A total overburden thickness of less than 36.6 meters (120 feet) except where multiple seams or a single thick seam exists.

A rough estimate of minable reserves was made in Colorado after reviewing the paper by Speltz, 1976, on surface minable resources, and discussing limiting factors with personnel of the U.S. Geological Survey. Data for New Mexico were estimated by Tabet, 1977, based on the ratio of land affected so far by coal surface mining to the tonnage of coal mined.

According to data compiled from interviewing reclamation managers with coal mining companies, only a small fraction of the land with the potential for being affected by surface coal mining had been excavated as of January 1, 1978. In Wyoming, 4,573 hectares (11,300 acres) are estimated to have been affected by pit excavation since significant surface coal mining began. This accounts for less than 8 percent of the total land area of 61,005 hectares (150,740 acres) with surface minable reserve deposits of coal (Thilenius and Glass, 1974). The land area affected by surface mining excavation totals 1,619 hectares (4,000 acres) in Colorado which is 1 percent of the estimated land area of 133,551 hectares (330,000 acres) with minable reserves. A small percentage, 2 percent, or 2,752 hectares (6,800 acres), has been affected in New Mexico out of a potential area of 125,861 hectares (311,000 acres). The total land area expected to be disturbed in these three states by surface coal mining during the next 20 to 30 years could be as high as 259 to 389 square kilometers (100 to 150 square miles).

Environmental Legislation

To a large degree the rapid development of surface coal mining in the American West has occurred coincidentally with an increased concern for the environment in the United States. There is reason to ascribe a portion of this rapid development to the low sulfur content of western coal, an important environmental consideration. Increased concern for the environment was initially reflected politically on a large scale in the United States with the passage of the National Environmental Policy Act of 1969. While this law did not deal specifically with surface coal mining it was a harbinger of future state and federal laws that did. The state of Wyoming enacted its Environmental Quality Act in 1973. This act, which established a Department of Environmental Quality including divisions for water quality, air quality, and land quality, was directed largely at the environmental effects of surface mining.

In 1976, the legislature of Colorado enacted the Mined Land Reclamation Act. This act established the Mined Land Reclamation Board which was given the authority to grant all mining permits. The New Mexico Coal Surface Mining Act was enacted in 1972. Among its provisions was the establishment of a Coal Surface Mining Commission. The New Mexico law is the only one in the three states addressed solely to surface coal mining. Generally, the state bodies established by these acts, whether they be a council, board, or commission, are charged with the responsibility of promulgating rules, regulations, and/or guidelines necessary for the administration of the acts and the granting or denying of mining permits.

Following the recent law-making activities of the states regarding surface mining reclamation, the National Congress enacted the Surface Mining Control and Reclamation Act of 1977. This act created the Office of Surface Mining Reclamation and Enforcement in the U.S. Department of the Interior. States have until February 1979 to present programs demonstrating that they have the capability of carrying out the provisions of the federal act. If state rules and regulations meet federal standards, then the states will have the authority to enforce the rules and regulations in their respective states. This will not prevent the states having rules and regulations of their own which are more stringent than the regulations under the federal act.

Laws, rules, regulations, and guidelines

The state laws pertaining to reclamation of surface mined lands in Wyoming, Colorado, and New Mexico are similar in many respects. The Wyoming act defines reclamation as the "process of reclaiming an area of land affected by mining to use for grazing, agricultural, recreation, wildlife purposes or any other purpose of equal or greater value." The Colorado law defines reclamation as "the employment during and after a mining operation of procedures reasonably designed to minimize as much as practicable the disruption from the mining operation and to provide for the rehabilitation of affected lands through the rehabilitation of plant cover, soil stability, water resources or other measures appropriate to the subsequent beneficial use of such mined and reclaimed land."

The New Mexico act of 1972 does not define reclamation but states that "affected areas must be graded and revegetated" and "reclamation shall be an integral part of the mining operation and shall be completed within reasonably prescribed time limits." Mining permits are required in all of these states. Receiving a permit requires approval of a submitted mining permit application. In general, these applications must include a mining plan and a reclamation plan, and they involve the posting of a performance bond for reclamation. After the permit is granted, the mine operator must submit an annual report of mining and reclamation activities during the previous year.

While the laws do not appear complex, most coal mine operators who have sought permits will be quick to point out that the state regulatory or permit-granting body council, board, or commission has a relatively wide discretion in interpreting the law. As a consequence, obtaining a permit requires the operator to be aware of the latest rules, regulations, or guidelines. Indeed, these regulations have been known to change two, three, or more times during the course of mine planning. While the state regulatory body apparently has the authority to accept variances of the regulations, the mine operators take some risk in causing a significant delay by deviating from the state regulations or guidelines. It is in the area of regulations for permit granting that the individual states vary considerably.

Differences in state requirements

These differences usually occur in the types or the detail of environmental baseline and reclamation information required. The detail of soils, overburden, and vegetation information desired for reclamation planning ranges widely (Table 36.2). Wyoming requirements are the most stringent of the three states compared, while Colorado appears to be the least. For example, the detail required in the mapping of soils to the level phase of a series, as stated in the Wyoming guidelines, is more specifically stated than in the rules and regulations of the other two states. "Series" designates a class in which soils are similar except for surface texture. The "phase" is a subdivision of series based on any characteristics or combination of characteristics potentially significant to land use or management of the soil; for example, slope, erosion, or stoniness. The mapping of soil is primarily carried out to locate sources of topsoil of different suitabilities for stripping in the land reclamation process.

Data on premining forage production measurements are used as a base to evaluate revegetation at the end of the reclamation process. In Wyoming enough plots within installed livestock exclosures must be clipped for each grass community so that a 10 percent reduction in total annual productivity can be detected with 90 percent confidence. This requirement requires exploratory field vegetation data to be gathered before the appropriate number of plots to be clipped can be determined. The regulatory agency in Wyoming (Division of Land Quality) is much more concerned about the chemical characteristics of overburden than the agencies in Colorado and New Mexico. Wyoming has recently increased the number of overburden chemical analyses required.

Topsoil replacement (topsoiling) is now a widely accepted and major integral step in the reclamation process. The Wyoming definition of topsoil for reclamation as the "total available soil depth that will support plant growth" suggests that:

1. Most soil material is superior to spoil as a root medium.
2. Topsoil stripping and replacement depth could be much deeper than 305 millimeters (12 inches), and reach perhaps 1.5 meters (5 feet).

Although detailed regulations interpreted from the 1977 Federal Surface Mining Act have not yet been forthcoming, it is possible that all requirements in Table 36.2 may remain the same except the reclamation completion period. The federal law states that in areas of less than 0.66 meters (2.17 feet) annual average precipitation, the operator assumes the responsibility for successful revegetation for 10 years after revegetation operations begin.

Table 36.2. Selected Environmental and Reclamation Requirements for Mining Permits

State	Soil mapping detail	Vegetation sampling	Overburden laboratory analyses	Topsoil replacement depth	Reclamation completion period
Wyoming	Soil phase	Forage productivity measurement to meet statistical re-	10-12 laboratory determinations	Total available soil depth that will support plant growth	In accordance with reclamation plan
Colorado	General soil description	Map & describe plant communities	none	6-12 in. if available	5 years
New Mexico	Soil association	Forage productivity measurement	none	6-12 in. if available	Not specified

Table 36.3. Land Areas Disturbed by Surface Coal Mining Activities and at Reclamation Stages January 1, 1978 (Acres)

States	Disturbed	Graded	Planted		
			Topsoil	Spoil	Total
Wyoming	18,790	7,350	4,470	1,380	5,850
Colorado	5,600	2,890	1,260	1,420	2,680
New Mexico	7,350	4,090	1,080	2,000	3,080

Note: Compiled from interviewing reclamation managers or reviewing annual reports to regulatory agencies for 20 mines in Wyoming, 6 mines in Colorado and 5 mines in New Mexico.

Reclamation Progress

Significant progress in the process of reclamation is being made on surface mined coal land in Wyoming, Colorado, and New Mexico. A portion of this activity was already underway at several mines before the state laws granting permits became effective. Reclamation laws in general apply to all lands disturbed or affected by mining activities and not just the mining pit. Land surfaces affected by roads, railroads, and construction of facilities are subject to reclamation. In Wyoming 7,604 hectares (18,790 acres) of land are estimated to have been disturbed by surface coal mining activities to January 1, 1978 (Table 36.3). Grading of spoil piles has been accomplished on 2,975 hectares (7,350 acres) or 39 percent of the disturbed area (80 percent of the graded area). Colorado estimates show 1,170 hectares (2,890 acres) or 52 percent of 2,266 hectares (5,600 acres) of disturbed land to be graded and planted. A slightly higher percentage of disturbed land, 56 percent, has been graded in New Mexico. The fraction of disturbed acreage in New Mexico which has been planted is estimated to be 42 or 75 percent of the graded acreage. Acreage planted on spoil generally reflects reclamation conducted before the states adopted permitting regulations. However, in New Mexico little or no topsoil has been available in significant areas: as a consequence other surface or overburden material has been used as a vegetation medium. The value of replacing topsoil on spoil where it is available is widely recognized. The capacity of topsoil to provide available plant nutrients, available water holding capacity, and favorable physical characteristics for root growth far exceeds the capacity of spoil to provide these advantages. In addition, some types of spoil exhibit toxic characteristics.

Reclamation of surface mined lands in the states of Wyoming, Colorado, and New Mexico has not progressed for a sufficiently long period to evaluate its success to completion as defined by the regulatory agencies. Complete bond release has occurred on only a small fraction of the surface mined coal lands in Colorado and apparently on none in Wyoming and New Mexico. Nevertheless, it is estimated that plant vegetation has been established to some degree on some 60 to 80 percent of the planted areas on noncultivated lands.

Value of technical detail in revegetation plans

The regulatory body which has authority to grant mining permits in Wyoming, Colorado, and New Mexico generally requires detailed revegetation plans and specifications. Among such specifications may be seedbed preparation operations, types and rates of mulch application, species, and rates and site areas for different planting mixtures. The regulatory bodies apparently feel that the requirements for such specifications are necessary so that the mining companies will carry them out step by step, and that they will make the mining companies become aware of the technical information necessary for revegetation. The overall result will presumably be better and earlier reclamation. A supplementary reason for this is that detailed planning facilitates the calculation of reclamation costs from which the regulatory body sets the performance bond. Some of the required planning detail does not appear justified because good evidence for the success of the specifications has not been obtained on site or in the area. Revegetation plans developed for the mining permit application frequently must be changed after the first year's field experience. Qualified agronomic or range management personnel who have been given the authority and support of the company to establish revegetation are the greatest assurance of a successful job. They are likely to be delayed by adhering to detailed plans and changes emanating from the state regulatory body. It would be desirable if regulatory bodies could give their major thrust to preventing essentially irretrievable actions which would reduce the probability of successful reclamation.

The major part of the performance bond covers the cost of stripping the topsoil, grading the spoil, and spreading the topsoil. These operations can be successfully accomplished with a high degree of certainty. However, the success of revegetation in the semi-arid and arid regions of the West is dependent upon unproven techniques for the most part. Many reclamation managers are doing applied research and testing on site. It is this approach which will result in more successful reclamation. Regulatory bodies have the authority to stop mining activities if reclamation is not proceeding satisfactorily, so it can be assumed that mining companies will, therefore, want to achieve reclamation goals.

Discussions with individuals in state regulatory bodies, and participation in mining permit application hearings, have revealed that some mining companies apparently feel that the reclamation plan, including the environmental baseline studies that were made on site, are "paper" requirements to obtain mining permits after which they really have little meaning. This is emphasized by the failure of some companies to incorporate collected environmental data for use as the basis for the reclamation decisions which are made. The intent of the regulatory bodies is that the reclamation should be based on environmental data which the mining companies must obtain.

Importance of reclamation goals

The achievement of good reclamation in a short time and at minimal cost depends on setting reclamation goals at the initiation of mine planning and pursuing those goals steadily to completion. This rationale is difficult to accept because although it will result in monetary savings, any reclamation adds to the cost of coal production.

Reclamation is a fairly complex process because it requires several widely different types of expertise to plan and manage. It is also constrained by regulations, weather, and mining plans. The following major steps are involved in reclamation:

1. Soils, overburden, hydrology, vegetation, and wildlife inventory surveys.
2. Mining permit application preparation.
3. Topsoil stripping.
4. Overburden placement.
5. Grading of overburden.
6. Topsoil replacement.
7. Revegetation.

Coordination of reclamation operations is essential because:

1. They must be done sequentially.
2. They begin prior to mining and continue until after it is finished.
3. People of various disciplines who use different technology and standards are involved.

Currently, mining companies are using several approaches for obtaining expertise in reclamation programs. Some larger companies have assembled reclamation groups or divisions within their organizations. Others rely on consultants for planning and developing mining permit applications and operate their own field reclamation programs. Still others prefer to have consultants for all reclamation planning and management. Consultants are also engaged by regulatory agencies to evaluate the success of reclamation.

Professionals from the fields of engineering, hydrology, geology, soil science, range science, agronomy, meteorology, and biology are involved in some phase of the permitting and reclamation process. The experience of Woodward-Clyde Consultants, which has personnel representing the above disciplines, has shown that successful and timely reclamation planning, permit application development, and reclamation management requires close coordination and strong leadership of the reclamation team working in concert with the mining project manager and the regulatory agency.

Summary

Requirements for mining permit applications in the states of Wyoming, Colorado, and New Mexico are likely to become more similar in the future because of the recently enacted Federal Surface Mining Control and Enforcement Act. While surface mining of coal is rapidly increasing in these three states, reclamation has been completed on a very small fraction of the disturbed land. It has been estimated that the land now planted, which is 30 to 50 percent of the disturbed area, will likely be considered reclaimed in the next 2 to 5 years.

Most of the unnecessary work in the total reclamation program can be avoided with improved coordination of such diverse activities as mine and reclamation planning, preparation of permit applications, and field operations. Climatic variability including unpredictable precipitation and shallow soils characteristic of many areas in Wyoming, Colorado, and New Mexico require that the reclamation manager has a good understanding of plant-soil-water relations for semi-arid and arid areas. Perhaps, with a few years of successful reclamation, regulatory bodies can leave the agronomic planning initiative to the reclamation manager.

References

Coal Surface Mining Act, 1972, State of New Mexico.

Colorado Mined Land Reclamation Act, 1976, State of Colorado.

Colorado Mined Land Reclamation Board, 1977, Rules and Regulations, Denver, Colorado.

New Mexico Coal Surface Mining Commission, 1973. Regulations of the State of New Mexico Coal Surface Mining Commission, Socorro, New Mexico.

Smith, J. B, M. F. Ayler, Clinton C. Knox, and Benjamin C. Pollard, 1972, "Strippable coal reserves of Wyoming; location, tonnage and characteristics of coal and overburden." U.S. Bureau of Mines, Information Circular 8538.

Speltz, C. N., 1976, "Strippable coal resources of Colorado; location, tonnage, and characteristics of coal and overburden." U.S. Bureau of Mines, Information Circular 8713.

Surface Mining Control and Reclamation Act of 1977, Public Law 95-87, 95th Congress, United States of America.

Tabet, David E., 1977, "Reclamation of Coal Surface Mined Lands in New Mexico." Annual Report, New Mexico Bureau of Mines, Mineral Resources.

Thilenius, J. F. and G. B. Glass, 1974, "Surface Coal Mining in Wyoming: Need for Research and Management." Journal of Range Management 27.

U.S. Bureau of Mines and U.S. Geological Survey, 1976, "Coal Resource Classification System of the U.S. Bureau of Mines and U.S. Geological Survey." Geological Survey Bulletin 1450-B, U.S. Government Printing Office, Washington, D.C.

Wyoming Department of Environmental Quality, Division of Land Quality, 1977 Guidelines, Cheyenne, Wyoming.

Wyoming Environmental Quality Act, 1973, Amended in 1974 and 1975, State of Wyoming.

Reclamation of Lands Disturbed by Coal Mining in British Columbia

37

J. D. McDonald, Senior Reclamation Inspector,
and J. C. Errington, Reclamation Inspector,
British Columbia Ministry of Mines and Petroleum Resources,
Victoria, British Columbia, Canada

Introduction

Coal has been mined in British Columbia almost continuously since the 1850s by conventional underground methods.

By modern standards, these mining operations caused little environmental disturbance when compared with current open pit operations.

The initiation of large open pit mining at Kaiser Resources operation in the East Kootenay during 1968 provided the impetus for present day B.C. mine reclamation legislation.

B.C. coal production is small by world standards and is currently around 9 million tons annually. Most of this total output is coking coal that goes to the Japanese market from the mines of Kaiser Resources Ltd., Fording Coal Ltd., and Coleman Collieries Ltd. Thermal coal is also shipped to Ontario Hydro from Byron Creek Collieries Ltd.

This chapter presents an overview of reclamation on lands disturbed by coal mining and coal exploration in British Columbia. Stability of waste dumps is an essential part of mine reclamation since the dumps must be both physically and biologically stable. In order to achieve stability for reclamation, it is essential that mine planning and mine reclamation be integrated in the total planning process.

Location of Coal Reserves

Coal is found at a large number of locations in British Columbia (Figure 37.1). At the present time, the three most important coal regions are the Crowsnest, Peace River, and Hat Creek coalfields. The Crowsnest Coalfield in the southeast corner of B.C. is presently the most important field in this province. It extends from the United States border northward a distance of 161 kilometers (100 miles), and has a maximum width near Fernie of 24 kilometers (15 miles). There are four companies currently mining in this area.

Figure 37.1. Map of British Columbia showing areas underlain by major coal-bearing formations (Coal Task Force, 1976).

The coal is classified as medium-volatile bituminous and is found in formations of Lower Cretaceous age. Coal is contained within the coal-bearing member of the Kootenay Formation. More than 20 significant coal seams, with a total thickness in excess of 45.7 meters (150 feet), are known within this member. Both thermal and coking coal are mined in this coalfield.

The Peace River Coalfield, located along the Rocky Mountain foothills in northeastern British Columbia, has received increased interest in recent years. The coal measures run from the Alberta border in a northwesterly direction to the Prophet River, a distance of approximately 483 kilometers (300 miles). To date, most exploration activity has been concentrated south of the Peace River.

The potentially minable coal seams vary considerably in this coalfield in both number and thickness. As many as 11 seams have been indicated in the Gething Formation and at least six in the Gates Member of the Commotion Formation. These seams vary in thickness from an absolute minimum now considered minable of 0.91 meter (3 feet) up to 9.1 meters (30 feet) or more. The classification of the coals is generally medium-volatile bituminous, usually with a low sulphur content and high calorific value. Many of the coals have excellent coking qualities.

The Hat Creek Coalfield, situated in south central B.C. near Cache Creek, has been assessed by B.C. Hydro in order to study the feasibility of its development for using the coal as fuel for thermal power generation. This deposit has a total stratigraphic thickness of at least 610 meters (2,000 feet).

Other deposits of coal in British Columbia are found in the Groundhog, Merritt, Similkameen, and Comox areas.

Legislation Governing Coal Mine Reclamation

Authority for regulation of coal mine reclamation is vested in the Ministry of Mines and Petroleum Resources by the Coal Mines Regulation Act, Section 8. The act requires that any person carrying out mining, exploration, or preparatory work for production must "carry out a programme for the protection and reclamation of the surface of the land and watercourses" and "leave the land and watercourses in a condition satisfactory to the Minister."

Before issuing a permit, the proposed program of environmental protection and reclamation must be filed with the minister, and Notice of Filing must be published in the Gazette as well as a local newspaper. The program is considered by other departments of the provincial government affected by, or interested in, the report through the Advisory Committee on Reclamation.

The act requires that the surface of the land be continually and progressively reclaimed and that a security be deposited not exceeding the sum of $1,000 per year for each acre of land used. This security may be used toward payment of the reclamation cost if the terms and conditions of the permit are not fulfilled.

Reports must be submitted annually to the Reclamation Section, Ministry of Mines and Petroleum Resources. At this time, permit conditions and bonding are reviewed.

Because major coal mining developments have a potential for large-scale impacts on the environmental, social, and economic conditions in the region of

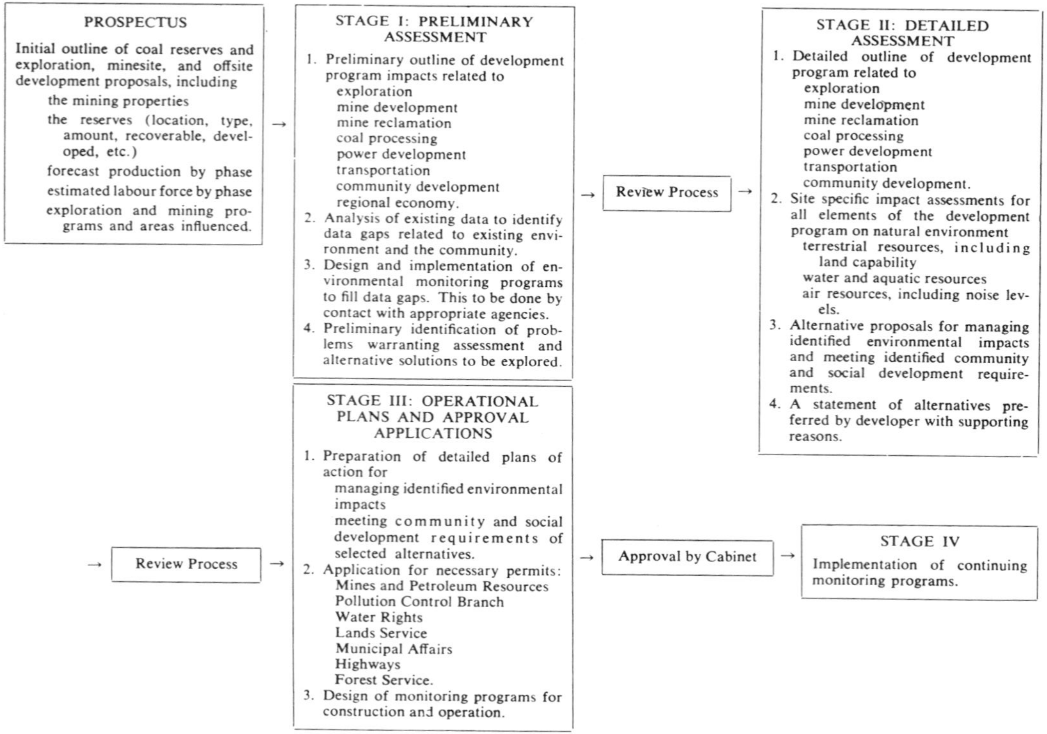

PROSPECTUS

Initial outline of coal reserves and exploration, minesite, and offsite development proposals, including

the mining properties

the reserves (location, type, amount, recoverable, developed, etc.)

forecast production by phase estimated labour force by phase exploration and mining programs and areas influenced.

→

STAGE I: PRELIMINARY ASSESSMENT

1. Preliminary outline of development program impacts related to

exploration
mine development
mine reclamation
coal processing
power development
transportation
community development
regional economy.

2. Analysis of existing data to identify data gaps related to existing environment and the community.

3. Design and implementation of environmental monitoring programs to fill data gaps. This to be done by contact with appropriate agencies.

4. Preliminary identification of problems warranting assessment and alternative solutions to be explored.

→ Review Process →

STAGE II: DETAILED ASSESSMENT

1. Detailed outline of development program related to

exploration
mine development
mine reclamation
coal processing
power development
transportation
community development.

2. Site specific impact assessments for all elements of the development program on natural environment

terrestrial resources, including land capability
water and aquatic resources
air resources, including noise levels.

3. Alternative proposals for managing identified environmental impacts and meeting identified community and social development requirements.

4. A statement of alternatives preferred by developer with supporting reasons.

STAGE III: OPERATIONAL PLANS AND APPROVAL APPLICATIONS

1. Preparation of detailed plans of action for

managing identified environmental impacts
meeting community and social development requirements of selected alternatives.

2. Application for necessary permits:
Mines and Petroleum Resources
Pollution Control Branch
Water Rights
Lands Service
Municipal Affairs
Highways
Forest Service.

3. Design of monitoring programs for construction and operation.

→ Review Process →

Approval by Cabinet →

STAGE IV

Implementation of continuing monitoring programs.

Figure 37.2. Coal Development Assessment Procedure (Environmental and Land Use Committee, 1976).

development the "Guidelines for Coal Development" were issued in March 1976 under the authority of the British Columbia Environment and Land Use Committee.

The guidelines identify the diversity of provincial government interests and public concerns associated with coal and related developments, and establish a procedure for the developer and government to assess and manage all major impacts and for these to be identified publicly. There are two parts to the guidelines. Part 1 describes the provincial government's four-stage project assessment process. Part 2 describes in more detail the information required by various government agencies in their review of coal and related developments.

No attempt is made to describe the guidelines in this chapter although an outline of the "Coal Development Assessment Procedure" is shown in Figure 37.2, and the "Procedure for Processing Impact Assessments of Proposed Coal Developments" is shown in Figure 37.3.

The staged assessment procedure plays an important part in identifying problems, engineering solutions, and planning for reclamation.

Revegetation Problems and Procedures

Crowsnest Coalfield

The Crowsnest Coalfield (southeast coal block) is situated in the Rocky Mountains in the southeast corner of British Columbia. It contains the only four companies currently operating coal mines in this province: Kaiser Resources Ltd., Fording Coal Ltd., Byron Creek Collieries Ltd., and Coleman Collieries Ltd. Coleman Collieries Ltd. is mainly an Alberta-based company but a small portion of their mining activity occurs within British Columbia.

The main mining and exploration areas fall within a biogeoclimatic zone described by Krajina, 1965, as the Engelmann spruce-alpine fir zone at elevations from 1,372 to 2,134 meters (4,500 to 7,000 feet). This zone is characterized by an overstory of Engelmann spruce, alpine fir, lodgepole pine, alpine larch, and whitebark pine forests on west to southeast aspects and, usually, grass-shrub communities on south and southeast aspects. These grass-shrub communities are important wintering areas for ungulate populations.

The entire coalfield is an important area for many big game species, including elk, moose, mountain goat, bighorn sheep, and deer (mule and white-tailed).

The slopes of this zone are generally very steep and the terrain rugged. Soils are regosolic, acid brown, and brown wooded and usually form a shallow mantle over bedrock or glacial till of varying depths. The acidity/alkalinity (pH) of surface soils ranges from 5.2 to 7.0.

The climate is characterized by warm, dry summers and cold winters. The mean annual temperature ranges from a minimum of minus 17.7 degrees Centigrade to a maximum of 25.9 degrees Centigrade. Annual precipitation amounts to 579 millimeters (22.8 inches) of which 219 millimeters (8.6 inches) falls as snow. The continuous frost free period averages 72.5 days with the growing season averaging five months in the Elk River Valley.

Factors determining revegetation in the Crowsnest Coalfield. Although there are many conditions that may inhibit revegetation of coal waste material,

PROSPECTUS (formulated by developer)
↓
Submission by Developer to Department of Mines and Petroleum Resources;
related discussions
↓
Circulation to Provincial Government line departments by Coal Development
Steering Committee*
↓
Receipt of agency comments by Steering Committee
↓
Consultation(s) with Developer's Representatives and (or) Consultants
↓

STAGE I: PRELIMINARY ASSESSMENT
(Contact as necessary with line departments throughout study stage)
↓
Submission of report to Mines Department
↓
Initial review by Steering Committee
↓
Circulation and review by line departments
↓
Consultation(s) between Developer's Representatives/Consultants and Steering Committee
↓
Integrated formal commentary to Developer's Representatives/Consultants by
Steering Committee
↓

STAGE II: DETAILED ASSESSMENT
(Contact as necessary with line departments throughout study stage)
↓
Submission of report to Mines Department
↓
Initial review by Steering Committee
↓
Circulation and review by line departments
↓
Consultations between Developer's Representatives/Consultants and Steering Committee
↓
Integrated formal commentary to Developer's Representatives/Consultants by
Steering Committee
↓

STAGE III: OPERATIONAL PLANS AND APPROVAL OF PERMIT APPLICATIONS
(Direct liaison with appropriate regulatory departments)
↓
Submission of detailed plans and analyses as required for statutory approvals
by line departments
↓
Successful projects granted necessary permit approvals

---

* Consists of representatives of Departments of Mines and Petroleum Resources and Economic Development
and the ELUC Secretariat.

*Figure 37.3. Procedure for Processing Impact Assessments of Proposed Coal
Developments (Environmental and Land Use Committee, 1976).*

there are no major barriers to the establishment of plant growth that cannot be overcome with adequate planning and amelioration techniques. Conditions that may inhibit revegetation in this area include steep slopes, low soil moisture, altered soil nutrient levels, and high elevations.

Steep slopes inclined at or near the natural angle of respose (37°) were recognized as a major barrier to plant survival. At the Kaiser Resources Ltd. mine a slope angle of 28°, sometimes referred to as the "biological angle of repose," was found to be the steepest gradient on which revegetation could be established easily. At Kaiser, a maximum slope of 26° was aimed at for all waste dump materials and this angle became the unofficial standard. Recently it has been found that vegetation can be established at steeper slopes; Kaiser and Fording Coal are both conducting research into the biological ramifications of increasing this angle, especially how this pertains to slope length and aspect.

Low soil moisture in critical summer periods is a major barrier to plant germination and survival under certain conditions. Soil moisture is affected by slope angle, aspect, permeability, and color. Coal waste material tends to be permeable and dark in color. Excessive dryness occurs on slopes with a south to southwesterly exposure that receive the maximum amount of sunlight and heat absorption. Examples of this effect can be seen on revegetated waste dumps as well as in the natural plant community.

Waste materials are different in their physical and chemical makeup from the natural soil medium. These differences do not prevent revegetation of waste material but determine to a large extent, the selection of species. Waste material has a pH ranging from 6.0 to 8.5 which is higher than the natural soil medium (5.2 to 7.0). This slightly elevated pH appears to be the primary cause for failures in many of the native tree and shrub plantings at the Kaiser operation. The most successful plantings have been black cottonwood (*Populus trichocarpa*), cut leaf birch (*Betula* sp) and European birch (*Betula* sp) which are species that are tolerant of higher pH conditions.

Waste materials are also deficient in nitrogen and phosphorus and require applications of fertilizer to support and maintain plant growth. There has been an emphasis on promoting nitrogen fixing species in order to establish a maintenance-free reclaimed area.

Mining occurs at elevations exceeding 2,134 meters (7,000 feet). When mining plans were announced there was great concern expressed regarding the establishment of vegetation at this elevation. Test plots have been established at the Kaiser Resources mine at 2,103 meters (6,900 feet), the Fording Coal mine at 2,225 meters (7,300 feet), and the Coleman Collieries mine at 2,103 meters (6,900 feet). Kaiser is now successfully conducting operational reclamation at these elevations, and a surprising number of species are growing well and producing viable seed.

Revegetation procedures. Although there are a wide number of procedures employed to revegetate disturbances in the Crowsnest Coalfield, the basic reclamation practice involves recontouring the waste dump to an angle of 26° then fertilizing and seeding with a grass and legume mixture. At Kaiser, heavy-duty harrows are then drawn across the slope to bury the seed and to create small terraces to prevent erosion and contain surface water. Coleman Collieries also buries the seed by running a Caterpillar D9 bulldozer up and down the finished slope and creating the necessary microrelief with the Caterpillar tracks.

Many other techniques have been attempted or are used regularly to tackle specific problems. Some of these are outlined below:

1. Hydroseeding is used on selected areas. Slopes that are too steep for conventional machinery and cannot be recontoured are reclaimed in this manner. The tailings lagoons at Kaiser Resources mine are also seeded by hydroseeding, because the seed drills or harrows mix up the surface of the tailings material causing it to dry out on the surface.
2. Maintenance fertilizer applications using a cyclone seeder mounted below a helicopter have been used successfully at Kaiser Resources mine during the past two years. Fertilizing by helicopter has reduced overall costs, is rapid, and provides more uniform distribution of fertilizer. Areas that have difficult ground access are also seeded and fertilized by helicopters.
3. The use of native tree and shrub species propagated in Kaiser Resources nursery is an integral part of their program.
4. Generally the salvage of surface soil material is not practiced in the East Kootenay coal mines. Tests are currently underway at the Fording Coal mine to assess the costs and benefits of topsoil salvage and surface dressing on waste dumps.

Peace River Coalfield

The Peace River Coalfield is located in the foothills on the eastern slopes of the Rocky Mountains. There are no active mines in this area but exploration activity has intensified in recent years and, should the economics of coking coal improve, there is promise of several major coal developments taking place in the near future.

This region lies much farther north than the Crowsnest Coalfield and consequently the climate is much more severe. For example, the treeline occurs at elevations of 1,524 meters (5,000 feet) in the Peace River Coalfield compared to 2,134 meters (7,000 feet) in the south.

The Peace River Coalfield lies in a remote wilderness setting with limited access. However, with incursions by coal interests as well as seismic petroleum and forestry industries, the character of the region is rapidly altering.

Most of the coal licenses fall within the Engelmann spruce-subalpine fir zone at lower elevations and the alpine tundra zone at higher elevations.

The region is excellent big game habitat and supports populations of moose, caribou, bighorn sheep, mountain goat, black and grizzly bear, elk, and deer.

Factors determining revegetation in the Peace River Coalfield. Experience in revegetation of disturbances in the Peace River Coalfield has been gained through reclamation during exploration programs, and from several field trials set out by mining companies and the Ministry of Mines and Petroleum Resources.

The major problems associated with revegetating disturbances caused by exploration occur above the treeline. Here the climate is severe and plant growth and survival are limited by an extremely short growing season, high wind velocities, and low nutrient levels.

The extreme climatic conditions above the treeline severely restrict the number of species available for use in reclamation programs. Although there are several grass species that perform adequately in the short term, no legume species has survived longer than one year. Because sites are generally low in nitrogen and phosphorus and nutrients cannot be supplied by legume growth, fertilizer additions are necessary for successful revegetation using grass species.

In areas below the treeline, revegetation of exploration disturbance presents no major problems. Many species of grass and legume grow successfully and are available for reclamation programs. Generally, sites have low nitrogen availability and rely on legume growth for several years in order to inject enough available nitrogen into the system to allow grass survival. Low soil moisture on coarse-textured sites occasionally limits legume growth, and affects the growth of the entire grass legume mixture. These areas require applications of fertilizer for successful establishment of a grass sward.

Revegetation procedures. Three broad types of seeding techniques have been tried in the Peace River Coalfield. These are broadcast seeding, harrowing after broadcast seeding, and hydroseeding.

Broadcast seeding, the most common method used, consists of scattering seed on the surface of the ground. Broadcast seeding has been accomplished by scattering seed by hand, by handheld cyclone seeders, or by cyclone seeders mounted on equipment with a 12-volt power takeoff. Aerial seeding using helicopter or fixed-wing aircraft has not been tried.

Broadcast seeding without fertilizer applications has resulted in satisfactory growth at most sites below the treeline. Above the treeline, this method of seed application has generally resulted in poor growth, but the failure may be attributed to lack of fertilizer rather than seeding technique. Results of plot trials above the treeline indicate that fertilizer applications must accompany seeding for successful growth and germination, and that, given adequate fertilizer, most of the grass species used in seeding programs will grow well.

Harrowing and fertilizing all areas above treeline have been recommended by the Ministry of Mines and Petroleum Resources since 1976. A variety of methods have been used for burying seed. The best method appears to be a harrow towed behind a small Caterpillar tractor and followed by seed and fertilizer applications and then reharrowing.

It is too early to see if methods involving harrowing have been successful; however, preliminary observations indicate that seed catch will be largely improved over former broadcasting techniques.

Hydroseeding was used in an experimental way in the alpine areas of Bullmoose/Chamberlain. Initial results indicate favorable growth over the first summer, and good survival is anticipated next year.

In conclusion, since the inception of mine reclamation legislation in 1969, coal mines in British Columbia have made major strides in reclaiming mine waste materials. As a result of this rapid progress in revegetating disturbances under extreme climatic conditions in mountainous terrain, the complete reclamation of future developments would seem assured. Planning, research, and operational reclamation will continue to overcome barriers to revegetation.

References

Coal Task Force, 1976, "Coal in British Columbia: A Technical Appraisal." Parliament Buildings, Victoria, B.C.

Environmental and Land Use Committee, 1976, "Guidelines for Coal Development." Parliament Buildings, Victoria, B.C.

Krajina, V. J., 1965, "Ecology of Western North America." Dept. of Botany, University of British Columbia.

Index